T. F. Bourdillon

Travancore Forests

T. F. Bourdillon

Travancore Forests

ISBN/EAN: 9783337211400

Printed in Europe, USA, Canada, Australia, Japan

Cover: Foto ©berggeist007 / pixelio.de

More available books at **www.hansebooks.com**

REPORT

ON THE

ﬀorests of ﬀrabancore

BY

T. F. BOURDILLON, F. L. S.,
Conservator of Forests, Travancore.

WITH 4 MAPS.

DATED DECEMBER, 29, 1892.

TREVANDRUM:

PRINTED AT THE TRAVANCORE GOVERNMENT PRESS,

1893.

INDEX.

ii

CHAPTER III.

PART II.

SUGGESTIONS FOR FUTURE MANAGEMENT.

APPENDICES.

INTRODUCTION.

One of the recommendations made by the Commission which was appointed in April 1884 to report "on the Forest Administration of Travancore and its defects," and to suggest remedies for the better conservancy of the forests, was that they should be carefully explored and reported on.

I was accordingly appointed in June 1886 to make a thorough examination of the forests of the State "to mark and define those tracts which should be permanently reserved," to submit a report on their resources, especially noting the condition and extent of the forests of teak, to prepare maps showing the character of the forest in different localities, and finally to bring to the notice of Government any points worthy of attention.

Before commencing the work it was necessary to determine how far my travels should extend, for it was obviously of little use visiting those parts of the country where the land was chiefly in the hands of the people, and where there was not much timber. I accordingly decided to confine my exploration to the interior hilly tracts, which occupy about one half of the State, and in which only about one per cent of the total population reside.

In most places an arbitrary but well-marked line separates these hilly tracts from the more thickly populated low country, but here and there, where the people have taken up small plots of land on the outskirts of the forests, its selection is not so easy.

In carrying out my explorations I was very much assisted by the excellent maps of the State which we have, which though prepared more than half a century ago are still wonderfully accurate, showing how little the country has altered within that time. As might be expected, there are some errors in them, the courses of some of the rivers in the interior forests are wrongly marked, ridges run in directions contrary to their true directions, and the names are often wrong or misleading. But on the whole, the errors are few, and the maps remain a standing testimony to the diligence and accuracy of Messers Ward and Conner who prepared them.

The errors are more numerous in North than in South Travancore, and the features of the country, especially about the boundary of Travancore and Cochin, are not always correctly marked, a fact which has caused much trouble in the demarcation of the line. The course of the Thēvi-ār, an affluent of the Periyār is wrongly given. In Central Travancore, the Kalār, one of the tributaries of the Acchankōvil river, is made to flow into the Rāni, a pardonable mistake seeing that there are 2 streams of the same name, one joining the Acchankōvil river at Thora, and the other falling into the Rāni at Kumarampārūr. Coming farther South, the Sāttār a tributary of the Kōtha is made to flow in a wrong direction. I am also informed by the officers of the Revenue Survey that the triangulation of some of the peaks is not correct.

Commencing my travels in South Travancore in June 1886, I worked my way as far North as Kovillūr, when in July I was summoned to Courtallum to meet Col. Campbell Walker, one of the Conservators of Forests of the Madras Presidency. After a month's absence, I resumed my work, but early in September I was placed on special duty in connection with the boundary between Travancore and Cochin, and was absent for 4 months. Returning in January, I continued my exploration till it became too hot to work in the low country, when I went up to the Peermade hills, but I had scarcely arrived there when I was recalled to Quilon to act for the Conservator for 3 months. In June 1887 I again resumed my work in spite of the wet weather, and steadily travelled through the country between Madattburai kāni and Mundakayam, reaching many places never before visited by a European. By December 1887 I had arrived at Peermade, and about 6 weeks were spent in exploring that neighbourhood, when I was recalled to the low country and appointed Deputy Conservator of Forests at Malayāttur. For a whole year my work was interrupted, but at the beginning of 1889 I was allowed to spend about 2 months on the Peermade plateau in exploration. In June of that year I was again placed on this special duty, and by February 1890 I had visited all the forests of North Travancore, including the Idiyara valley, the Cardamom Hills, the High Range and the Anjināda valley.

The time actually expended in exploration did not exceed 21 months, spread over more than 3½ years, while the time allowed me was 3 years. After I had completed my travels, I was invalided to England for 6 months, and since my return I have had the charge of the Forest Department, and I have only been able to find odd moments while attending to other work for the completion of this Report.

In some ways it is unfortunate that so long a time has elapsed between the initiation of my work and its completion, but in other respects this is not a disadvantage. Every year shows more clearly that some measures are urgently required if our forests are to be properly conserved, and Government are more likely to take the matter in hand now than they would have been two or three years ago.

In the course of my travels I must have covered a distance of about 7,000 miles, some of it by boat, some on horseback, but the greater part on foot. I had also during my residence for several years in South Travancore become familiarized with the character of the forests there. In addition to this, I have lost no opportunity of obtaining information from Hillmen and others as to parts of the country which I could not myself reach. This Report represents the result of my work.

Throughout this Report I have spelt the names of places on the Hunterian system, endeavouring to transliterate each name according to its pronunciation into the exact English equivalent, and the result is that some familiar names have become hardly recognisable. Nevertheless this is undoubtedly the proper method, exception only being made in the case of those names whose orthography has already been settled by the Post Office, such as Travancore, Nagercoil, Cottayam &c.

In the preparation of a Report of this kind one of the chief difficulties to be met with is to know how far to descend to details, and where to write briefly and generally. Perhaps I might with advantage have made this paper shorter, but there is this to be said, that there is hardly any detail or fact which, if carefully ascertained to be true, is not worthy of record. The extent and condition of our forests, the character of the timber, the numbers of the population and their characteristics, traces of former inhabitants, and the present state of forest roads and rivers are matters of interest to ourselves, and will be doubly so to our successors. How invaluable are the remarks let fall by Messrs Ward and Conner, writing 70 years ago, on the system of collecting cardamoms at that time, on the seeding of bamboos, on the extent and condition of the forests and the work carried on in them then, and what would we not now give for fuller and more complete details. If then this Report may seem too prolix, I offer this apology for it. All my statements I have endeavoured carefully to verify.

In the first chapter of the first part I have given a general description of the country, its geography, population, physical features and climate.

In the second chapter I have treated of the forests of the State. I have described the forest area river by river, showing the character of the forests, and the population resident in each. I have referred to the system of hill-cultivation, to the damage done by grass-fires, and I have given some particulars of the principal timber trees.

In the third chapter I have described the past and present management of the forests, showing what steps have been taken to conserve them, the arrangements in force for the felling, delivery and sale of timber, the miscellaneous work done by the staff of the Forest Department, and the offences against the rules and regulations of the Department.

In the second part I have thrown out suggestions as to the future working of the Department, showing where improvements are required and where the present system should be maintained.

Finally, in the Appendices I have collected information on the cardamom monopoly, on the Hillmen, on the trees of Travancore and other matters.

Quilon,   T F. BOURDILLON,
December 29th 1892.   Conservator of Forests.

REPORT

ON THE FORESTS OF TRAVANCORE.

PART I.

THE FORESTS AND THEIR MANAGEMENT IN THE PAST.

CHAPTER I.

GENERAL DESCRIPTION OF TRAVANCORE.

1. The State of Travancore is situated in the South of the Indian Peninsula, and lies between the parallels of North latitude 8° 3′ and 10° 22′: in *Geographical position.* longitude 76° 14′ and 77° 38′ East of Greenwich.

2. Its shape is triangular with the apex towards the South, and its two sides, of which the Western is rather the longer, run in a Northerly and *Shape.* North Westerly direction. Its greatest length from North to South is 174 miles, and its greatest width, near the Northern boundary is 75 miles.

3. It is bounded on the West and South by the Arabian Sea, and by portions of the State of Cochin, which run down in a narrow strip and *Boundaries.* intervene between Travancore and the Sea. On the East by the British Collectorates of Tinnevelly and Madura, and on the North by the Collectorate of Coimbatore and the State of Cochin.

4. The area comprised within these limits is generally estimated at 6,731 square miles, but in reality it is considerably more, and the correct *Area.* amount is probably over 7,000 square miles.* Compared with Ceylon, Cochin and the adjacent Collectorates, Travancore is about one fourth the size of Ceylon, (24,702 square miles), five times the size of Cochin, (1,361½ square miles) rather larger than Tinnevelly, (5,381 square miles) and Malabar, (6,002 square miles) and smaller than Coimbatore (7,804 square miles.).

5. For administrative purposes the State is portioned out into Four Divisions, and thirty two Taluqs, the latter varying in extent from 612 square *Divisions.* miles (Thodupura) to 47 square miles (Paravūr).

6. The population at the Census of 1881 was 24,01,158 which, distributed over 6,731 square miles of country gives 356 to the square mile, *Population.* compared with 109 in Ceylon, 441 in Cochin, 310 in Tinnevelly, and 410 in Malabar. The density of the population varies greatly in the different Taluqs. In the smallest (Paravūr) it is 1,318 to the square mile, while in Thodupura it is only 40. The population is much denser all along the sea coast, and for some distance inland, for here the conditions of existence are more favourable, communication is easier, and the coconut tree, which represents the chief wealth of the country, thrives best. As we proceed inland the density of the population rapidly decreases, except along the chief rivers, and the lines of communication with Tinnevelly and Madura, and on the hill plateaux it is very low.

7. The first Census of Travancore was taken by Messrs. Ward and Conner at *The different* the time of their survey of the State in the years 1816–20, and *Censuses.* though it was necessarily more inaccurate than latter ones, especially in all remote districts, yet it may be assumed to have been fairly correct. §The total population was at that time 9,06,587. §Another Census was taken in 1836 showing a population of 12,80,668, and another in 1854, at which time, if the figures are to be relied on, the total population was less by 18,000 than it was eighteen years

---

* The above figures are taken from the memoir of Ward and Conner's survey mentioned above but the survey excludes from Travancore, (1) the whole of the Anjanāda Valley together with a large portion of the High Range, aggregating about 229 square miles, (2), The Idiyara Valley, about 42 square miles, and (3) a portion of Forest of about 6 square miles near the Alvākkurichi gap, which was under dispute until a few years ago, when it was handed over to Travancore. Further, Travancore gained considerably when the unexplored portion of the boundary between this State and Tinnevelly was surveyed. The boundary is the watershed, (except in one or two places) and it was found that this was usually much nearer the Eastern side of the hill plateaux than Lieut Ward had calculated, for, owing to the difficulty of the ground he was unable to undertake a regular survey there. On the other hand Travancore may have lost a little in the adjustment of the line between Pannımāda Kuttha and Parutthamala which divides it from Coimbatore.

§ Census Report of 1881.

befoic. From that time the population has steadily increased, and was 23,11,379 in 1875, and 24,01,158 in 1881. Considering the large area of waste land available to the people, this slow increase is a matter for surprise, more especially as there is practically no emigration from Travancore, while there is a certain amount of immigration. In other parts of the world where land is easily obtainable, the increase of population is very rapid, even leaving out of consideration the increase by the influx of foreigners; in Canada for instance families of 14 and 15 children are by no means uncommon among the French "habitants," but in Travancore, for what reason I do not know, large families are unusual.

8. The population is divided according to Religion as *Religion.* follows :—

Hindus ...	...	...	17,55,610
Christians	...	...	4,98,542
Mahomedans	...	...	1,46,909
Jews	..	...	97
			24,01,158

9. The language chiefly spoken is Malayalam which claims 80·69 per cent of the people, Tamil comes next with 18·31 per cent and other *Language.* languages make up a total of 1 per cent. The Malayalam spoken in Central and South Travancore is largely intermixed with Tamil, and is very mongrel, but in North Travancore, where the Tamil influence is very slight, it is pure.

10. As regards the occupations of the people, some inter- *Occupation.* esting statistics may be obtained from the Census Report.

11. Out of the 11,97,134 males in Travancore only 6,83,759 are returned as having occupation or admitting that they have none, the other *Occupied males* 5,13,375 are of course mostly children.

12. . Of the persons having occupations the following per-*Percentage of occupa-* centages are given under different classes of work. *tions*

Professional class	...	...	...	... 5
Domestic class	...	...	..	... 2
Commercial class	...	...	...	... 9
Agricultural class	...	...	...	...37
Industrial class	...	...	...	...18
Indefinite and nonproductive class	...			...29
				100

13. From this it will be seen how large a proportion of the population is engag-
More than 50 per cent ed in agriculture. The members of the Industrial class though *of the population engaged* numbering 18 per cent, on account of the want of labour-saving *in agriculture.* apparatus and machinery, show a comparatively small outturn of work in proportion to the labour expended. The sixth or indefinite class includes 1,95,537 labourers, the greater part of whom must be engaged in tilling the fields or in other branches of agriculture, so that, adding even half of these to the agricultural class, we find that over 50 % of the working male population are engaged in raising produce from the land.

14. * The Revenue, which has increased rapidly within the last few years, reached the high figure of 77,64,494 Rs. in the year 1064 M. E.—
Revenue 1888-89 A. D. the expenditure being Rs. 70,03,048. The chief items in the Revenue were

Land Revenue ...	...	...	...20 lacs.
Salt	...	...	...	...17 lacs.
Tobacco ...	...	..	...	... 9 lacs.
Customs	... ·	...	...	.. 6 lacs.
Forest ...	...	...	...	... 5 lacs.
Abkarry and Opium	...	...	... 5 lacs &c.

* Administration Report for $\frac{1888-9}{1064}$.

15. The value of exports in the same year was Rs. 1,24,82,640, and of imports
Exports
Rs. 78,67,341 showing a difference of 46 lacs of Rs. which go to
pay the land tax, salt tax, customs duties &c. and to supply money
for saving.

16. The chief article of export is "copra" the kernel of the coconut, and
this and other products of the coconut tree yield more than 54
Different articles of export.
per cent of the total value of the exports. Garden produce, such as
arecanuts and ginger, gives 10 lacs or 8 per cent, pepper contri-
butes 9 lacs or 7·2 per cent, jaggery (the produce of the palmyra palm) 5 lacs,
and so on. Travancore exports no rice.

17. The chief articles of import are tobacco to the value of 25½ lacs = 32·4 per
Imports.
cent of the total imports. Cotton, thread, and piece goods 20¼
lacs = 26 per cent, and rice and paddy 12¼ lacs = 15·6 per cent.

18. Although rice is the staple food of the people, and is grown in large
quantities, the amount is not sufficient for their requirements, and
The Rice grown in the country not suffi-cient for the popula-tion.
as already stated, the imports of this article are large and amount
to some 700,000 cwts a year. The reason seems to be that it is
more profitable and certainly requires very much less labour to
grow coconuts, pepper or arecanuts than rice.

19. The yield too of rice, except in the Nānjināda where the fields are heavily
manured, is small, and does not exceed 10 or 12 per cent in the
Yield of rice fields
ordinary paddy fields; as a dry crop grown on the uplands after
a long fallow, rice gives a very variable yield, fluctuating between 20 and 200 fold,
but the usual crop, if the season is favourable and the weather propitious, is 40 fold.

20. The area of assessed rice lands has been estimated at 3,61,632 acres paying
an average tax of 4 Rs. but this varies between ½ and 35 Rs.
Area of assessed lands.
per acre. *

21. The area under coconuts, jacks and other fruit trees, known as garden
lands, has never been exactly estimated, the tax being levied on
Area of garden lands
the number of the fruit-bearing trees themselves and not on the
area under cultivation, except when the gardens contained no taxable fruit-bearing
trees. The most important taxable trees are (1) coconut (2) areca (3) jack (4) palmyra
rated at from ½ to 4 chuckrams each. §

22. Writing in 1820 Captain Conner divided the country
Distribution of the surface of the country
according to its features and its suitability for various products as
follows—

Distribution of the surface.

Rice lands	741 sq. Miles.
Slopes available for the temporary cultivation of rice and various dry grains about	1000 do.
Suparee and coconut topes chiefly along the coast	356¾ do.
Sandy extent covered with palmyra trees chiefly to the South	115 do.
Lakes, rivers, tanks.	157½ do.
Sites occupied by buildings of every description about...	20 do.
Pasturage and superficies occupied by low chains of hills about	1961 do.
Hills and forests scarcely any part of which is improvable.	2379½ do.
Total ..	6730¾ do.

and he thus proceeds :—

" It is thus seen that, on a subtraction of the mountainous, woody, and "
" watery parts, about two-thirds only remains applicable to the purpose of pro- "
" fitable cultivation or pasturage ; indeed it may generally be said that the whole "
" riches, population, and cultivation of Travancore, are confined to a contracted "
" strip along the beach."

---
* Memorandum on the Revenue Survey and Settlement—1885.
§ One chuckram—½½ or nearly ¾ of an anna

23. In addition to the products mentioned above which form the staple products of the country, a certain quantity of coffee, cinchona and tea is also grown and exported. Coffee had been grown in the low country under the heavy shade of jacks and other trees for many years before the success of the enterprise in Ceylon turned the attention of some Ceylon planters to the suitability of the hills of Travancore to the cultivation of that product. General Cullen, a former Resident, had already opened experimental gardens at Ashambu and the Vélimala hills, and had been successful, so that when Messrs. Grant and Fraser came over from Ceylon in 1864 or 65, the future success of the enterprise seemed quite assured.

Coffee.

24. The Travancore Government anxious to welcome the introduction of capital into the country, at first gave away land on the hills, and afterwards sold it at an upset price of 1 R. per acre, which was raised in 1873 to 10 Rs. but so great was the rush for land, and such the anxiety to secure it, that in 1874, the last year when auctions were held, some lots were sold for over 80 Rs. an acre. From first to last 39,400 acres of land, mostly virgin Forest, were sold to Europeans and a few natives, and then the Government decided to stop any further sales. *

Commencement of the coffee enterprise.

25. Meanwhile a great part of the lots sold had been cleared and planted. many Europeans had arrived in the country and the exports of coffee rose to 45,700 cwts in 1879–80.§

Increased Exports of coffee.

26. But about the year 1872 a fungoid disease (*Hemileia Vastatrix*), which had attacked the leaves of the coffee plant in Ceylon some 3 years previously, began to appear in Travancore, and very rapidly spread through every estate in the country. As however the price of coffee remained high, and "the leaf disease," as it was called, seemed to yield to manure and careful management, extension of cultivation still went on.

Coffee leaf disease.

27. After 1877 prices began to fall, until they reached a figure which could only pay those owners of estates, who were getting large crops, and this the disease prevented. Under these combined circumstances, one estate after another was abandoned, and at the present moment almost the only estates still cultivated with coffee, are those which had been planted on exceptionally fertile soil, or which have other advantages enabling them to yield good crops. The exports of coffee last year amounted to only 6,546 cwts. Prices of coffee are again much higher now, owing to short crops in Brazil, and it is possible that the area under that product in Travancore may be again slightly increased, but it is unlikely that there will ever be any great extension.

Causes of the abandonment of the coffee estates.

28. Although leaf disease and low prices together combined to ruin the coffee enterprise, other causes contributed to bring about that result. Much of the land planted was steep, and the heavy rain washed the soil away, other parts were exposed to wind, or contained poor soil, while nearly everywhere the coffee plants were hurriedly planted out without a proper rejection of the weak and sickly specimens. Indeed those who have studied the question are generally agreed that the leaf-disease was a result rather than a cause, and that it would never have done the damage it did, unless the coffee plants had been already in a sickly condition, a belief that is supported by the fact that in Coorg where the soil is good and cultivation receives due attention, leaf-disease is little feared.

Leaf disease not the only cause.

29. As coffee began to fail the European planters turned their attention to other products, and cinchona was largely planted, chiefly through the coffee, or in company with tea, and the number of plants put out during the last ten years must be very large. Cinchona grows very well in Travancore above 2,500 feet elevation, but the valuable kinds of crown bark (*C. Officinalis* and varieties) only thrive at 5000 feet and upwards. On the High Range of Travancore with a fine soil, light rainfall, and absence of wind, cinchona has succeeded probably better than any where else in

Commencement of other enterprises. Cinchona.

* Statistics supplied from the Huzur.

§ Administration Reports. The figures for 1875–6 and 1876–7 are not available, as in those years there was no export duty.

India, and if prices rise again, the planters there who are now only collecting enough bark to keep their Estates going should realize large profits.* On Peermerd in Central and South Travancore *Cinchona Succirubra* has been planted and though the growth is more rapid than at a higher elevation, the bark contains less Quinine, and the tree is capricious, failing in some places and succeeding in others.

30. The tea plant was introduced into Travancore by General Cullen many years ago, but it was not till 1876 or 77 that any efforts were made to extend its cultivation. The first attempts proved so successful, that a large area has now been planted up, and some extensions are made every year. The annual out-turn must be nearly a million lbs. of made tea, but there are no statistics on the subject available. The area under tea is probably about 3,000 acres.

Tea.

31. The tea plant is found to thrive equally well at sea level and at 6,000 ft. In the former situation it is difficult to start on account of the drought; and the greater heat causes it to winter more thoroughly, but during the monsoon months the growth is remarkably rapid, and fully makes up for its cessation in the dry weather. At the higher elevations the growth is more regular, but less rapid, and the yield less, but the quality is better, so that the advantages of either situation are about equally balanced. Probably the best elevation is midway between the two at about 3,000 ft. It was here that the first success was secured, and here a very considerable yield can be obtained, while the climate is more favourable for Europeans than in the low country.

Success of tea cultivation.

32. The present position of the tea industry in Travancore is decidedly encouraging. The yield is good and the class of tea is on a level with that of Ceylon. The mistake of planting steep land, made in the case of coffee has been avoided, and the plants have been singularly free from diseases or attacks of insects. Only one or two of the estates at the lower elevations have suffered from the tea mosquito (*Helopeltis antonii*), which punctures the leaves and prevents the plants flushing. The low prices now ruling, which are the result of extended cultivation, do not admit of large profits, but it is nevertheless possible to obtain a very fair return on the capital expended.

Present position of the tea industry.

33. Although the number of European planters in the country is only about half what it was ten years ago, there are some 30 still remaining. These together with the labourers they employ and the magisterial and engineering staff required for their wants make up together a very considerable population, the whole of which resides on the hills and within the forest area. The coolies who amount to 4,000 or 5,000 in number are recruited chiefly from the neighbouring districts of Tinnevelly and Madura. These are the best coolies on the estates, as many of them have been to Ceylon, and have learnt their work there, and they are also thoroughly acclimatised to the hills. The Malayalies as a rule detest the hills and do not thrive there, but a few Pulayans are employed, and some hundreds of Shānars and native Christians from the neighbourhood of Nagercoil.

Population living on the hills.

34. Until quite within recent times the only means of communication between one place and another was by water, or by narrow foot-paths leading from village to village. Travancore is fortunate in having in the Backwater a system of rivers and canals an easy means of travelling in the neighbourhood of the sea. The Backwater consists of a series of fresh-water lagoons running parallel to the sea and separated from it by a strip of land varying in width from a few hundred yards to several miles. Where these lagoons were not continuous, canals were cut to connect them, so that it has always been easy to travel by boat from Trevandrum to Paravūr on the Northern boundary of the State, or inland up the chief rivers, if the water has not run so low as to prevent the passage of boats.

Means of communication.
The Backwater.

35. But in South Travancore, where there is no Backwater or canal, roads were a necessity, and though at the time of Lieut Ward's visit, the villages there were connected by such means of communication, yet these roads were hardly passable for wheeled carriages. A vast improvement has taken place within the last 30 years, and now South Travancore is inter-

Ronds.

---

* Prices are at present lower than they have ever been, the unit being quoted at less than 2d.

sected by a whole net work of excellent cart roads. Many others have also been cut in all directions through Central and Northern Travancore, some of them, as at Shencottah and Kumili, joining similar roads in British territory and offering a means of exit from the country.[*]

36. **North Travancore the most backward.** North Travancore as being farther from the capital, more thinly populated in its inland Taluqs, and possessing good means of communication in its many fine rivers, was the last part of the country to be opened by roads, but within recent years many cart roads have been cut there and others are now under construction, so that in a few years it will be possible to travel to any part of the low country, and to many portions of the hills by cart.

37. **Bridle paths.** Where the steepness of the ground or the smallness of the traffic made it undesirable to open cart roads, the Travancore Government has cut a great many bridle paths, and there is scarcely an estate on the hills that is not connected with a cart road by one of these.

38. **Railway.** Although there is at present no railway in the country, it is under contemplation to connect Quilon, or Trevandrum, by rail with the South Indian Railway at Tinnevelly, and Travancore will then be in railway communication with any part of India where the iron horse runs.

39. **Chief ports.** The chief ports of Travancore are Alleppey and Quilon, the former accessible at all times, the latter closed during the monsoon. A great deal of produce passes out of the country through the Arrukkutti chowkey in the North and is exported from Cochin, as also through Arammula in the South and Shencottah in central Travancore. Colachel is the chief port of export for jaggery, but there is not much other produce shipped there.

Physical features.

40. **General character of the Boundaries of Travancore.** As I have already stated, Travancore is bounded on the South and West by the sea, or by a narrow strip of land which separates it from it. On the North it is bounded partly by Cochin Territory lying at a low elevation, and partly by hills and high ground belonging to Cochin, or to the British Government. On the East the boundary is, with three small exceptions, the lofty mountain-range which forms the back-bone of the Peninsula. These exceptions are (1) the Anjinada valley, about 130 square miles (2) the Shencottah Taluq about 65 square miles and (3) the Eastern slopes of the Mahenthragiri Hills about 6 square miles, all of which drain to the East, the drainage of the rest of Travancore is to the Westward.

41. **The Highest mountains and peaks.** The lofty mountain ridge to which I have alluded reaches its highest elevation at Anamudi (8840 ft.) in the extreme North-East of Travancore. Round it are clustered numerous other peaks which attain a height not much less. The chief of these are Eravimala, (7880 ft.) Kūttamala, (7800 ft.) Chenthavara, (7664 ft.) Kumarikkal, (7540 ft.) Karinkulam, (7500 ft.) and Thēvimala (7200 ft.) which run in a horse shoe shape with the opening facing towards the North East. These lofty hills, with the lower ground connecting them, form the elevated plateau, extending over about 100 square miles, known to Europeans as the High Range.

42. **The High Range.** Properly speaking this tract can be hardly called a plateau : it is rather a succession of high hills, with deep valleys between, running down to a depth of 2,000 to 3,000 ft. below them. The greater part of this area is covered with beautiful, short grass, with stretches of heavy forest on the lower ground, and before any estates were opened here it was a famous place for game of all kinds. Even now, thanks to the system of preservation adopted by the planters, the Neilgherry ibex (*Hemitragus hylocrius*) is very plentiful on some of the mountains, though the Muthuvans and people from Anjinada have exterminated the animal in all places easily accessible to them.

[*] A canal has been partially cut from Trevandrum Southwards to Colrchel, but I believe no part of it is ever used.

43. From the High Range, the land slopes steeply down in three directions :
North East to the Anjinâda Valley, to the West into the
The slopes of the High Range. Valleys of the Kandam Pâra, Parishakkuthu and Idiyara
rivers, all of which fall into the Periyâr, and Southward to the
Cardamom Hills and Peermerd.

44. Anjinâda lies in a secluded valley between the lofty peaks of the High
Range. In its upper part it is called the Thallayâr Valley,
The Anjinâda. and here the elevation is above 4,000 ft. Sloping gradually to
tho North East, it opens out into the Anjinâda Valley proper,
a level terrace 2 or 3 miles wide and 5 miles long lying at an elevation of 3,000 ft.
The soil of the Anjinâda is very fine, and the climate is quite different to that of any
other part of Travancore, being as dry as in Shencottah or Thôvâla, and cooler on
account of the higher elevation.

45. Below and to the East of Anjinâda, the land slopes
Country below the Anjinâda. rapidly down to the British frontier probably at about 1,500
or 2,000 ft., and this tract is very feverish, and contains no
resident population.

46. To the west of Anamudi the slopes are exceedingly steep, almost precipitous, and descend to a greater depth than on any other side.
Western slopes of Ana-mudi. The Kandampâra, Parishakkuthu and Idiyara rivers start at an
elevation of from 400 to 800 ft., from the base of the Anamudi
slopes, though they are fed by streams running down from the higher elevations.
They are separated by strong ridges rising to over 3,000 ft. in some instances, and
running out towards the sea in a Westerly and North-westerly direction.

47. To the South of the High Range the land spreads out into a hill-plateau
of considerable width. Sixty miles South of Dêvikulam, the
The southern slopes of the High Range, and the Peermerd plateau. Periyâr, by far the largest river in Travancore, takes its rise in
an extensive forest lying at over 3,000 ft. elevation. From
here it runs in a general direction almost due North but inclin-
ing to the West, winding however a great deal in its course, until it has reached a
point close under the High Range. Here its waters are joined by those of the
Mûnâ, or Mothirapura River, which takes its source on the slopes of Anamudi, and
turning to the West the Periyâr plunges down between immense cliffs of rock, and
finally falls into the sea at Cochin and Paravûr. Thus for the first 40 or 50 miles
of its course the Periyâr traverses a hill-plateau 60 miles long and 20 broad lying at
a general elevation of from 3,000 to 3,500 ft. with peaks and hills running up to
4,000 and even 5,000 ft. These are the Cardamom Hills and Peermerd.

48. South of Peermerd the Mountain Range, or backbone, of which I have
been speaking, is of no breadth except at a place called Muthu-
The mountain backbone south of Peermerd. kuri Vayal in South Travancore, where the Kôtha River, like
the Periyâr, runs for the first part of its course almost due North
across a plateau at an elevation of about 4,000 ft., and measuring some 10 miles long
by 6 broad.

49. For the rest of its length the mountain backbone is a mere ridge sloping
down on either side, and running N. N. W. and S. S. E. at an
The Range in the ex-treme south. elevation of about 4,000 ft., with isolated peaks rising here and
there to 5,000 and even 6,000 ft. Agasthiar peak (6,200 ft.)
Mahêntbragiri (5,500 ft.)

50. From the main range or axis, and from the western watershed of the
Peermerd plateau and the High Range, rocky spurs run out
Mountain spurs run-ning out from the main range. to the West and North-west, extending in some instances to
within a short distance of the sea, and forming a series of
parallel valleys drained by numerous rivers.

51. Speaking generally the plateaux and higher hills are covered with grass,
or consist of bare rock, the forest being confined to the hollows,
Vegetation on the hills. but there are exceptions in the case of the tracts where the
Periyâr and Kôtha River take their sources, which are covered
with heavy forest matted with reeds of various kinds, which it is difficult to penetrate.

52. The slopes of·the hills on the other hand are covered, or were originally covered before they were cleared for cultivation, with heavy moist forest containing a great, variety of trees.

Vegetation on the Hill-slopes.

53. At the foot of the Hills, the low country commences and extends to the sea coast, varying in width from 15 to 30 miles, and consisting chiefly of flat land interspersed with rolling hills which seldom rise to 200 ft. above their surrounding valleys. It is here that the greater part of the population resides, as I have already shown (paras 6 and 22), its density increasing in the neighbourhood of the sea.

The low country.

54. Travancore is watered by eight principal rivers and many minor ones. During the wet weather these are filled from bank to bank with a large volume of water rolling in a strong current to the sea,. but in the dry weather the water runs very low, especially in those streams which take their sources on the outer slopes of the Hills, or in the low country itself. In the larger rivers—the Periyār, the Rāni and the Manimalla—there is always a considerable amount of water, due no doubt to the fact that the heavy forest at their sources does not allow the rain which falls in the wet weather to run off too rapidly. Thanks to the heavy rain-fall thus stored up and conserved, almost every part of Travancore is amply supplied with water, and that too of an excellent character.

The Rivers of Travan- core.

55. The Geology of Travancore has been studied by Dr. King of the Geological Survey, from whose Report this description is taken.*

The Geology.

56. The Hills so far as they have been examined are composed entirely of igneous rocks of the gneiss series.

Composition of the Hills.

"These gneisses are generally of the massive grey section of the series and" "are generally quartzose rocks. So quartzose are they, that there are locally, fre-" "quent thin beds of nearly pure quartz rock which are at times very like reefs of" "vein quartz."

57. It was to examine these thin beds of quartz, in the hope that they might contain gold, that the services of Dr. King were obtained from the British Government. Dr. King's report goes on to say that on the area inspected by him there are "practically no" "auriferous quartz-reefs though the quartz itself gives the very" "faintest traces of gold" and that it is unlikely that gold in paying quantities will be found.

No prospect of gold oc- curring in paying quanti- ties.

58. In several places and especially in the neighbourhood of Mlāppāra on the Peermerd plateau, near Kouniyūr and in other localities quartz is found abundantly lying about the hill- side in blocks.

Abundance of scatter- ed quarts blocks.

59. The gneiss rocks of South Travancore are remarkable for the occurrence in them of large numbers of garnets of very fine colour, but of small size.

The Garnets of South Travancore.

60. "The great feature about the gneisses in Travancore, and indeed also in" "Cochin and Malabar, is their extraordinary tendency to wea-" "ther or decompose, generally into white, yellow, or reddish" "felspathic clayey rocks, which in many places, and often very extensively, ulti-" "mately become what is here always called laterite."

Laterite.

61. The question of the origin of laterite is still under dispute, but whatever the decision may be as regards all descriptions of this substance, it is certain that some kinds have been formed by the decom- position of rocks, and this process of formation can often be seen, in cuttings on the road side, partially completed.

Its origin still under dispute.

62. Laterite is largely used in the construction of buildings, as, though soft when dug out of the quarries, it hardens with exposure, and be- comes like brick, but for high walls it is a dangerous material as it has a tendency to crush under great pressure. In different localities, it varies

Its uses.

---

* Logan's Malabar.

much in quality. The laterite obtained near Trevandrum is very variegated in appearance, being streaked with pink and white, and is poor in quality, but in North Travancore it is homogeneous and more uniform in colour.

63. In addition to the igneous rocks referred to, the only *Quilon and Warkalla beds.* other formations worth noticing are the Eocene rocks near Quilon, and the Warkalla beds, supposed to be of upper Tertiary age. Both are of small extent.

64. Plumbago is found in some parts of Travancore, as for instance, in the *Plumbago.* neighbourhood of Neduvengaud, but it is said to be mixed with impurities, and to be therefore unable to compete with the same article from Ceylon.

65. There is a very curious absence of lime-stone from all parts of the State, the more curious, because it is found in considerable quantities *Absence of lime and lime-stone.* in Tinnevelly and Madura within a very short distance of the Travancore boundary. Most of the soil too contains but a small proportion of this substance but in one of the estates in the Periyâr valley, the soil of which I had analysed some years ago, the percentage of lime in the sample was considerable.

66. The soils of Travancore vary considerably in character through the country. *The soils of Travancore* The best consist of a chocolate coloured loam with a certain quantity of sand intermixed : others again are gravelly, with laterite at a short distance below the surface, and others are very pale, almost white, with a large quantity of decaying felspar in them.

67. Taken all together they cannot be considered good. It is to the climate that Travancore is indebted for its luxuriant vegetation. The *As a rule they are not good.* soil on the High Range, where the rainfall is light and the risk from wash is less, is good, and it is here, if any where in Travancore, that lime-stone will probably be found. The most fertile soils are those of the valley of the Periyâr, especially on Peermerd. On the banks of the other large rivers, the Muvâttupura river, the Konniyûr river and others, there is a great depth of alluvial deposit, but organic matter is wanting.

6?. This absence of humus is more marked on the steep slopes of the higher ground, probably because the heavy rains wash it down into the *Absence of humus.* ravines; and this is a strong argument for the retention of forest on such slopes, for when covered with trees and vegetation the loss by wash in such positions is infinitesimal.

69. Mr. Vincent notices the same absence of organic *Absent also in Ceylon* matter in Ceylon.[*]

70. By far the poorest bit of soil in Travancore is found to the North of Nedumangâd in the open grass land at the foot of the Hills. *Great poverty of the soil North of Neduman-gâd.* The only parts of this country fit to be cultivated are the swampy hollows lying between the grassy ridges ; here the soil, improved by the ashes of the burnt grass washed down for generations, is able to yield small crops of paddy, which do a little more than pay for the cost of raising them. The hills around are covered with long grass or stunted trees, which show, as well as the appearance of the soil itself, its miserable character.

Climate &c.

71. The temperature of the low country of Travancore varies little all the year round. Thanks to the cool breezes which blow up from the *Temperature in the low country.* sea regularly every day during the hot weather, the thermometer seldom shows a greater heat than 90°, and in the coldest season it rarely falls below 70°. At the foot of the Hills, and in places out of the influence of the sea breeze, the thermometer rises 5° or 6° higher and sinks probably as much lower. Messrs Ward and Conner, indeed, register a heat of 107° and 108° experienced by them in North Travancore in 1819, but such temperatures must have been quite exceptional, and were recorded when in camp under canvas, so the figures I have given above show the maximum that may be expected.

---

[*] Report on the Forests of Ceylon.

72. On the Hills the temperature varies with the elevation. In the Periyâr valley near Peermerd, where the country is completely shut in by high land, the extreme range of temperature is very great, and varies from little more than 45° to over 90°, or upwards of 50° in the twenty four hours.

Temperature on the Hills.

73. At the higher elevations the air is naturally much cooler, and on the High Range the climate is quite that of a temperate region, the thermometer falling to 25°, or less, at night, and rising to 50° or 60° in the day.

Temperature at the higher elevations.

74. This equable character of the temperature would make the climate exactly similar all the year round, were it not marked at different times by wet or dry weather, which, rather than an occurrence of cold or heat, indicates the seasons.

Seasons marked by wet or dry weather.

75. The wet weather commences about the beginning of June, and daily rains, often very heavy, may be expected till the end of August. In September and October the rainfall becomes much lighter, and under the influence of bright sunshine and occasional showers vegetation is at its height. Towards the end of October the North East monsoon begins to assert itself, and though the mornings are usually fine all through November, a heavy shower often amounting to several inches, may be expected each after-noon. In December the weather becomes more settled, and with the new year the dry season may be said to have commenced, and this continues, the temperature gradually increasing, until the showers at the end of April and beginning of May cool the air. At intervals through the dry weather showers may be expected, and it is seldom that two months elapse without a little rain falling. March is the hottest month in the year.

The wet weather.

The dry weather.

76. During the continuance of the " Rains" the air is saturated with moisture, and at Allepey I have frequently noticed a difference of only half a degree between the wet and dry bulb thermometers. In the dry weather the ground is no doubt very dry, but the drought cannot be compared in intensity to what is experienced in other parts of India at the same season.

Saturation of the air in the wet weather.

77. As I have shown, the greater part of the rain registered in Travancore is brought by the South West monsoon winds, and falls between the middle of May and the end of August. The amount varies considerably, being less in South Travancore, and it gradually increases along the sea board to its Northern limit. The range of Hills too, which rises inland, offers an effectual barrier to the passage of the rain clouds farther East, and causes the precipitation of a greater quantity there than falls in the low country. A short distance beyond the edge of the Ghauts the rainfall sensibly decreases, and in the Peermerd plateau and the High Range the South West rainfall is very slight on their Eastern edges.

The rain that falls in Travancore is chiefly brought by the South West monsoon.

78. The rainfall of those parts of Travancore is almost entirely due to the North East monsoon, and the character of the rain that falls at this time of year is that it descends in sudden and very heavy showers, which often cause high floods and much damage to buildings and cultivation.

The North East monsoon supplies the rain on the North Eastern parts.

79. Rainfall on the coast or in its neighbourhood going from South to North.

Distribution of rain.

Rajâkkamangalam	...	...	...	40·7 inches.
Palpmanâbapuram	...	...	...	54·0 do.
Trivandrum	...	...	...	67·9 do.
Quilon	...	...	...	87·8 do.
Allepey	...	...	...	114·2 do.

* Administration Report 1888-9 Average of ten years.

Rainfall on the Western face of the Hills proceeding from South to North.

South Ashambu	...	...	...	(about) 100 inches.
* Central Ashambu	...	...	...	...(2,000 ft.) 130
* North Ashambu	...	...	...	...(2,600 ft.) 169
† Agasthier Peak	...	...	...	...(6,200 ft.) 199
* Pon Mudi Hills	...	...	...	...(2,600 ft.) 170
§ Peermerd Hills	...	...	...	...(3,300 ft.) 210

Rainfall in the Periyar Valley—Peermerd some miles East of the Ghauts (2,800 ft.) 60-100 inches.

Rainfall on the Eastern edge of the Ghauts on the Peermerd plateau (3,000 ft.) 40-80 inches (probably.)

Rainfall on the High Range (5,500-7,000 ft.) 40-100 inches.

80. The result of this almost perennial moisture, combined with a high temperature, on the vegetation is to stimulate it to an amazing

Heavy rainfall and heat very favourable to growth. degree. Probably nowhere in the world are the conditions of growth so favourable as in Travancore, and consequently we find the ground completely covered with trees, or shrubs, wherever it is not cleared for cultivation. No sooner has this cultivation ceased than the land becomes again covered with a dense growth of bushes. But for this fortunate state of things, the system of hill-cultivation, so wasteful in its processes, would have long ago reduced the greater part of Travancore to the condition of a desert.

81. In spite of the only moderate fertility of the soil, the trees attain a great height, while the Flora contains an unusual number of species.

Great size of the trees and extraordinary variety of species Advantages and disadvantages of this. Instead of finding 4 or 5 chief trees and 6 or 8 less abundant ones, as would be the case in a European forest, our forests often contain over a hundred different species varying in every conceivable manner. Such a variety is in one way an advantage, because a greater amount of timber can be grown on a given area, if the species are different, than if all the trees are the same, but it is a disadvantage in this way that, in all probability, only a small proportion of the timbers is known, or used by the neighbouring population, leaving a great many valueless species occupying land to the exclusion of their betters.

82. Owing to its rich vegetation Travancore presents a great contrast in appearance to the neighbouring district of Tinnevelly. Standing

Appearance of the country contrasted with that of Tinnevelly. on any of the higher peaks that mark the boundary between Travancore and Tinnevelly, it is possible to see at the same time these two portions of the Peninsula. To the East will be seen the dry level plains of Tinnevelly, unmarked, if the visit happens to be in the hot season, by any cultivation, and dotted here and there at long intervals with melancholy looking palmyras. But to the West, in spite of the drought, the forests will be seen to stretch from the summit of the hills to the low country in one unbroken sheet of evergreen trees, if they have not been cleared for cultivation. Along the foot of the Hills stretches a belt of deciduous forest with long grass between the trees, and beyond that again will be seen gardens, planted with mangoes, jacks, and other fruit trees, with belts of coconuts becoming more numerous in the direction of the coast.

83. Throughout Travancore the climatic conditions of the forests in the low country are so similar that the forests themselves do not differ

The low country forests very similar all through the State. materially, though certain trees are found in one part that are not found in another, or are replaced by others. Some curious instances of this occur, where a tree is found in great abundance in one or two valleys, and no where else in Travancore, or the world. In other places a tree or shrub will be found, which is very common in other parts of India, but is extremely rare with us. These anomalies are probably due to some slight difference in the character of the soil, which favours the growth of one tree and is unsuitable to another.

---

* Private registers.
† Observations taken by J. A. Brown for the Travancore Government average of 3 years
‡ Administration Report 1888-9 Average of ten years.

84. Of the illnesses most fatal in Travancore fever claims the greatest number of victims. Yet the kind of fever common among the people is not of a virulent type, except in certain places, and it proves fatal, more from the want of stamina of the patient, or from his inability to get a change, than from its dangerous character. The climate of the different parts of Travancore is, as I have shown, so similar that in order to get a complete change it is necessary to make a considerable journey, and this the sufferer has frequently neither the means nor the inclination to undertake. The worst parts of Travancore for fever are (1) Shencottah (2) The Periyâr valley on Peermerd, and all the country to the East of the Periyâr, including the Cardamom Hills. (3) The lower part of the Anjinâda valley. (4) The small part of the Kôttashêri valley belonging to Travancore. (5) The Idiyâra valley. These comprise in fact all those parts of Travancore which the hot weather sea breezes do not reach. In many other parts of the forests it is unwise to stay more than a fortnight without a change to the low country, but the danger of fever is not so great.

Most prevalent forms of sickness. Fever by far the most fatal. Its character.

85. The most feverish months are April, May and June, and in the low country jungles December, when the showers are beginning to cease, is considered very unhealthy.

The feverish months.

86. Cholera is not nearly so fatal in Travancore as it is in the drier districts of the Madras Presidency. This may be attributed to the fact that the heavy rains wash away its germs, and fill the tanks and streams with pure water. Nevertheless cholera is always more or less prevalent in Travancore, chiefly in the dry weather in the more open parts of the country, and among the fish-eating population of the sea coast.

Cholera less prevalent than in parts of the Presidency

87. Smallpox attacks a fairly large number of people every year, but it is seldom fatal, except where vaccination has not been properly carried out. Among the Hill people smallpox does much harm, as none of them are ever vaccinated, and when once this epidemic appears it often sweeps away a whole village, as these people do not know how to treat it, and having a greater dread of it than of any other illness, they usually abandon the sick to their fate, leave their huts, and go into camp in the jungle till all danger of infection is past.

Smallpox fatal chiefly among the Hillmen

88. Dysentery is prevalent in the wet weather on the hills, where the cold monsoon winds blowing on the body, when wetted by the rain, often produce diarrhœa and fever, which develope into dysentery. In the low country, owing to the high temperature, this disease is less common.

Dysentery common on the Hills.

89. It will be seen from what I have said, that the conditions of existence, owing to the high temperature and constant moisture, which encourage the growth of all kinds of vegetation, are very favourable. A man need only scratch the ground and sow paddy, or plant a few mangoes, jacks, and coconut trees, and for nine months in the year he can remain idle. The standard of comfort too is very low, as the wants of life are so easily supplied and consequently, though there is a great deal of poverty, there is no suffering. The lower classes lead a hand-to-mouth existence, spending each day what they earn, and keeping nothing for the morrow, but they never starve.

Conditions of existence very favourable, & standard of comfort low

90. The ease with which a living is made reacts upon the character of the people, and makes them most unenterprising. As workmen, the labouring classes of Travancore, those of the south excepted, are far inferior in intelligence to the corresponding classes in Tinnevelly and Madura, and though they work well for a time, they are less capable of continuous and sustained labour.

Effect upon the character of the people.

91. Travancore, in spite of its advantages, cannot be called a rich country in the sense of having a large accumulation of wealth. The property of even the most wealthy consists in fields and gardens, and though it is at times considerable, the whole of it is invested in immovable property leaving no spare capital for investment or for the

Travancore is not a rich country.

development of the resources of the country. Thus it becomes the duty of Government itself to support, if it does not also originate, every new enterprise.

92. Nothing can be truer than the remark made by the late Governor of Madras Sir M. E. Grant Duff that it is the "first duty of a landlord to be rich"; meaning that though agriculture affords a good investment for capital, it is speculative on account of the uncertainty of its returns, and therefore that there should always be a large capital in reserve to help the agriculturist in bad seasons. The advantage of such a provision is especially apparent in the case of new products and new enterprises, which seldom realise, at first, the expectations promised, however carefully the anticipated profits may be calculated.

Capital required for developing the resources of a country.

93. In such a state of things, when there is a large area of waste land, and when the amount of accumulated capital in the country is small, Government assistance and the retention of monopolies become a necessity, because they enable many persons possessing only small means to invest in agriculture, and to make a living with the support of Government capital, when they would otherwise be unable to invest at all, or by doing so would become the slaves of the money-lender. Withdraw the support of Government suddenly, and the land cultivated by the smaller ryots will either be abandoned or be absorbed in the property of the larger capitalists.

In the absence of capital Government has to supply it, or to retain a system of monopolies.

94. Until the year 1860 the Travancore Government held a monopoly of pepper, and the effect of its abolition would have been disastrous did not pepper cultivation occupy a peculiar position, in as much as the vines are grown in gardens in combination with other products, and therefore a large capital is not required for the production of the spice. Cardamoms are still a monopoly, which it has been suggested to abolish, and I shall have something to say about this later on. Tobacco, salt &c. are also monopolies, but in a different sense, and lastly we have timber, and here more than in any other country it is imperative in Government to preserve and improve its forests.

The pepper, cardamom and other monopolies, including timber.

95. Forest trees take long to grow, and it is probably half a century before the money sunk in planting them is finally recovered. In countries where there is a large amount of capital seeking investment at very low rates, it may be worth the while of the capitalist to expend some of his money in the planting of trees, but where there is no available capital, and where the rate of interest is 12 per cent, it is certainly the duty of Government to look to the future timber supply of the country, or the inevitable destruction of the forests, which must follow neglect, will prove a national calamity.

Forests do not offer an investment to the capitalist, and it is the duty of Government to protect them.

96. Plantations, when the trees are eventually felled, often prove very profitable, but the planting of trees, or preservation of forests should not be regarded by Government merely as investments; the result of a neglect of such works should be taken into consideration. At the present time we can sell teak at a profit, at 10 Rs. a candy : if it had to be imported from abroad the purchaser would have to pay at least 30 Rs. for the same quantity.

Plantations often very profitable, but they should not be regarded merely as objects for investment.

97. The character of the country, as I have described it, thus lends itself to a division suggested by the varying density of the population, the races that compose it, and differences of climate and vegetation in its two portions. In the low country the people are more or less crowded together, while on the hills they are scattered thinly over large areas : in the former the Malayalies predominate, in the latter there are few Malayalies, but more Europeans and Tamils and a certain number of hillmen, and even in the sub-alpine jungles, Mahomedans take the place of Hindus. In the low country again the climate is hotter and more equable and the rainfall less, while on the hills the air is cooler, the range of temperature wider, and the rainfall more copious. Lastly, the flora in these two parts of the country is more or less distinct.

The character of the country suggests a division into forests, and low country.

4

98. In the hill country I include the whole of the hills and their slopes, together with a fringe of forest land at the foot of them, which *The extent of the forest area just half the total area of the State.* has either not been cultivated at all, or is cultivated only at intervals: This may be called the forest area, and it extends over 3,544 square miles. The low country includes all the rest of Travancore; and, if my estimate of 7,000 square miles for the total area of the State is correct, it amounts to nearly 3,500 square miles, the forest line thus dividing the country almost exactly in two.

99. Hitherto I have been speaking of Travancore as a whole: in the next chapter and in future, my remarks will be confined to the forest *My remarks in future will be confined to the forest area.* area only, except where I have to refer incidentally to other portions of the country.

CHAPTER II.

THE FORESTS OF TRAVANCORE.

100. The forests of Travancore, judging by the character of the soil, the old trees still standing, and the decayed logs sometimes dug up in *The forests once extensive.* the low country, must have been much more extensive than they are now. Most probably the whole country was at one time covered with trees.

101. But, assuming this to have been the case at our period, it is quite impossible to ascertain the condition or extent of the forests at *Impossible to ascertain their extent at any given time.* any given date, because no regular examination has ever been made of them until the present time.

102. Messrs. Ward and Conner, as they passed through Travancore surveying it, noticed in their diaries that such and such places were cover-*No information obtainable from Messrs. Ward and Conner's memoir.* ed with forest, and that the country north of Trivandrum, for instance, was densely wooded, but they do not expressly mention the size of the timber, and the height of the trees, nor whether the ground was merely covered, as it is now in many places, with a thick scrub of bushes 4 or 5 feet high. Their descriptions therefore do not help us much.

103. But of one thing there can be no doubt, and that is, that even when Travancore was covered with trees, the forests in the north *Forests in the north and the South different.* must have been very different from those of the south.[*]

104. Travelling through the country, examining the flora and the soil as I went along, I could not help noticing a great difference between *Former condition of the forests of the south.* those of the north and south. The low country in the southern part of Travancore was at one time chiefly covered with long grass, with large trees scattered through it, as is the case all about Pālōda, Madathurakāni and so on, and land covered with this sort of forest being unsuitable for cultivation, the condition of these forests is very much what it has always been, except that they are thinner from having had the best trees cut out of them, while the grass itself is more rank for the same reason, that the stronger light encourages its growth.

105. But in north Travancore from the Rāni river northwards, as well as about Trivandrum and some other parts in the south, long *And of the forests in the north* grass is very scarce, the hills, rounded and less rocky here, being covered with scrub, indicating a better and moister soil and a wetter climate. At one time then, all this land must have been covered with a dense forest of large trees growing close together, with no grass under them. and this sort of land being suitable for cultivation, the whole of it in all easily accessible places has been cleared, and has been under cultivation at one time or another. Very old trees of this primeval forest may still be seen dotted about North Travancore, of such species as Aval (*Holoptelea integrifolia*), Ara-anjeli (*Antiaris toxicaria*), Peru (*Ailantus Malabaricus*), and so on, all of them trees which grow naturally only in moist forest.

---

[*] The classes of forest are different, but within each class the trees are the same. See para 125.

106. In some places the ground will be seen covered with short grass, looking
Exhausted land covered with short grass. beautifully green and fresh in the wet weather, but completely burnt up in the dry months. This grass is very different from the coarse grass of the hills, and shows that the ground has been cultivated over and over again till it possesses no fertility whatever. Land covered with this scant herbage may be seen all about Quilon, Adūr, Malayattūr and many other places.

107. When Travancore was first inhabited there can be no doubt that the
Probable extent of the forests in former times population was collected along the sea-board, or was scattered along the banks of the principal rivers, and communication must have been chiefly by boat. The rest of the country was a wilderness of trees. As the population increased, paths were made through the forest from village to village, and also over to Madura, and Tinnevelly, whence produce was brought into the country, and as these lines of communication became more used, villages sprang up along the main routes, but probably up to the end of last century all the rest of Travancore was covered with forest.

108. And here it is necessary to notice the occurrence of ruins of old buildings,
Frequent occurrence of old ruins in the forests. which are so often met with in travelling through the forests, especially of Central and North Travancore. As these places are quite uninhabited now, the natural inference is that the population in former times was much larger than it is at present, and this seems to have been the opinion of Messrs Ward and Conner, who state that in their time the wild animals were gradually, but surely, driving back the people to the sea coast, as since the disarming of the population at the beginning of the century they were quite powerless to keep the animals and especially the elephants in check.[*]

109. The disarming spoken of must be the disbandment of the Carnatic brigade
The occupation of these ruins must date back to a very distant period. in 1805, and of the local militia after the attack on Col. Macaulay in 1808,[§] and doubtless the effect of depriving the people of fire arms must have been to place them at the mercy of the wild beasts. There were however probably some other reasons which induced the population to leave the forests, indeed Lieut. Ward himself mentions in one part of his diary, that at Nellikkal, near the Shabari-mala pagoda, there were at the time of his visit numerous remains of buildings, but that the population was said to have deserted the place 300 years before.[†] We must therefore seek some further explanation of this and similar migrations.

110. I have shown that at one time the population was scattered along the sea
Reasons why the population should at one time have been more scattered than it is now. board, the banks of rivers, and the lines of communication with Tinnevelly and Madura, and that the rest of the country was covered with heavy forest. It seems to me very probable that many people preferred to go out into the jungle and form colonies there, which held very little communication with the rest of the State. In such places they would find a fertile virgin soil close to their doors, which would give them very large returns for a small expenditure of labour. Owing also to their isolated position, these villages would probably escape the taxgatherer, and in an unsettled state of the country, would avoid being mixed up in the petty wars and faction fights that no doubt occurred then. They must have suffered a great deal from fever, and from the invasions of elephants, in fact even at the present time the villages of the inland show signs of having been built so as to be out of the reach of these animals. Thus at Chēttakal, at Nūruvakkāda and at Rāni itself, the compounds surrounding the houses are protected by high stone revetments, rising 8 or 10 ft. above the path ways, and the houses themselves are only accessible by single pieces of stone let into the revetment, or by wooden bridges from one compound to another.

111. But, as population increased, the people in these remote parts began to
Population tends to concentrate under a more settled Government. lose the advantages of their secure position. First of all they would not be able to get land for clearing so near home: then they would find that the tax collector came for his share of the produce, and should their crops fail it would be difficult to get

---

[*] Travancore Records, Ward and Conner's memoir Vol. I p. p. 41.
[§] Shangoony Menon's History of Travancore.
[†] Lieut. Ward's diary October 8, 1818.

food from elsewhere, while in years of plenty they found that they could not dispose of their rice. Thus it yearly became less advantageous to live in these remote places.

112. Then again, in early times there were very many tracks through Travancore to Tinnevelly and Madura ; some of them must have been exceedingly bad, and none of them very good, but even in Lieut. Ward's time the traffic had begun to converge to a few good routes. At the present time, all the traffic of Travancore with the East passes along three roads the Arammula, the Shencottah, and the Peermerd ; these are cart roads, and even though it may be much longer to go, say from Kambam to Thodupura viâ; Mundakayam, yet the road itself is so much better than the old track, that transport by that way is cheaper.

Similarly, traffic has a tendency to converge to the best routes, and population would shift with it.

113. In Lieut. Ward's time there were many more routes than at present and these were available for pack bullocks, but several of them, like the pass from Ariyanāda to Pāpanāsam and from Udumbannūr to Thevārām were much less used than formerly, and the people settled along these lines of communication would leave them, as the traffic along them declined.

Some of the worst routes abandoned even in Lieut. Ward's time.

114. Again, as the paths and means of communication all about the country improved, the people would find that there was no object in living in feverish jungles for twelve months, when they could live in the large villages for 9 months, and for the other 3 could go out in the morning and clear land, or reap the crop, and return to their homes in the evening as they do now.

As means of communication improve the necessity of living in the jungles would become less.

115. Thus there would be a tendency to leave the forests as places of residence, and to settle in the villages. This process is still going on now, for instance, the small town of Thodupurai is (I am told) much more populous than it was 40 years ago, while outlying villages, like Velliamattam near it, are completely abandoned.

This migration from the forests to the more civilised parts still going on.

116. Lastly, many of the ruins found in the forests are those of shrines, and we have no evidence to show that they were ever visited except for perhaps one or two days in the year.

Many of the ruins those of shrines.

117. I do not think, therefore, that we are justified in concluding from the occurrence of ruins in the forest, that Travancore was ever more thickly populated than it was in 1818, when Lieut. Ward estimated that it contained nearly one million persons. That the population has changed its distribution, and converged to the villages and small towns there can be no doubt.

Population in former times was more scattered, but not necessarily larger.

118. I have endeavoured, by comparing the figures of the last census, with those given by Ward and Conner to ascertain if these conclusions can be supported by them, but I have not been able to arrive at any satisfactory result, because I do not know if the areas of the Taluqs are the same now as they were in 1818. In some cases I believe that alterations have been made. It must also be borne in mind that the change of residence has been rather from a remote and small village to a larger village in the same Taluq, and not from one Taluq to another.

No information to be obtained from statistics.

119. Up to the end of last century the quantity of timber exported from our forests, teak excepted, must have been very small. There has always been a steady demand for this timber, chiefly indeed for local consumption, but some of it also for export. At the time of the survey in 1816–20, a Deputy Conservator was stationed at Rāni, whose duty it was to see teak felled and brought down to Allepey, where it was sold by the Commercial Agent, who was also at that time the Conservator of forests. A little later than this, and perhaps then too, the Commercial Agent used to hold his office at Kōthamangalam in the wet weather, so as to properly supervise the felling of this valuable timber, and the collection of cardamoms, and in the dry weather the season's collections of both sorts of produce were sold at Allepey.

The demand for timber, teak excepted, was very small up to the end of the century.

17

120. In the last 40 years the population has doubled, and the people have made great advances in civilization, and, as a consequence, the demand for other sorts of timber has very considerably increased, while at the same time very large areas of land have been cleared for cultivation. While then the people have left the more remote forests, and come down to the civilised parts, the cultivated area has extended towards the East, and thus it is that the line of division between the cultivated area and the forests is so marked.

The more recent demand for timber and its results.

121. The dividing line which I have selected, and which, generally speaking, follows the limits of cultivation runs as follows :—

Definition of the forest line.

From Mahēnthragiri peak it runs East and then South, coinciding with the boundary between Travancore and Tinnevelly. After travelling South for about 10 miles, it leaves Tinnevelly and turns West and South-west following for 5 or 6 miles the foot of the rocky hills North of the Arammula pass. Still following the foot of this rocky spur (Tandaga mallay in the map), it turns North-east and runs for about 5 miles until within a short distance of Ananthapuram (Ananta waram).* From here it runs 4 or 5 miles North West, skirting the cultivation, and then 3 miles South West to near Mēlapputhūr. From here it runs about 7 miles West slightly North past Vadakkūr to near Ponmana. Thence it runs more to the North, and passes over the rocky ridge overhanging Kulashēkkrapuram (Molagaddy malay), and leaving Killyūr Kōnam on the left, strikes the Kōtha river a short distance above Mayitunni. From this point it follows the Kōtha for 2 miles down to the Arianāda dam. The line then turns North West, and runs nearly 7 miles in a stright line to Kōvilūr. From here it turns rather more to the North, and runs direct for some 8 miles to Parathippulli (Coeeodi), following the bandy road. Here it leaves the bandy road, and keeping to the East of it, crosses the Ariyanāda river one mile above the town of that name. It then runs about 12 miles North West, passing Enāthi at the distance of a mile, and leaving it on the left hand, it runs past Bombayikōnam at the distance of 2 miles, and then turning due North, crosses the Vāmanapuram river 3 miles above the town. It then proceeds due North through Kummil, and at a distance of 2 miles from Kodakkal, and to the East of it ; then North slightly East and crosses the Erūr–Kulathuppura road, 2 miles East of the former place ; then 5 miles North Northwest to the Kulatthupura river, and then for 2 miles the river is the boundary. The line then turns Northwest, and leaving Chālakkara on the left, runs to the letter " C " in Aunaycolum, a place rather to the South of Kōnniyūr. From here it goes North, and crosses the Acchankōvil river above Kalleli, then Northwest 5 miles till it touches the Kalār river which it follows down to its junction with the Rāni river at Kumarampērūr. From here it goes 10 miles Northwest to the Karuvālikkāda hill near Alappāra, and from here 5 miles almost due North to the Karuppalli hill, and on to the Mani mala river. The line then follows the river up to Kānnyirappalli. From this place to Eerāttapētta the new cart road is the boundary. From Eerāttapētta the direction is North Northwest for six miles to Kayūr, and thence Northwest 9 miles to where the bandy road crosses the ridge dividing the Pālāyi and Thodupura valleys near Kilanthara. After crossing this ridge the line turns Northeast and passing through Mrāla (Mirthala), strikes Udumbannūr at a distance of 11 miles, it then runs Northwest for 10 miles, and crosses the Vadakkan Ar 2 miles above its junction with the Shangaruppilla thōda. Here the direction is North for 4 miles, the line cutting across the entrance of the Mullaringāda valley. From this point the forest line runs West Northwest for some 20 miles, parallel to the Periyār, and at a distance of 4 miles from it, till it is within 3 miles Southeast of Malayāttūr. It then turns North, and crosses the Periyār 1¼ miles above Malayāttūr, and proceeds up the Illi thoda, the boundary between Travancore and Cochin. It then follows the same boundary round to the West for about 3 miles, and then proceeds to the Kōttasshēri river, which it strikes at Erāttanukkam on the Travancore–Cochin boundary after following a winding course for some miles. The Northern boundary of the forest area corresponds with the boundary between Travancore and Cochin, and Travancore and Coimbatore. Its Eastern boundary is identical with that between Travancore and British Territory except that a portion of the Shencottah Taluq is omitted, as it contains no forest land.

---

* The names in brakets (re those given in the map of Travancore,

122. The mountainous region enclosed within this boundary measures 3,544 square miles, as already stated, and is watered by 18 rivers of different sizes, which take their origin on these hills, and flow away to the low country and the sea.

Total area within the boundaries.

123. Owing to the character of this part of the country, it will be seen that it is completely protected on the North and East from the operations of timber smugglers, with the exception of the Eastern slopes of Mahēnthragiri, which contain very little good timber, and the Anjināda valley, which belongs to the Pūnniyātta chief, and which is not of large extent.* Every where else the natural outlet is to the West, and in transporting our timber to the coast the large rivers of North Travancore are an immense assistance.

The natural outlet for the timber is all to the West.

124. In describing the forests within the above boundaries I cannot do better than divide the whole region according to the catchment area of each river, a method which will enable me to describe the forests and the rivers themselves in detail, and afterwards to sum up the results of my explorations. It must not be supposed that this area includes all the waste land, or even, all the forests in Travancore. Outside this area, there is a considerable extent of land suitable only for forests, or grazing, such as the Vēlimala hills near Nagercoil, but these are isolated blocks which may be left out of consideration, until the rest of the forest area has been put under a thorough system of conservancy.

The Forest area to be divided into its river basins, for purposes of description.

125. From what has been already said, and from the analogy of other countries, it may be inferred that the forests of this region do not all present the same characteristics, and I have for convenience sake divided them into four classes.

Division of the forests into 4 classes according to their characters.

(1) Heavy moist forests of evergreen trees.

(2) Land originally covered with moist forest, but now overspread with scrub of various ages, the resulting growth after being abandoned by hill cultivators.

(3) Deciduous forest, with grass growing under the trees.

(4) Rock, and land covered with short grass, and useless for any purpose except pasture.

126. The first class of forest at one time extended all over the low country of North Travancore, but as it covered the best soils, it has been gradually cut down there, and is now confined to the slopes of the Hills, and to perhaps one-third of the upper hill plateaux. The trees composing it grow very close together, and exhibit an extraordinary variety of species, and owing to the absence of grass and to the fact that the trees themselves are evergreen, forest fires do very little harm here. In spite of the great choice of woods they offer, these forests are, as a rule, less valuable than the deciduous forests of class 3, the greater part of the timber being unknown; nevertheless some of the trees command a high price.

flora of the dense moist forest.

The following are the most important :—

Ebony—Diospyros melanoxylon. ?	Pāthangkŏlli—Pœciloneuron Iudicum.
Kambogam—Hopea parviflora.	Cotton—Bombax Malabaricum.
Anjili—Artocarpus hirsuta.	Chīni—Tetrameles nudiflora.
Jack—A. integrifolia	Enna—Dipterocarpus turbinatus.
White Cedar—Dysoxylum Malabaricum.	Payini—Vateria Indica.
	Vedi pilāva—Cullenia excelsa.
Red Cedar—Cedrela tuna.	Pōla—Sterculia alata.
Punna—Calophyllum tomentosum.	Mala-ūram—Pterospermum rubiginosum, and 2 other kinds.
Nānga—Mesua ferrea.	
Gamboge—Garcinia Cambogia and 4 other species.	Olan kāra—Elœocarpus serratus,and 2 other kinds.

* The Shencottah Depôt is supplied with timber carried over the ghauts from near Kallatthuppura, but the natural outlet of all these forests is to the West, and the smuggling of timber over to the East is not great on account of the difficulties of carriage.

Peru—Ailantus Malabaricus.
Dammer—Canarium strictum.
Vengkotta—Lophopetalum wighti-
            anum.
Kadapilāvu—Karrimia paniculata.
Mango—Mangifera Indica.
Redwood—Gluta Travancorica.
Thēn chēra—Semecarpus Travan-
            corica.
Ambalam—Spondias mangifera.
Shurali—Hardwickia pinnata.
Malam puli—Dialium ovoideum.
Kurangādi—Acrocarpus fraxinifo-
            lius.
Mūtta Kongu—Pygeum wightia-
            num.
Nāval—Eugenia, several species.

Mani marutha—Lagerstrœmia flos reginœ.
Kadamba—Adina cordifolia.
Pāla—Chrysophyllum roxburghianum.
Do.—Dichopsis elliptica.
Karin thuvara—Diospyros microphylla
    and insignis, and 2 other species.
Erilappāla—Alstonia scholaris (often
    found also in grass land.)
Nutmeg—Myristica, laurifolia and 4
            other species.
Cinnamon—Cinnamomum Zeylanicum
            and others.
Kola māva—Machilus macrantha.
Thōndi—Bischofia javanica.
Aval—Holoptelia integrifolia.
Ara ānjili—Antiaris toxicaria.
Fig—Ficus 9 or 10 species.

127. The second class of forest contains no timber of any value except Vāga
(Albizzia procera), as the bushes and scrub springing up after

Flora of the second class.

a burn are of useless kinds of trees, such as Ama—Trema
orientalis, Vattakanni—Macaranga Roxburghii, Clerodendron
infortunatum, and Mallotus albus. Vettilapatta—Callicarpa lanata, and so on.

All of these bushes or small trees are short lived, and after growing for perhaps 10 years give place to better kinds of trees.

In this class are included all lands cleared for cultivation of any sort, whether
for coffee, tea, rice, rāgi or other produce.

128. The third class consists chiefly of forest growing on poorish land lying at the
foot of the hills, and is very abundant in South Travancore.

Flora of the third class of deciduous forests

These grass forests are found also covering the ridges and higher ground, where the soil is too dry for the moist forest to grow.
A small part of the hill plateaux also is covered with forest of this description.

The deciduous forests contain a much smaller number of species of trees than
the moist forests, but their value is greater. The most important of these are .—

Teak—Tectona grandis.
Blackwood—Dalbergia latifolia.
Sandalwood—Santalum album.
Irūl—Xylia dolabriformis.
Vēnga—Pterocarpus marsupium.
Thēmbāva—Terminalia tomentosa.
Ven Teak—Lagerstrœmia lanceolata.
Meili—Vitex altissima.
Pūvan—Schleichera trijuga.
Vekkāli—Anogeissus latifolius.
Mullu Vēnga—Bridelia retusa.
Ven Marutha—Terminalia panicu-
            lata.
Thāni—T. belerica.
Nay Thēkka—Dillenia pentagyna.

Nux Vomica—Strychnos nux vomica.
Gallnut—Terminalia chebula.
Uthi—Odina wodier.
Kumbil—Gmelina arborea.
Pēra—Carya arborea.
Nelli—Phyllanthus emblica.
Vāga—Albizzia procera.
Mūra—Buchanania latifolia.
Chinna Kadamba—Stephegyne parvi-
            folia.
Mala uthi—Stereospermum xylocar-
            pum.
Murukka—Erythrina indica, and some
            others.

These forests suffer much from grass-fires which become yearly more intense
as the trees are felled, and the grass is thus encouraged to grow, while the branches
and tops of the logs left about feed the flames.

Fourth class.

129. The fourth class of forest is of course worthless, as
far as timber is concerned.

Rivers taken in order from the South.

130. Commencing at the Mahēnthragiri peak, from whence
I began to trace the forest line, the first portion of the forests to
be described is the drainage basin of the Hanamannathi river.

(1) *The Hanamannathi river.*

131. This river rises in the moist forest near the Mahēnthragiri peak at about 5,000 ft., and joined by several lesser streams, whose sources *Its sources.* are o.i the Eastern slopes ot the Mahēnthragiri hills, flows into British territory, and is there largely used for irrigating the paddy fields.

132. The Mahēntbragiri hills, comprise a considerable extent of forest land estimated by Mr. Hayne of the British Forest Department at *Area of its drainage* 6½ square miles. This seems to me a low estimate, and at the *basin.* time I visited the spot I put down the area at 15 square miles, or 10,000 acres, being 10 miles long by 3 at its widest part, with an average width of 1½ miles. In my estimate I probably include some land omitted by Mr. Hayne.

133. This tract is bounded on the West by the mountain backbone already referred to, which runs from 5,500 ft. to about 3,000 ft., dipping *Divisions of its area.* down at one place called the Chūralvari gap to 600 ft. It is divided into 6 sections, called Kūttāmpuli : Chūralvari : Kāttādi : Kānumpuli : Karumpāra : and Ilanthayādi ; taken in order from South to North.

134. The greater part of these hills is steep and rugged, and slopes directly from the ridge to the plain. They are divided by numerous *The slope of the land.* ravines which run parallel to each other in the direction of Panagudi, the streams uniting outside Travancore territory.

135. The soil is very poor over the major portion of this tract, the only excep *Its soil.* tion being on the coffee estates belonging to the late Mīrānji Miya Sahib situated in its North Western corner. It contains a great deal of rock and stone, and the slopes exhibit evidence of suffering severely from wash, the hill sides being deeply scored in many places.

136. The climate is dry, and the total rainfall probably does not exceed 40 inches, nearly all of which falls during the North-East monsoon. *Climate and rainfall.* At the time when the rest of Travancore is drenched in the heavy rain and thick mist of the South-West monsoon, the Mahēnthragiri hills do not get one drop of moisture, the only indication of the monsoon that reaches them, being the violent wind, that blows steadily down from the hills to the plain below. So strong is this wind that the forest trees growing on the slopes are stunted and twisted into abnormal shapes, and, where they are fully exposed to its force, never grow to any height.

137. Of the 15 square miles that constitute this forest tract, the largest portion, amounting to about 8 square miles, consists of rock, short grass, *The area divided accor* and worthless land, but it is possible that if this area was closed *ding to the 4 classes of* *forest.* and cattle were excluded, some of it might soon be clothed with scrub and bushes. About 6 square miles are covered with grass with trees and bushes growing through it (class 3). and about one square mile was originally forest, but has now all been cleared for coffee with the exception of a quar ter of the area which was too steep to be felled.

138. The most impoitant trees growing here are the tamarinds, which seem to have sprung up from seed dropped by herdsmen and cattle, *Its chief trees. The* and, in spite of the very high wind, grow and bear large crops, *tamarinds.* which are sold by auction, and yield a revenue of some Rs. 200 per annum.

139. Next in importance is the teak tree, which is not plentiful, and is only found between 800 and 1,500 ft. elevation. It suffers severely *Teak, its character.* from the high wind. The trees, however, though not growing large, are sufficiently so to yield timber for furniture, cart wheels, and other small articles of this kind.

140. Other trees and bushes found here are, *Dalbergia latifolia, Strychnos* *Other trees.* *nux vomica, Anogeissus latifolius, Givotia rottleriformis, Grewia* *tiliæfolia, Pongamia glabra, Acacia latronum, Euphorbia antiquorum, Cassia auriculata, Mundulea suberosa,* in the drier parts, and *Stereospermum chelonoides, Bischofia javanica, Lagerstræmia lanceolata, Scolopia crenata, Mallotus albus, Eugenia,* several species, *Albizzia* 2 species, *Citrus,* and *Glochidion,* in the moister forests.

141. As already stated, there are 2 or 3 coffee estates in the North-western corner of these hills, which used at one time to yield very heavy crops, but owing to leaf disease, and one thing and another, they had, at the time of my visit, all the appearance of having been abandoned, and were quite deserted. The older parts were planted by a former Collector of Tinnevelly, upwards of 40 years ago, and they form the best portions of the estates. The newer clearings seem to have been more hurriedly opened, and to have suffered severely from the wind. Many parts are planted with cloves, nutmegs, pomeloes, and other fruit trees which seem to thrive well. The elevation of the estate is about 2,500 ft. above sea level.

Estates within the area.

142. A bungalow was built by the Forest Department about 5 years ago close to the Chûralvari gap, at an elevation of about 500 ft., but the choice of the site was unfortunate, as the wind whistles through the gap with frightful violence during June and July, and at the time of my visit, it had blown off a great portion of the roofing, which consists of tiles. As there is a very excellent chattram at Panagudi only four miles away, this bungalow has never been, and is little likely ever to be used, so its destruction is not a matter of great importance as regards accomodation for a Forest Officer inspecting these hills.

The forest bungalow at Chûralvari.

143. With the exception of the maistrees and coolies employed on the coffee estates, and at the time of my visit there were none, there is no population resident within the area of the Mahênthragiri hills.

No resident population.

144. These hills are under the charge of an Aminadar stationed at Panagudi who has under him, a Gumastha, 6 Peons and 2 Watchmen, and the total pay of the staff amounted in 1886 to Rs. 70 a month. The returns are about Rs. 900 a year, realised from the sale of tamarinds, fire wood, leaves for manure, and cattle fees, so that this portion of the forests is self supporting. In addition to the above the Aminadar has charge of certain forests on the western slopes of the mountain back bone, and for these, 2 Watchmen are allowed but this land yields no return, firewood and grazing being permitted free to Travancore subjects.

Management of the area.

145. The chief path that traverses this area is a very rugged foot path running from Ananthapuram through the Chûralvari gap to Panagudi, a distance of about 8 miles. Another path leads from Panagudi up to the coffee estates, and from there it passed on to Ashambu over the hill, but, since the abandonment of the coffee estates in the neighbourhood of Mahênthragiri, it has been closed. Another path runs from the same coffee estates above Panagudi in a Southerly direction, and crossing the ridge descends to Ananthapuram, through the Kunimuttha Chôla estate. This is very rough and steep.

Paths through this area.

146. Owing to the light rainfall in this part of the country, the streams contain but little water, except during the North East monsoon, when the greater part of the rain falls, and as these hill slopes are very bare, the rain that falls on them rapidly runs off, much to the chagrin of the villagers in the plain below, whom a better and more regular supply of water would greatly benefit.

Owing to the denudation of the slopes, the rain that falls soon runs off.

147. The Collector of Tinnevelly, aware of this state of things, and anxious to improve the water supply of the southern portion of his district, proposed to the Travancore Government that these hills should be rented to him for Rs. 1,000 a year, and in November 1881 the late Dewan offered him a lease of the Mahênthragiri hills above Panagudi for 3 years at Rs. 1,200 a year. This offer the Collector was disposed to accept, but the Advocate General at Madras, to whom the matter was referred for his opinion, said that as long as the land belonged to Travancore, all offences and crimes committed there would have to be tried under Travancore law, and as we had not, and have not still, any law corresponding to the Madras Forest Act for punishing forest offences, the negotiations fell through, the Travancore Government promising however to protect the Mahênthragiri slopes as best they could.

Proposal by the Collector of Tinnevelly to rent these slopes. The project abandoned.

148. At that time these hills were under the control of the Dewan Peishcar of
the south, and they were then transferred from his authority,
and placed in the charge of the Conservator of Forests. Previous to the transfer, the trees had been largely felled on these
hills, though indeed the class of timber, in consequence of the
dry climate, and very high winds prevailing there, could never have been very good.
When the Conservator took charge of this tract all felling of timber was stopped, as
was also the system of cultivating the lower slopes, and the burning of wood for
charcoal.

Changes in the manage-ment with a view to pre-serving the scrub and re-aforesting the slopes.

149. Beyond this very little seems to have been done,
and the villagers of the plain below come up and collect firewood
and carry away loads of leaves for manure, or burn the grass,
and bring their cattle to graze as they did formerly.

The villagers come up much as they always did.

150. I had been informed before I visited these hills that I should find the
bare slopes covered with small plants of many species, as the
Aminadar had received orders to sow the seeds of indigenous
trees there, but up to the date of my visit in June 1886, these
orders had not been carried out, indeed the Aminadar pleaded
ignorance of any such orders.

The proposed sowing of the slopes with the seeds of trees never carried out.

151. When I met Col. Campbell Walker, one of the Conservators of Forests of
the Madras Presidency, at Courtallum in 1886, he complained
very much that no steps had been taken for the proper conservancy of these hill forests, and said they had very much deterio-
rated within his recollection.

Complaints made by British Forest officers.

152. The great objection to the closing of these forests is that it would cost
money, as we should have to increase the number of guards, and
to pay for this extra staff, we should get no immediate return. The
proximity of British territory, where all the people who collect
forest produce from these hills live, and whither they can easily escape, makes the
difficulty of capture, and of punishing offences against any forest law very great. At
the present time, the villagers from Panagudi graze their cattle on the borders of
Travancore, and within it, and as soon as they see any one coming, they drive their
animals across the line, and so avoid the payment of fees. I am told also, that they
have even been known to go in numbers, and forcibly release cattle which had been
impounded for grazing in Travancore without due payment.

Difficulties in the way of proper conservancy.

153. These people depend entirely on the Mahēnthragiri hill forests for their
firewood, for leaf manure, and for grazing. It would therefore
be impossible to close the whole area at once. I think however
that it would not be difficult to shut off some of the higher land and refuse admittance
to any one there, until the scrub had grown up to a good height. Grass fires should
be rigorously excluded. When the growth on the reserved portion had reached a
fair height, firewood collectors might be cautiously admitted, and another part of these
hills should be closed, but we should always endeavour to prevent grazing and the
collection of leaf manure (both of which habits very rapidly ruin the bushes,) from all
the land that has been once closed for conservancy.

Suggestions offered.

154. The boundary between Travancore and Tinnevelly should be always kept
open. In consequence of disputes some years ago it was carefully surveyed and demarcated, but many of the stones have
fallen down, and the line has never been cleared again, and the
consequence is that when I visited these hills again in 1888 I found that the villagers in British territory laid claim to land which, judging by the map, seemed to me
to certainly belong to Travancore.

Necessity for keeping the boundary clear.

(2) The Palli or Vadasheri river.

155. Crossing from the Eastern to the Western side of the Mahēnthragiri
ridge, I passed from the basin of the Hanuman Nathi river to
that of the Palli or Vadashēri, the most southerly river in Travancore.

The Palli river.

156. If the streams that combine to form this river, one takes its rise not far
from the Mahēnthragiri peak and South of it, at an elevation of
Its sources and course. more than 3,000 ft., and passing down a steep gorge with coffee
estates on each side of it, reaches the low country a short distance to the west of
Ananthapuram : another rises in the Kunimuthu Chōla Estate : another drains
Mr. Cox's Estate " Black Rock " : another the same gentleman's Estate " Olivers, "
and the level ground along the foot of it, while two smaller ones drain the rocky spur
called Poyuga Mala (Tandaga Mullay.)

157. All these streams pass out of the forest area before uniting to form the
main river, and the land drained by them is of small extent,
Its area divided accor- and amounts to only 38 square miles. This may be portioned
ding to the classes of out among the 4 classes of forests as follows (1) heavy forest
forest on it. in small detached pieces, nowhere in large blocks, 2 square
miles, (2) Coffee Estates, plantain gardens, and secondary growth after the clearings of Hill-men, 6 square miles, (3) Dry grass forest with trees scattered through
it, 20 square miles, and (4) Rock and short grass, 10 square miles.

158. The chief trees growing within this area are teak and blackwood. The
former is very abundant but of small size, and sometime extends
Teak growing within over large areas to the exclusion of every other tree. This,
this area, and its chai- the teak seems to·be able to do, and to survive where no other
acter. tree will grow, from its ability to stand prolonged drought. In
such circumstances the trees seldom grow more than 20 ft. in height and 6 inches in
diameter and begin to branch within 5 feet of the ground. This "Kōl teak,"·as·it is
called, is very abundant all along the edge of the cultivation about Ananthapuram,
on Poyuga Mala, and on the slopes above Aragiyapannyapuram. It also occurs as
the chief tree near the Vīrappuli timber depot, and where the soil is good it attains a
fair height and girth.

159. The zone of the blackwood tree, as pointed out by Dr. Cleghorn in the
description of his trip to the Ana mala hills, is above that of the
Blackwood. teak, and the former tree is found in considerable abundance,
and of excellent quality, about half way up the hills, being especially common on
some of the coffee estates in the Olakkara valley, where it has been left for shade.[*]
It is also found but more rarely on the Poyuga Mala range of rocky hills, and in the
drier forests near the depot.

160. The Vēnga *(Pterocarpus marsupium)* is met with on the grass-hills above
the depot, and attains a good girth here, but owing to the wind,
Vēnga. its natural tendency to branch low is aggravated, and the available quantity of this fine timber is small.

161. Vekkāli *(Anogeissus latijolius)* occurs in great abundance on the slopes of
Poyuga Mala, but does not attain a large size, on account of the
Vekkāli. poverty of the soil. In other parts of the grass forests it is also
found but less commonly.

162. Thēmbāva *(Terminalia tomentosa)* occurs in the grass forest in the neighbourhood of the depot, but it has been largely felled, and is now
Thēmbāva far from common.

163. The valuable Kongu *(Hopea parviflora)* is extremely rare within the area
I am describing. Growing in the moist forest, nearly all of
Kongu. which has been cleared for cultivation, and felled on every
opportunity, it is now found only on the banks of rivers, or in places where access is
difficult.

164. Of the other and more useful trees *Lagerstrœmia lanceolata, Adina
cordifolia, Vitex altissima, Stereospermum chelonoides, Bridelia
Other trees. retusa*, are not uncommon. *Terminalia belerica*, a tree of very
large size, is conspicuous by its height so much over-topping the trees around it.

---

[*] Cleghorn's Gardens and Forests of South India.

165. The presence of *Filicium decipiens* on the Kunimuttha Cherᵣ-ᵥ ate is noteworthy, as it is a tree that grows only in a dry climate, and is not found elsewhere in Travancore except to the East of the Periyār on the Peermerd plateau.

Filicium decipiens.

166. Parts of the forests within this area, on account of the poverty of the soil, and the abundance of rock, as on Poyugá Mala are poor, the trees being stunted. In the better portions near the depot, where the soil is richer, the timber has been largely felled, so that the quantity available from these hills, Kōl Teak excepted, is not large.

Character of the forests

167. The rocky hills which form the dividing ridge between the basin of the Palliār, and the slopes above Panagudi, and those to the North of Black Rock run up to about 3,500 ft. (Wudda Mullay) is a rocky hill about 3,000 ft. high. (Tandaga Mullay) about 2,500 ft. (Vallaut Mullay) a bare rock some 3 miles to the North West of the Virappuli depot, and a conspicuous point, is 600 ft. These slopes are for the most part steep and very rocky. In the more level portions the soil is decidedly good, but elsewhere the gneiss crops too much to the surface, and stones are very abundant. It errs on the side of being too sandy rather than too clayey.

The highest hills and the slopes in this area.

168. The rainfall is light, and ranges from 30 to 80 inches, well distributed through the year. During the monsoons and especially in the months of January and February, the wind blows with extreme violence at the higher elevations, and does much damage to the coffee estates exposed to its force.

Rainfall and wind.

169. There are no villages within this area, with the exception of the ones belonging to the Hillmen or Kānikkār, who do not remain in any fixed place, but move about, changing the site of their habitations every few years. As a rule, these people do not lay any claim to the land they cultivate, beyond a prescriptive right drawn from a frequent occupancy of the same hill slopes extending over a very long time, but in the case of these particular Hillmen, they claim to have a title to certain valleys, which was given to them by a former Rajah, and on the strength of this they have sold portions of their land, and on others claim a sum of one rupee per acre a year from any one who likes to cultivate it. What the limits of their territory are no one knows, and it is very probable that they have no valid title to any but a small area of land, and that they get money for the lease of certain portions which do not really belong to them. The one colony of these people numbers about 30 persons.

The number of the Hillmen located here.

170. Besides these Kānikkār, there is a resident population of perhaps 200 persons employed on the coffee estates, a number which is considerably augmented during crop time, and may then be treblal or quadrupled. A few villagers live on the plantain gardens at the foot of the hill slopes, but they return to their homes when the crop is gathered, and their number fluctuates greatly. Altogether we may take the average population of this tract at 350 souls.

Population living in the coffee estates.

171. The most important estates are Black Rock and Olivers belonging to Mr. Cox. The former is one of the oldest in the country, and portions of it still bear well. Besides these I must mention the experimental garden opened by General Cullen at Ashambu, which subsequently passed into the hands of H. H. the late Maharajah. The older part of it, a very small area, used to give very heavy crops of coffee, but the portions more recently opened suffer from wind, and have not been so successful. There are also 8 or 10 native estates, most of them abandoned, but one or two, where shade was retained, still do well.

Coffee estates

172. These forests are partly under the Mahēnthragiri Aminadar, and partly under the Aminadar in charge of the Virappuli depot. The boundary between them is the stream coming down from the Ashambu and Victoria estates. The southern portion under the former Aminadar is not worked for timber, but fuel gatherers and cattle are allowed into it and the returns are nil.

Management of these forests.

173. The Vīrappuli depôt used, until about 4 years ago, to be under the Peishcar of the Southern Division. It is now placed under the
Recent change in the supervision of the assistant Conservator of Forests stationed at
control. Nagercoil. The pay of the establishment here was at the time
of my visit Rs. 43 a month, the staff consisting of an aminadar, a gumastha and 4
peons.

174. The forest on the Parali ār controlled by this aminadar has been heavily
worked in former years, and most of the timber brought to the
These forests hard work- depôt is felled beyond its limits. In 1886 the custom was for
ed the Conservator to engage 2 or 3 contractors to fell, and saw up
timber in the forests, and to bring it to the depot, where it was sold at fixed rates,
and I believe this method is still in force. The returns vary between Rs. 9,000 and
Rs. 12,000 a year, and the total expenses amount to about one half the receipts. The
timbers mostly brought to the depot are teak, kongu, vēnga and thēmbāva. A
tax was levied on fuel at the time of my visit, but I believe this has since been abolished.

175. Part of this forest, so much of it in fact as is contained between the cart
road from Vīrappuli to the hills, and from the same place to
Part of these forests to Chorlakkōda, I have recommended to be reserved. This forest
be reserved. is called Valang-kunna, and consists of undulating grass land
with numerous trees, and especially a large number of young teak, growing up
through it.

176. The large quantity of Kōl-teak found on these hills deserves some attention. It would very much benefit the trees that are left if some
Large quantity of Kōl- of them were thinned out. A ready sale would be found for
teak deserves attention these thinnings which answer admirably for furniture or cart
wheels. Unfortunately these patches of Kōl-teak are much scattered, and it would require some consideration to determine how they should best be worked, and looked
after. The Assistant Conservator might be asked to report on this subject.

177. Besides the tracks mentioned in describing the forests of the Mahēnthia-
giri slopes, good bridle paths run up to the Ashambu, and Black
Roads and paths. rock estates, and a cart road connects the Olivers estate with the
large bandy road running to the hills. This passes the Vīrappuli depôt in a Northerly direction towards the hills, and Southwards to Nagercoil. The more open parts
of the forest are easily accessible for foot passengers, and do not require made paths
to inspect them.

(3) *The Parali or Thāmravarani river.*

178. Proceeding in a Northerly direction the next river which drains part of
the forest area is the Parali or Thāmravarni. The main branch
Sources and course of of this river rises at 4,000 ft. on the Southern slopes of the
the Thāmravarni river. plateau of Muthukuri Vayal, and collecting the numerous
streams that water the large valley called Shāmbakkal, below Valiya mala, it descends
rapidly to Kālikkayam, where the more level land is reached ; two miles farther on
it is joined at Māsappidi by its other branch formed by the junction of 3 large streams
which descend from the Mahēnthragiri, the Rora (Swāmikurichchi) and the Corric-
mony (Māra mala) valleys. The Parali river then runs West with a slight inclination to the South, and passes out of the forest area a mile above Ponmana. At this
point an ancient dam, which has been recently repaired, crosses the river, and diverts
a large portion of its waters into a channel which runs for 3 or 4 miles to the Palli ār
which river it largely increases, and enables to irrigate an extensive area of paddy
fields, which it could not otherwise do. The undiverted part of the river flows down
its old bed, and uniting with the Kōtha river above Kuritthura falls into the sea at
Thēngāpattanam. This river is too small either for navigation, or the floating of
timber.

179. The area watered by the Parali ār, which is included within the forest line, amounts to 71 square miles, which may be divided as
The division of its area follows (1) moist forest, 10 square miles, in small isolated blocks
according to the class of at the high elevations, or consisting of land too steep or too in-
forest on it. accessible for coffee cultivation (2) secondary forest, coffee estates,

Hillmens' clearings, &c. 16 square miles (3) grass forest, with large trees growing through it, 30 square miles, and (4) rock and grass land, 15 square miles.

180. Within this area are found some medium sized teak in moderate quantity,
Principal trees found here. both on the level land and on the lower slopes of the hills. Blackwood of large dimensions is found, being especially fine at about 1,800 ft. below the Corriemony coffee estate. At the same place vênga of good size occurs, and was being felled by the contractors at the time of my visit. On the other branch of the Parali är in the neighbourhood of Kälikkayam, there was a considerable quantity of very fine kongu growing at an elevation of about 1,400 ft., but the contractors have been working here extensively for a long time, and the supply must be getting short. Some excellent thêmbäva of large size, and straight bole, was to be found near the Mäsappidi bridge, and again in the direction of the depôt, but there has been a considerable demand for this timber lately, and it may well be getting less common. Of other trees I may mention *Mesua ferrea*, *Diospyrus melanoxylon ?*, *Bridelia retusa*, *Adina cordifolia*, *Vitex altissima*, *Stereospermum chelonoides*, *Logerstrœmia lanceolata*, and *Artocarpus hirsuta*.

181. In this valley there is a good deal of cultivation by Hillmen, who know-
Damage done to teak and blackwood by the Hillmen. ing that they may not fell the teak and blackwood, nevertheless when they find them in their clearings, lop the trees very severe-ly, that they may not shade the soil, nor cause drip. This does much harm to them.

182. The high hills, from which the streams that form the Parali är flow,
The chief peaks and hills. reach their greatest elevation to the North and North-east of this area, and culminate in Valia mala (Wallia malley) about 4,700 ft., a pointed pinnacle of rock, which rises at the upper end of the Shämbakkal valley, and in Mahênthragiri 5,500 ft. The mountain back-bone here runs at a considerable altitude, and seldom falls below 4,000 ft. From it very strong rocky spurs, covered with grass, run out towards the low country. Be-tween them the land slopes steeply at first, and more gently lower down, opening out into valleys, with only a slight incline, which were very well adapted for cultivation
Large extent taken up for coffee cultivation. with coffee. Completely covered with virgin forest 30 years ago, these valleys seemed to the pioneers of that enterprise ex-actly suited to their requirements, and the whole of the suitable land lying between 1,500 and 3,000 ft. was rapidly bought up. Altogether about 6,000 acres were sold by Government for coffee cultivation within this small area, and the greater part of the land was opened and planted. Of the 10 or 12 fine es-tates which once gave employment to so many coolies, only portions of 5 or 6 remain and these contain but a small area under coffee, the greater part being planted with tea.

183. The soil of the upper part of these valleys is free and good, but has been
The soil is good. much washed : and in the lower valleys and at the foot of the hills it is even better, as is testified by the fact that the Hill-men are often able to cultivate the same land for 6 or 7 years in succession, whereas in other places a move is made, at the least, every 2 or 3 years.

184. The rainfall is from 80–120 inches, and is well distributed. The climate
Rainfall, climate and wind. is the same as in the adjoining tract of country, and the wind is almost if not quite as bad. One or two of the coffee estates however, on account of their position, manage to escape it. Chorlakkôda at the foot of the hills, and the forests in the neighbourhood of Mäsa-ppidi bore at one time a bad reputation for fever, but this part of the country seems to be much healthier now.

185. The resident population consists of about 270 Hillmen, who are located
The resident population. chiefly in the long valley which runs down to the Parali är on its North side, and lies to the West of the Kuru mala hill (Moodamanuddy mullay), and some 500 maistries and coolies employed on the estates, making a total of about 800 souls.

186. In the Shämbakkal valley, which was some 12 years ago the head quar-
Objects of interest. ters of the coffee enterprise, a Post Office was opened, and a hos-pital was built by subscription in memory of John Grant one of

the first pioneers. These are still kept up. In the valley to the West of Kuru mala a keddah for capturing elephants was constructed many years ago, but I have never been able to discover anything about it.

187. The forest area included within the drainage basin of the Parali ār is

Management of the forests here.

partly under the Vīrappuli aminadar, whose staff I have already described, and partly under the Kalkolam aminadar, the boundary between their respective divisions being the main branch of the river, and the village of Ponmana. The latter functionary had under him in 1886 four peons whose salaries added to his own came to Rs, 35½ a month.

188. At the time of my visit the Kalkolam aminadar was not working the

Timber operations.

forests for timber, his duties being confined to the collection of beeswax and cardamoms from the hills, hence his small staff. Any timber felled within the area supervised by him was taken to the Vīrappuli depôt, but since that time I understand that a depôt has been opened at Kula-shēkkrapuram.

189. As stated above, the forests near the Vīrappuli depôt have been regularly

These forests have been worked for a long time, and new tracks are required to get the timber out.

worked, and yield about 1,000 candies per annum. The difficulty of getting timber becomes greater every year, as the contractors have to go further for it, and at the same time we have no means of knowing whether we are overcutting or undercutting the forests. By making some new bandy tracks we could certainly bring more timber to market, but the forests require to be more carefully examined in detail than I had time to do, so as to see how these tracks should be taken, and what amount of timber may be expected when they are completed.

190. Four years ago I recommended that the area enclosed between the Vīra-

Part of these forests recommended for reservation.

ppuli-Māsappidi road, the Vīrappuli-Choralakkōda road, the Pandian channel, and the Parali ār aggregating about 10 square miles, partly within the drainage basin of the Palli, and partly within that of the Parali river, should be reserved, but I have not heard that any steps have been taken to declare it.

191. The catchment area of the Parali ār is traversed by an excellent cart road,

Roads and paths. The main road from Nagercoil to the hills.

which starting from Nagercoil, passes the Vīrappuli depôt, and crosses the river at the Māsappidi bridge shortly above the junction of its two branches. Here the road forks, and one portion running along the left hand side of the Northern branch for 4 miles, crosses it by a ford at Kālikkayam, and ascends through the forest to the Balamore estate at 1,500 ft. where there is a rest-house. Beyond this, numerous bridle paths run in various directions, one of them continuing the ascent of the hi l as far as the plateau of Muthukuri vayal. The other portion of the cart road runs for several miles in the direction of the Māra mala estate, while a bridle path connects the Māsappidi bridge with the estates at Swami kuricchi.

192. From Vīrappuli another cart road runs to Chorlakkōda, and from thence

The Chorlakkōda road.

to Ponmana and Kulashēkkarapuram, skirting the forest area. At the time of my visit in 1886 a trace for a bandy road was being made from Ponmana to Kālikkayam, so as to avoid crossing and recrossing the river, but I have not heard that this trace has ever been converted into a road of the usual width.

(4) *The Kotha river.*

193. The Kōtha river takes its rise on the Southern extremity of the Muthu-

The sources and course of the Kōtha river.

kuri vayal plateau, and to the east of the Valiya mala (Walliamullay) peak, at an elevation of about 4,500 ft. It then runs sluggishly at a general altitude of 4,200 ft. and in a Northerly direction for 6 or 8 miles traversing the undulating table land, called by the above name. It then begins to descend, slowly at first, and spreading out at intervals into large still reaches, and then more rapidly, rushing down over large boulders. Passing under very steep banks it reaches an elevation of 1,800 ft. in the Mottacchi valley, after flowing for 14 miles. From here it continues to descend with great rapidity, tumbling over falls 30 ft. high, and eddying among huge boulders with which the river is studded. The banks on each side of the river are very steep and rocky and unsuited for cultivation. After descending 1,500 ft. in little more than two miles it passes out of the valley, but still continues to fall rapidly till an elevation of 250 ft. is reached. From here it begins to flow in a more leisurely manner, and its waters are joined by the Katār and Sātthār, which take their rise on the Motavan potta, and Thaccha mala hills to the South, and affect a junction before meeting the river. These streams are scarcely knee deep in the dry weather, but during the rains they become rapid rivers often impassable for days.

194. Proceeding in a southerly direction the Kōtha is joined on the North by
Its course on the borders of the forest area and outside it
a tributary which takes its rise in a narrow valley East of Klāmala (Callanmullay). About a mile below this, at a point where the river begins to fall again, are found the remains of the Ariyanāda dam, constructed of huge blocks of granite. The native engineers who planned this work, made the mistake of building it with an angle pointing down the river so as to take advantage of an island in the middle of the stream, but the water soon found the weak part in the building, and at some period before the recollection of the present generation the wall of huge granite slabs was breached. This dam was built with the intention of diverting some of the water into the Parali river above the Pāndian dam, and so eventually into the Palliār, whose stream is so largely used for irrigating the paddy fields of the Nānjināda. At the time of Lieut. Ward's visit the dam itself seems to have been in good order, but the object for which it was built had not been attained, and, from what I have since learned, I believe that this was because it was constructed at a point too low down the Kōtha to enable the water to run into the Parali ār. *After passing the Ariyanāda dam the Kōtha, here about 150 ft. above the sea, leaves the forest area. About 4 miles lower down where it is near Kulithura river 150 yards wide, it is precipitated over the Thrippārpu (Tirpanippoo) fall, the height of which is about 50 ft. This is considered a very sacred place, and there is a large pagoda, and bathing places of cut stone here. Continuing in a southerly direction the Kōtha is joined above Kulitthura (Cooletoray) by the Parali ār, and a few miles farther on falls into the sea at Thēngā pattanam (Thaingaputnum).

195. Below Thrippārpu small boats may sometimes be used for navigation du-
The Kōtha not suited for navigation
ring the wet weather, but in the dry months the water is confined to a small stream in the middle of a dry sandy bed, and the large girder bridge across the Kōtha near Kulithura would seem ridiculous, were it not known that the water some times rises 40 and 50 ft. high.

196. Portions of the river, as between the Ariyanāda dam and Thrippārpu, and
Nor suited for floating timber.
that place and the sea, may sometimes be used for floating timber, but it cannot be said to be either suitable for navigation or for the conveyance of logs.

197. The area drained by the Kōtha ār, which falls within the forest limits,
Area of the forests on the Kōtha river divided according to different classes.
may be estimated at about 99 square miles, to be divided as follows :— (1) moist forest, most of it on the Muthukuri plateau, 30 square miles. (2) Coffee estates, Hillmen's clearing, paddy fields and secondary growth, 20 square miles. (3) Parkland forest with large trees and grass, 30 square miles. (4) Useless land, rock and short grass, 19 square miles.

198. The Muthukuri vaynl plateau consists of an elevated basin covered with
The Muthukuri vayal plateau.
low hills, of no great height, and lying between two parallel ridges rising not more than 500 ft. above the level of the lowest part. It is some 7 miles long by 3 broad, and is almost entirely clothed with moist forest. The forest however is not so large, and does not contain trees of the same size as are found in the low country. Small open glades frequently occur, and the trees are often matted together with a dense growth of reeds or rattans. The trees themselves are not of any value, though several of them, such as the Iluppa (*Dichopsis elliptica*), supply useful building timbers.

---

* Lieut. Ward's diary January, 26, 1817.

199. At the South-western corner of this plateau there are several open patches, or vayals of short grass land, aggregating perhaps 1,000

Its suitability for a sanitarium. acres, and often frequented by elephants or bison. Here at an elevation of rather over 4,000 ft. His Highness the late Maha Rajah had a bungalow built, and there are besides this one or two small bungalows belonging to the L. M. S. Missionaries living at Nagercoil. From its easy accessibility, and the level nature of the ground, this place would make an excellent sanitarium, but it would be necessary, in order to make it more fit for habitation, to clear away some of the wood.

200. The plateau takes its name from certain pits to be seen in the largest

Origin of the name. patch of grass land, which according to tradition were dug for precious stones, though it is most probable that they are merely old elephant pits.

201. In the Mottacchi valley, and on the outer slopes of the rocky ridge that

Very little original forest on the slopes. encloses the Kōtha on its western side. most of the timber has been felled, and very little original forest now remains except at high elevations. At one time kongu and anjili must have been abundant, and of very large size, the few trees still remaining attesting this.

202. The forests of the level land at the foot of the hills and on their lower

Character of the forests at the foot of the slopes. Teak. slopes, contain for the most part deciduous trees, and belong to the third class. Here we find the teak tree, which was at one time abundant near Kākkacchāl, but is now met with in ones and twos, or small clumps, and not of large size. It occurs chiefly in the valley of the Sāthār, and on the outer slopes of the Thaccham mala ridge and sometimes attains a diameter of 18 inches. The total quantity of this timber is not large, for immediately on crossing the Kōtha to its western bank teak is no longer seen.

203. Blackwood is said to be abundant in one place on the western bank of the

Blackwood found in these forests. Kōtha, and in different places on the slopes of the hills it attains a large size. About a dozen years ago 70 or 80 logs of good girth were felled in various parts of the forests, but there was no demand for the timber, and when I visited these parts in 1886 I saw the logs lying about near where they had been felled. I believe they have since been brought to the depôt.

204. Kongu is found growing on the banks of the Kōtha, and before the es-

Kongu scarce. tates were opened it was very numerous in the moist forests. It is getting very scarce now.

205. Vēnga, once abundant in the sub-alpine forests has

Vēnga has been largely felled. been so largely felled, that it is scarce in all easily accessible places, but is found at a distance from the roads.

206. Thēmbāva is common, and grows particularly tall and straight. The soil

Thēmbāva and other trees seems exactly suited to it. Of other trees *Dysoxylum malabaricum, Diospyros melanoxylon, Mesua ferrea, Adina cordifolia,* and others are found in the lower moist forests, and *Lagerstrœmia lanceolata,* and *Terminalia chebula* in the grass land.

207. I have already described the country where the Kōtha takes its rise.

More important hills and high land to the East of the Kōtha. Leaving the Muthukuri plateau, it finds itself in the Mottacchi valley, in a very rugged country : on its left is an immense bastion of rock, with a sheer fall of fully 2,000 ft. In the centre of the valley is the Mottacchi peak, a bare rock 4,500 ft., and at its northern end, whence a strong stream runs, the peaks rise to over 5,000 ft. From this elevated land a very high ridge of grass and rock extends in a south easterly direction shutting out the sea and the low country from view, the grass is short and gives excellent feeding to large herds of ibex. This ridge which is 5,000 ft. at its upper end is precipitous to the west. Below it lies a long valley sloping to the Kōtha and covered. mostly with grass. Beyond this again, and between it and Klā mala (Cullan mullay), is a grassy plateau, with an elevation of 3,000 ft.

8

called Vengalam mala. Klá mala itself is a razor edged rock precipitous on both
sides and rising to 2,500 ft. Outside it and nearer the sea is yet another hill of lesser elevation, called by the Hillmen Kuniccha mala.

208. To the East of the Kōtha the most conspicuous hill is Thaecham mala
(Sucha mullay), which rises to a rough grassy peak, between
Hills to the East of the river. the Kalár and Sathár, to an elevation of 2,000 ft. To the north-
west of this hill, but still on the East of the river, is Mōthira
mala (Mothan mullay), a lower grass hill with a conical summit.

209. The ridge that divides the basin of the Kōtha from that of the Parali ár
consists chiefly of grass. Running out from the main range
The dividing ridge between the Kōtha and Parali. it ends at Kal Padava (Mauran mullay) 2,700 ft. Outside it
and detached from it is the conical hill called Kuru mala or
crown hill (Moodamanuddy mullay).

210. The slopes of these hills are all more or less steep, but the soil to the East
of the Kōtha is uniformly good. To the West of that river it
Great difference of soil on the two sides of the Kōtha. is exceedingly poor, as may be seen all along the road to Kōvilūr (Coviloor), and the timber is stunted, the trees being of the
same kinds as are found in the poor tract near Nedumangáda
and to the North of it.

211. The rainfall here is from 80–170 inches, being rather more than on the
hills draining into the Parali ár. The climate is the same as
Rainfall and climate. in that valley, and the wind blows at the same seasons, and
with equal force.

212. About 5,000 acres of forest land were sold for coffee cultivation within
the basin of the Kōtha. Not an acre of this land, the greater
Cultivation of coffee now abandoned. The Hill- men living on the slopes of the hills. part of which was in cultivation ten years ago, is now kept up,
and there is therefore no resident population on the upper hill
slopes. About 400 Hillmen, broken up into 15 or 16 villages,
inhabit this country. They are divided into Kānikár, who live in the interior and
Vālenmár who are confined to the outer portions of the hills: the latter mix more
with the low country people and are perhaps more civilised, but to an ordinary observer the points of distinction are minute. In the neighbourhood of Klá mala, the valleys are often planted with arecanut and jack trees, with oranges, citrons, and limes,
and also with the sappan (*Cæsalpinia sappan*), and here these hill races seem inclined
to settle down more, a tendency which is by all means to be encouraged.

213. Besides the Hillmen there are a few Náyars and Mahomedans settled at
Kannimámmudi (Cunnemamode), and Karinchira (Currinjur-
Resident population at the foot of the hills. ray), whose numbers may amount to 100, so that the total population of this basin may be set down at 500 persons. Once a week
a market is held at Alacchōla, a place near the two last named spots, but the people
who collect to sell or buy produce disperse by the evening and there is no resident
population there.

214. For about 3 years before my visit in 1886 the forests at the foot of the
hills near Mayilūnni (Mileoony) had been worked by Mr.
Timber operations in these forests. Change in the position of the depot. Chisholm, and a native, the timber being taken to Virappuli,
but soon afterwards a depôt was started at Kulashōkharapuram
(Colashagrapooram), and timber was sold there. This has now
been closed and a depôt has been opened at Nagercoil. Just outside the forest area,
and at a distance of 3 or 4 miles from Kōvilūr is Panicchamūda, a small village,
where there was a watch station at the time of my visit. Since then a small depôt
was opened there, but this has been closed, and all the wood felled in the neighbourhood is taken either to Trivandrum or Nagercoil for sale.

Rates of timber. 215. In the Appendix will be seen the rates paid to the
contractors, and by purchasers for the different kinds of wood.

216. A small quantity of wax and a few thoolams of cardamoms are annually
collected on these hills. The latter kind of produce is very
Wax and cardamoms. The latter much stolen. much stolen by the natives of Tinnevelly, who come up by paths
known only to themselves, and help themselves to this spice

without any one being a bit the wiser. Nor is it easy to know how to stop them. Cardamoms are scattered in small patches all through the forest, and it is quite out of the question to protect the whole of them. But as I shall discuss the subject of the cardamom monopoly later on, I will not pursue it further now.

217. I have not suggested the reservation of any land within the basin of the Kōtha for the reason that the forest is much intersected by the cultivation of Hillmen and it would not be easy to find a block sufficiently large to warrant the maintenance of a staff to protect it. The reserves which I have recommended cover a large area of ground, and will occupy the attention of the Department for some time in demarcating and guarding them. But the assistant Conservator of the South, who will have little of this work to do, might be directed to make a closer examination of these parts, with a view to reporting further on their capabilities.

No reserves recommended in these forests.

218. Forest fires are very harmful in the deciduous forests of this river, and should be prevented.

Prevalence of grass fires.

219. The valley of the Kōtha is well supplied with roads. A cart road runs from Kulithura past Kulashēkharapuram, and one branch continues on for 5 or 6 miles up to the Sātthār valley to the foot of Kalpadava estate. Another crosses the Sātthār and proceeds for 3 miles to the Kallār, which comes down from Mutavanpatta estate. From the end of these roads numerous bridle paths run up to the estates, others again intersecting and joining them. Since the abandonment of these properties the paths have doubless become much overgrown, but they are there, and would only need clearing to make them passable again. A bridle path known as the District road was cut at a level of about 2,000 ft., running along from estate to estate, but this, like the other, has been abandoned.

Roads and paths.

220. A cart road from Kulashēkharapuram crossing the Kōtha at Thripparpu runs to Kōvilūr along the edge of the forest area, and, from different points on it, tracks run into the forest, but a guide is always required to show whither they go.

The cart road to the west of the Kōtha.

(5) *The Ney Ar.*

221. North-west of the Kōtha river the Ney ār drains the forest area. Its largest branch rises on the slopes of the Agasthiar peak at an elevation of nearly 6,000 ft., and descending with great rapidity often falling over rocks 100 or 200 ft. high, it reaches the foot of the hills in about 4 miles, and continuing 5 or 6 miles further in a Southerly direction it is joined by the smaller branch, sometimes called the Kalār which rises in the Mala ād valley near the Nūruthōdapāra (Nooratodia bluff rock), at an altitude considerably over 5,000 ft., and passing downwards forms a large waterfall, 300 ft. high, visible from Trivandrum, and below what was the Mala ād coffee estate.

Its sources and course on the hills.

222. The two streams together form a small river, which proceeds in a Westerly direction, and winding about through the low hills, which occur here, passes out of the forest area shortly above Mūnnara (Moonurray), a small village of no importance.

Its course in the low country.

223. For some distance above this place the Ney ār flows so smoothly that timber can be floated down it, but there is none of any size, or of weight light enough to permit of its conveyance by water. Neither is there any traffic by this small river, whose waters sink so low in the dry weather that they are passed almost dry shod.

Suitable for floating timber.

224. The area drained by the Ney ār extends to 60 square miles of forest land. Only a very small portion of this land is covered with moist forest, a few thousand acres close under the highest peaks comprising its whole extent, which may be set down at 4 square miles. The area covered by secondary forest, abandoned coffee estates, and Hillmen's clearings, the last of which are very extensive, amounts to 20 square miles. The third class, comprising grass land with large trees, may be set down at 24 square miles, while the useless land and rock cover fully half that area.

The Ney ār basin divided into different classes of forest.

32

225. Of trees the most important growing in the moist forests are the jack, red-wood (*Gluta Travancorica*), and enna (*Dipterocarpus turbinatus*), which yields the gurjun oil largely used in Burmah for rheu-matism. Kongu is not found on the hills but is scattered spa-ringly along the river sides. In the grass land teak, ven teak, and vekkāli (*Anageissus latifolius*) are not found, and blackwood is very rare. The most useful are Vēnga, which here grows to a large size, Thēmbāva (*Terminalia tomentosa*), and mullu vēnga (*Bridelia retusa*), but the latter two only attain a small size compared to the dimen-sions they reach elsewhere.

The most important trees growing here.

226. The most important peaks and hills here are Agasthiar 6,200 ft., a cone of solid rock visible at sea from a great distance off, and held very sacred by the Hindus, the Nūruthōdapāra (Nooratodia bluff rock) and other points, which must be upwards of 5,500 ft. high, overhanging the Mala ād valley ; Vengalam mala is a grassy plateau already referred to, on the Western edge of which is a pointed bluff rock rising to over 2,000 ft. and visible from near Trivandrum (Chenkullu mullay). The other hills and ridges Pēkkulattha mala (Awe mullay), Kāttādi mala (Kautadi mullay), Kuru mala (Cooroo mullay), and so on, rise to 1,200 or 1,500 ft. above sea level, and are clothed with long grass.

The highest peaks and hills.

227. The higher slopes of the hills are very steep, and the soil is poor and stony with a stiff under stratum. On the more level ground the soil of the ridges, which are rounded, and not abrupt, contains much clay and laterite, and is decidedly poor, but the lower lands which have been fertilised for centuries by the ashes of the grass washed down from the slopes, are more suited for cultivation, and the Hillmen have settled here in large numbers, and that apparently in quite modern times, as Lieut. Ward in his diary says, that at the time of his visit, they were afraid to occupy these parts for fear of wild elephants.*

Soil on the hill slopes and in the valleys.

228. The rainfall is rather heavier than on the hills a little farther to the South, and varies from 120–220 inches. The climate is similar to what prevails there, while the wind is even worse. The North-East monsoon winds of the Agasthiar peak being notorious throughout the country. The valley of the Ney ār, from its open character and proximity to the sea is not unhealthy, and fever is but little prevalent here.

Rainfall and climate.

229. About 2,000 acres of forest land were sold for coffee cultivation, and several coffee estates were opened below the Agasthiar peak, and in the adjoining valley of Malayāda and again in the small valley under Thagara mala, still farther South. These have all been abandoned. There are no buildings or objects of interest here, and the population is confined to the Hillmen, who are more numerous in this valley than in any other, and number about 500 persons divided among 22 villages.

The coffee estates and the large number of Hill-men.

230. The forests of the Ney ār contain little timber of any value except the vēnga, and as there are no roads into them, I believe that no trees are felled here. The presence of these Kānikār, whose vil-lages dot the hill sides in every direction, would interfere with the selection of reserves, and the character of the forests themselves is not encou-raging, though they could be much improved by the suppression of grass fires, a work which cannot be attempted until the more extensive forests have received attention.

These forests very poor and will not be reserved.

231. On the upper hill forests cardamoms are found in some quantities, but the greater part of these are smuggled over to Tinnevelly, by men who cross the hills by narrow foot paths ; one of these leads over close by the Agasthiar peak, while another comes up from Pāpanāsam rather farther North.

Cardamoms.

* Diary February 12, 1817.

232. The district road or bridle path already mentioned ran through the upper
part of the hills, and connected the larger estates, but this is
Absence of roads. quite adandoned. The only other road that is found in this
part is the cart road from Kōvilūr to Munnara, which skirts the forest area.

(6) *The Karamana river.*

233. The sources of the Karamana river are on the ridge to the North of the
Agasthiar peak, and on the outlying spur which terminates in
The sources and course the Sāsthānkōtta rock, and separates the valley of this river
of this river. from that of the Pālōda river. The Kāvilr, the main branch of
the Karamana, drains the " valley of the four winds " (Nāngkātta) which faces North-
west, and ., is joined by other streams from valleys to the North, all combining to
form a small river with a rocky bed often impassable in the wet weather. This is
marked in the map as the Mylaur, a name not known now. Proceeding 3 or 4 miles
farther it is joined by streams on both sides, and having reached the more level
ground, flows along in a more leisurely manner. From this point the Karamana can
be used for floating timber, and has been so used during the wet weather at different
times, some quite recently. This river leaves the forest area shortly above Ariyanāda,
a village of from 300~500 inhabitants, and at one time of some importance. It is
here a stream 50 yards wide, with high banks, and a sandy bed.

234. The catchment area of the Karamana is approximately 72 square miles,
the greater part being covered with grass forest. Of the first
The catchment area class, or moist forest, there are probably 5 square miles, all of it
divided out according to at the highest elevations. Of the secondary forest there are
the classes of forest. probably 20 square miles. Of the grass forest 35, and 12
square miles of rock and useless land.

235. The country here resembles that of the adjoining valley, and contains the
same trees, of which vēnga is the most common. Thēmbāva is
Timber found here. found in some places, but not in others. Teak is quite absent,
except where it has been planted, as at Arianāda, where General Cullen made a
plantation, which has not succeeded well. The soil of all this country seems too stiff
for that tree. Blackwood is scarce, but is occasionly met with.

236. In other respects, in rainfall, climate, soil, and in the general slope and
condition of the land, the valley of the Karamana closely resem-
Rainfall, climate &c. bles that of the Ney ār, and I need not describe it further.

237. The range of hills forming the boundary between Travancore and Tinnevelly runs at an elevation of 4,000 ft. with no conspicuous peak.
Elevations of the high- Sāsthānkōtta already alluded to is a bluff rock looking towards
er hills. the sea, and about 3,000 ft in height. Uthal mala (Wodala
mullay) is over 1,400 ft. high, its slopes being largely cultivated.

238. About 3,200 acres of land were sold for coffee cultivation in this valley,
and 7 large estates were opened. At the present time there
Coffee cultivation and are about 300 acres under tea on 3 estates, and all the coffee is
the resident population. abandoned. The population on these properties is perhaps 300,
and there are 300 Hillmen, scattered in 15 or 16 villages, making a total of 600
individuals, to which must be added 120 people living at Kōtūr (Mamalikeray), and
30 at Kīlpālūr (Keelpaloor), or 750 souls in all. The last named are either Nāyars or
Mahomedans engaged in cultivation or in trade.

239. The Southern part of these forests has been worked for thēmbāva which
was sawn up and carted out, but the vēnga at the time of my
Timber operations. visit had not been touched. Through the grass enormous numbers of seedlings of both vēnga and blackwood were visible, showing that the latter
must at one time have been plentiful, though scarce now. The Northern part of these
forests has been worked from Vithara (Viddiryaloor), a bandy track running for a
considerable distance due South from the 23rd mile stone, and both vēnga and
thēmbāva have been cut.

240. None of this land comes within the reserves which I have suggested, for the reason already stated that it is much broken up by Hill-

None of this land to be reserved. men's clearings. Grass fires do much damage, but until we can take the matter in hand, and a force sufficiently large to suppress them is maintained, nothing can be done to prevent them.

241. A bandy road from Mūnnara runs along the edge of the forest arca, and

Chief roads. is available for working it. A branch runs from Paratthipulli to Kōtūr 2 or 3 miles, and from thence a bridle path used to proceed up to the Agasthiar coffee estates, but this has long been abandoned. The same fate has fallen upon the District road, which connected all the estates, the only parts of it still kept up being a few miles at the Northern end of this valley.

242. The Hillmen have foot paths leading from one village to another, but

Paths on the upper slopes of the hills. these are changed as their cultivation is shifted from place to place. In olden times a path from Ariyanāda ran up to the top of the hill near Orakānnumpāra (Oorcannoompaurae) and was continued down into Tinnevelly. By this tract goods were bought on pack cattle, and others taken in return, and there seems to have been at one period a considerable traffic with the Eastern side, but in Lieut. Ward's time, Ariyanāda had fallen in importance, and the path was but little used. At present the only people who pass along it are cardamom smugglers.

243. The Killiyār is a small river which rises in the neighbourhood of Nedu-

The Killiyār. mangāda, but the area drained by it within forest limits is so insignificant that I have included it in the basin of the Karamana river, which it joins not far from Trivandrum.

(7) The Pālōda river.

244. The sources of the Pālōda river are on the lofty peak of Chemmūnji,

Sources and course of the Pālōda river. North-east of Trivandrum, and on the strong spur which runs out from the main range, and ends in the rocky cliffs of Ponmudi. Rising at an altitude of nearly 4,000 ft. the streams that form its main branch, descend with great rapidity and unite at the foot of the hills in the Kalār, a river which rises at times to a great height, and is never quite empty of water, thanks to the extent of forest which clothes the higher elevations. From this point the Kalār or Kavadiār, here spanned by a strong wooden bridge, runs with a more tranquil stream, but with a current still very strong in flood time, for 7 or 8 miles and then passes the hamlet of Pālōda, where it is crossed by another bridge. A few miles below this point, the other branch of the river, which drains the valley West of the Ponmudi peak joins it, and the waters, now largely augmented, run for several miles South, and then turning West pass out of the forest area as a river 80 yards wide, 3 miles above the village of Vāmanapuram.

245. The Pālōda or the Vāmanapuram river is used for floating timber and

Used for floating timber, and in places suitable for navigation. bamboos from as far up as Pālōda or even a little higher, but the quantity of logs conveyed by this route is not large, owing to the timber being as a rule too heavy for floating, and to the fact that it is more convenient to saw the logs up and convey the wood by cart to Trivandrum. Boats can ascend in the monsoon above Vāmanapuram, but owing to the strong current and the time occupied in working up against the stream, this mode of travelling is not often adopted.

246. The area drained by this river is 143 square miles, of which 23 are covered

The area divided according to the forest in it. with moist forest, chiefly at the higer elevations, 40 with secondary forest, hill cultivation, paddy fields &c. 70 with grass forest growing under large trees, and 10 square miles is rock and useless land.

247. The trees growing within this area are almost identical with those found

Character of the timber in these forests. in the basins of the neighbouring rivers. Here, as on the Ney ār and Karamana ār, teak is absent, or is found towards the West as a stunted tree of no height. Vēnga, and to a much less degree, the thēmbāva are the most abundant trees of any value, where the forests have not been touched, but the white marutha *(Terminalia paniculata)* is infinitely more common, though its timber is but little appreciated by the people. The frequent occurrence of the white maruthu is very characteristic of this region : in the open forests at the foot of the hills it often reaches 66 per cent of the whole number of trees, and attains a large size while the other kinds do not seem to thrive. A sample of the forests in these parts may be interesting.*

Distribution of trees in some of the grass forests. 248. Percentage of trees in untouched forest, average of several observations near Vithara January, 22, 1887.

Terminalia paniculata	50
Dillenia pentagyna	20
(Vēnga) Pterocarpus marsupium...	15
Careya arborea	5
Phyllanthus emblica	3
Buchanania latifolia and others	7

}100

The only saleable tree here is vēnga, the other woods being used only for agricultural instruments, rough buildings, and firewood.

249. The kongu tree is not found on the hills in such numbers as it is in South *Kongu becoming very scarce.* Travancore, its place being taken by *Hopea racophlea*, an allied species, possessing timber of equal value, but not attaining the same size. The more valuable kongu is met with along the sides of rivers, and must at one time have been abundant but the demand for it in Trivandrum has been so great during the last few years that it has been nearly exterminated from all accessible places.

Blackwood not abundant. 250. Blackwood is found occasionally, and grows to a fair size, but it is never so numerous as to form a percentage of the trees.

251. A very curious tree occuring here for the first time is the *Humboldtia alata*, remarkable not for the value of its timber, or so much *A curious endemic tree.* for the curious appearance of its pale winged leaves, as for the fact that it is confined exclusively to the valleys of the Palōda and Kulatthûppura rivers, and is found no where else in the world.

252. The most important peaks and hills in the basin of the Pâlōda river are *Peaks and hills in this basin.* the Chemmûnji peak, about 4,000 ft. high, at the head of the valley, Peramu Kôtta station (3,240 ft.), to the North of the river : Pallipāra a rounded hill near Chemmunji, and about 3,500 ft. high ; the several peaks of Ponmudi on the Northern spur overhanging the valley, which rise to over 3,000 ft. : Kakkâda mala, a small hill above Vithara, and South of it (600 ft.) : and Verhyan kunna, an isolated ridge of about the same height, not far from Kummil.

253. The soil on the upper slopes of the hills is better than in the neighbour- *Soil.* hood of the Agasthiar mala, but cannot be considered good. In the plain below, it is very poor as I have already stated.

Climate and rainfall. 254. The climate and rainfall of this basin present no differences to those of the adjoining country, and need not be discussed.

255. The heavy forest which clothes the upper hills has been partly sold for coffee cultivation, and what remains is too steep and too in- *Character of the forests.* accessible to be worked for timber. Small patches of forest occur in the plains, chiefly about swamps or on river sides, but most of the land of this kind that was once covered with forest has been cleared, and converted into plantain gardens, or paddy fields. By far the greatest part of the forest area in the basin of the Vāmanapuram river is occupied by grass forest, in which trees of the ordinary kind affecting such situations grow, but in consequence of the extreme poverty of the soil, to which I have already alluded, they are small and stunted. A large acreage is under cultivation, or has been cleared for it at one time.

---

‡ A very convenient way of calculating the number and kinds of the trees of a forest, is to note down the names of all growing within a chain or half a chain on each side of the path, as the observer moves along, measuring 10 chains in length. Adding these up, the total number of trees and of each species growing within one or two acres, as the case may be, is obtained.

or another. Besides land opened in coffee on the higher slopes, the Hillmen have made considerable inroads into the jungle, and in the plains a large area is occupied by paddyfields and gardens at Pālōda, Paccha, Vithara, Mathiri, Mundōṇi Kara, Kummil and other places.

256. The quantity of land sold for coffee cultivation within the basin of the *The coffee estates and the resident population on the hill-slopes.* Pālōda river was about 3,500 acres, comprising some 8 estates. Of these, portions of only 3 or 4 are kept up, but the coffee has disappeared and has been replaced by tea, which gives occupation to perhaps 400 maistries and coolies. In the lower land, there are 8 colonies of Hillmen, whose numbers amount to about 700 persons. In addition to these people, there is a considerable population living at Pālōda, Paccha and other places, the total number of which may reach 2,000, making 2,700 in all resident within the basin of the Pālōda river.

257. The chief villages are Kummil, the head quarter of a proverthy, with a *The chief villages and distribution of the population at the foot of the hills.* population of 300, : Paccha, a scattered hamlet with a very large extent of paddy land, and a population of 300 or 400, chiefly Nāyars and Pulayars : Pālōda, a place of similar size ; and Vithara on the cart road to the hills, at one time simply the residence of one enterprising Mahomedan, but now, since the opening of the coffee estates, a village of 200 or 300 persons, with a bazaar and a Police station. Mathiri, Mundōni Kara and Muthira are all small places, consisting of a few paddy fields, the owners of which live with their servants in gardens shaded by jack trees on the edge of their properties. Besides these places which I have named, the swampy valleys between the hills are often occupied by single families of Nāyars or Mahomedans, who live entirely by themselves, holding little communication with their neighbours.

258. The selection of the forest line in this neighbourhood was by no means *Selection of the forest line not at all easy.* easy, because in some places comparatvely far inland, I found the population fairly numerous, while outside them and nearer the coast, the area was uninhabited. I decided in the end that it was better to make a simple boundary, and to include a good area of cultivated land, than to exclude forest which might prove of value, or to lay down a line that would have been difficult to follow.

259. Twenty years ago the forests in the neighbourhood of Pālōda and Vithara *Timber operations. These forests have been very heavily worked during the last 20 years, and the timber is becoming very scarce.* must have contained much valuable timber, but the construction, about 1870, of the road from Shencottah to Trivandrum, which passes through Pālōda, and of a branch to Vithara to serve the coffee estates, made them easily accessible, and timber merchants began to fell very largely here. When I was speaking to some ryots at Pālōda in 1887, one of them said to me " Before the Shencottah road was cut vēnga, kongu and thēmbāva were so abun- " " dant that we could not see the sky for the branches of the trees, but now we can- " " not get any of these woods, they have all been cut down and taken away to " " Trivandrum. " The forests in this basin have been denuded of their best timber. in all the more accessible places. Rough cart tracks run into them from numerous points on the main road, and when I was travelling here in 1887 I found that the merchants who wished to deliver kongu in Trivandrum, had to get it from Motha a small hamlet on the Ayyūr–Madatthura road, 31 miles from the capital. No doubt by this time all the kongu in that neighbourhood has been worked out, and the timber men have to go farther North and East. At the same time the contractors from Quilon and from Shencottah have been extending in the same direction, so that in a few years they will all be felling in the same locality, and there will then be great difficulty in procuring good timber at all.

260. It is very desirable to select a reserve near Trivandrum, where the de- *Necessity of selecting a reserve, but difficulty of doing so. Distribution of trees in the grass forests.* mand for timber is considerable, and rapidly increasing, but this is a matter of no small difficulty, on account of the occurrence of numerous patches of cultivation in the forest, and also because of the poverty of the timber itself, but it is quite necessary to do something, or these forests will become poorer and poorer every year. The distribution of the trees in forest within the Pālōda basin, which has been worked, is as follows :—

Terminalia paniculata ...		42	
Dillenia pentagyna ..		19	
Careya arborea		17	
Phyllanthus emblica ...		14	} 100
(Vēnga)	Pterocarpus marsupium...	6	
	(Small and twisted.)		
	Other kinds	2	

Jan. 24 1887.

Here the only saleable tree is vēnga, which in this place reaches only 6 small trees per hundred.

261. Comparing this with the list already given, we can see how felling, without conservancy destroys a forest. The chances of vēnga for-

Very disastrous effect of timber operations if not carried out with caution, and in combination with conservancy.

ming a considerable proportion of the trees in this forest are smaller than previous to the commencement of felling operations here, and each year decreases those chances, as the grass fires kill out the young seedlings, which would take the place of the trees felled. Thus a mixed forest containing valuable trees, if not conserved on a proper system can very easily be ruined, and converted into a forest of nothing but worthless trees, and no amount of closing, unless assisted by artificial reproduction, can ever place it again in its original condition. Blocks of forest suitable for reserves might be chosen either near Pālōda, or South of Vithara, but for reasons already stated, they would be small.

. 262. The forests of the Pālōda basin as well as those of the Ney ār, the Kara-

Management of these forests.

mana, and the Kulatthuppura river, South of Punalūr, are nominally under the Aminadar of Nedumangāda, but at the time of my examination of them the duties of this officer kept him entirely in Trivandrum superintending the timber depôt; and he never visited, nor was he expected to visit the forests of the interior. The Vijārippukāren at Kadakkal is entrusted with the collection of wax, cardamoms, and honey, from all these parts, but he has no control over the forests themselves.

263. The timber is felled by contractors who cut down the trees without the least supervision from the Forest Department, saw them up,

The felling and delivery of the timber not supervised in the least.

and cart the wood into Trivandrum where it is delivered at the depôt and sold again at fixed rates. Kongu, vēnga, and thēmbāva are almost the only trees operated on; occasionally a few logs of blackwood are brought in.

264. The basin of the Pālōda river is very well provided with roads. The

Roads and paths.

main cart road from Trivandrum to Shencottah enters the area near Enāthi, and traversing it in a Northerly direction, passes through Paccha, and Pālōda, and leaves it again at Madatthurakāni. Another cart road runs from Nedumangāda up to Vāmanapuram, and Nelamayilam, and, though it does not pass through the forest, is largely used for conveying timber to Trivandrum. From Enāthi on the Shencottah road, a branch runs to Vithara, thence to Kalār, and from there it ascends the hills for about 10 miles, reaching within a short distance of the sanitarium built for Europeans near the Peramukōtta station, which it was cut to serve. From Vithara another cart road runs for a few miles in an easterly direction to the foot of the hill below Bonaccord estate, and a trace for a cart road leaves Vamanapuram, and passing through Mathiri, strikes the Shencottah road 3 miles North of Pālōda. Besides these larger roads, numerous bridle paths run in all directions to the different estates, and the district road which was cut to connect them all is kept up, the rivers being passed by excellent wooden bridges. One good path runs from the Kal ār up to the Oaklands estate, and, passing through it, continues up to the top of the hill, where it meets a similar bridle path, cut by the British Forest Department for the examination of their forests, coming from Pāpanāsam.

265. The cardamoms found on these hills are all of the "long" kind, and, as no

Cardamoms and wax.

gardens have been opened here, they are much smuggled by thieves from Tinnevelly. The total production of this basin may be perhaps 100 thulams a year, only about one-third of which, and this too only within the last few years, ever reaches Quilon. Wax is obtained in small quantities, and is delivered by the hillmen to the Vijārippukāran at Kadakkal.

10

(8) The Itthikkara river.

266. The Ayyūr or Itthikkara river drains obout 39 square miles of land included within the forest line. It takes its rise among the low hills
that are situated near Madatthurakāni, and those to the South-
west of Kulatthuppura and is formed by the junction of several
streams the Kalār, the Vēngōdathōda, the Karambanthōda, and the Kilparambu-
thōda.

Area of this river basin, and its sources.

267. The main river, created by the combination of these streams, leaves the
forest area at Manarkōda, and proceeding in a North-westerly
and Westerly direction for 7 or 8 miles, is joined by another
large stream, whose sources are within the forest area at Perangkāda (Perungat).
From thence the Itthikkara flows South-west and West, and falls into the sea at
Pāravūr, a short distance South of Quilon.

Its course.

268. In the wet weather these streams rise very rapidly, and the river can then
be, and is, very considerably used for the floating of timber, as
there are no obstructions in it, but in the dry weather the quantity of water in them is insignificant.

Used for the floating of timber.

269. The area of forest land included within the basin of this river, and amounting to 39 square miles may be divided as follows :—(1) Heavy
forest in detached blocks, and occupying the low lying land by
the river sides, sometimes extending for a good distance, 12
square miles. (2) Secondary forest, mostly paddy fields and old plantain gardens,
3 square miles, and (3) Grass forest with large trees, 24 square miles. There is no
useless land here.

*Area divided into differ-
ent classes of forest.*

270. The forests of the Itthikkara river are much better than those of the
Pālōda river, the soil being moister, and the proportion of heavy
jungle to secondary jungle being larger. Teak is absent or is
very rarely found, and blackwood is scarce and small. Thēm-
bāva and vēnga are found, the former in good quantity, as may be seen from the following figures.

*Character of the for-
ests.*

271. Distribution of trees in partially worked * forest
some miles to the East of Kadakkal. June, 3, 1887.

*Distribution of trees in
the grass-forests.*

(Thēmbāva)	Terminalia tomentosa	...	32		T. tomentosa and
	Dillenia pentagyna	...	22		T. Paniculata good.
	Terminalia paniculata	...	20		Bad leaf cover, only
	Careya arborea	...	10	} 100	ten trees to the acre.
	Lophopetalum wightianum		5		Vēnga worked out.
(Ven Teak)	Lagerstrœmia lanceolata		3		Average height 90
(Vēnga)	Pterocarpus marsupium	...	2		ft.
	Other kinds	...	..	6		

272. In the moist forest kongu and ānjili are found, and
in the more open parts of it, ven teak and meili (Vitex altissima).

*Trees in the moist for-
est.*

273. The small tree, kodagappāla (Holarrhena antidysenterica), whose seeds
are used medicinally, is found in these forests in considerable
abundance. Farther south it does not occur, but from this
point up to the northern boundary of Travancore it is constantly
seen.

*" Holarrhena antidysen-
terica."*

274. The highest hills within this basin do not rise more
than 500 ft. above sea level, and all the land has a very easy
slope.

The chief hills.

275. The rainfall and climate do not call for any special
remark. They are the same as in the adjoining river basins.

Rainfall and climate.

---

* This forest had been worked either for Quilon or Sheucottah; at both places thémbáva is not valued
nearly as highly as kongu and vénga, so it was not felled when the last mentioned trees were cut.

276. The population residing within this area is small and does not excee I
100 persons, living at Shedara, Motha, and Irakuli, where a
few Nâyars or Mahomedans have paddy fields.   There are no
coffee estates here, and no Hillmen, which largely accounts for
the existence of so much moist forest.

Population residing here.

277. Although it is not within the forest area, the large village of Kadakkal,
about one mile to the west of the forest line, is deserving of
notice.   The large number of chathrams, tanks, and shrines
seen here show that it was once a place of some importance.
It now contains 500–1000 inhabitants, and stands back from the
main road about a mile.   An elephant cage and Police station are located here,
and the Forest Vijârippukâran, whom I have already mentioned, makes this place
his head quarters.   His duties are to collect the wax, cardamoms, honey, dammer,
and lac found between the Punalur river and the Neyâr, and to assist him he has
a staff of 2 Watchmen and 4 or 5 Peons.   None of these articles are obtained in
the basin of the Itthikara river.   They are collected almost entirely by the Hillmen,
the only exception being in the case of cardamoms which are often gathered on
contract by some of the Planters.   About 24 thoolams of wax are collected annually
by this Vijârippukâran, and forwarded to Quilon and from thence to Alleppey.

The village of Kadak-kal the head quarters of a Forest Vijârippukâran.

278. The forest of the Itthikara basin is nominally under the charge of the
Nedumangâda Amindar, who resides at Trivandrum, as already
stated, and no Forest Officer ever comes to look after the
felling of the timber, so that the contractors do just what they
like.

Management of these forests.

279. The timber is felled mostly by contractors, who have to deliver it at
the Quilon or Shencottah depôts at certain fixed rates.   They
only fell kongu, ânjili, venteak, vênga and sometimes thêmbâvu,
and these are usually taken to the depôt by cart with the exception of the venteak,
which is floated down the river.   Watch stations on the road at Ayyûr and Nelamayilam maintain some check on the passage of timber.   Persons requiring other
woods than those kinds in demand at the depôts are allowed to cut down the trees
on obtaining permits to fell, and paying certain fixed rates per candy of wood
required.

Timber operations.

280. The only road passing through this basin is the cart road from Nelamayilam to Madatthurakâni, and all the timber that goes by
road is conveyed by this route.   The forests on either side are
untapped by any path but tracks for dragging timber, and it is not easy to penetrate
them.

Roads.

(9)   The Kalleda River.

281. The Kulathuppum or Kalleda river is the 3rd largest in Travancore,
and drains an area of 293 square miles, with in the forest limits.
It is formed by the junction of 4 large streams which issue from
as many distinct valleys.   The main branch rises in the most
southerly of these valleys, and is composed of numerous streams
whose sources are on the elevated plateau that stretches from the Alvâkurucchi
peak southwards to Chemmûnji, or on its slopes.   Descending, rapidly at first, these
streams fall into the main channels, running towards the west, the two most
northerly of which, when united, are called the Vellâr, the other the Pinanaâr.
Meeting in the bottom of the valley at an elevation of about 600 ft., and continuing
in a westerly direction, the river is joined after a couple of miles by a large stream
descending from the Churutta hill on its right, and, after proceeding for another 7
or 8 miles, by another stream which flows into it from the left or south side.   Here
the river leaves the Kulathuppura valley proper, and running over a rocky bed for
about 5 mile , it passes the village of the same name, which is situated on its left
bank.   At this point it is about 80 yards wide with a powerful stream in the wet
weather, and even in the hot season it is well supplied with water.   The elevation
is about 300 ft.

Sources and course of the upper part of the Kallada river.

282. About 3 miles below Kulatthuppura, the river, which has continued flowing nearly due North, is precipitated over the Minmutti fall or cataract, the water rushing with immense velocity among the rocks with which the channel is filled. Some 2 miles below this fall the river is joined on the right by another which descends from the Chenthrôni valley, and is called by that name.

Its course near the village of Kulatthupura.

283. The upper part of the Chenthrôni river is formed by the junction of the Umiyâr (Manimuttâr), which rises on the slopes of Alvâkurucchi, with another stream flowing from the lofty peak to the east of the valley. These two streams meet at an elevation of about 300 ft., and proceeding in a North-westerly direction, their united waters are augmented by those of the Parappâr on the left, and another small river of equal size on the right. Two miles below this the Kalâr joins them from the south, and after running for 6 or 7 miles North-west the Chenthrôni joins the Kulatthuppura river, as already stated, at an elevation of perhaps 200 ft. The two rivers then flow some 2 miles nearly due west, till they are joined by the Karuthâ-urutti which drains the valley of Arienkâvu.

The course of the Chenthrôni river.

284. This river is not of any great length, but has numerous tributaries rising on the high ground which forms a semicircle facing towards the south.

The Arienkâvu river.

285. The three branches, now united together in the Kalleda river, run West-north-west in almost a straight line for some 10 miles, and pass out of the forest area 3 miles above the town of Punalûr.

Course of the main river.

286. Here the river turns abruptly to the North, and then bends round again to the North west passing Patthanâpuram. About 2 miles below Punalûr it is joined on the right hand by the Châlakkara river which drains a considerable valley.

Its junction with the Châlakkara river.

287. It then proceeds, winding very much, first in a westerly direction, and then South-west until it falls into the sea a short distance North of Quilon, leaving Kalleda, from which it takes its name, on the left hand.

Its course below Punalûr.

288. Some 3 miles below the junction of the Karutha-urutti with it, the Kalleda river, now a fine stream between 100 and 200 yards wide, pours over a cataract at Ottakkal, but beyond this it is not obstructed, except about 3 miles below Punalûr, where a reef of rocks crosses the river, and where a paper mill is being erected.

The Ottakkal cataract and rocks at Punalûr.

289. The Kalleda river can only be used for navigation as far up as this reef of rocks last mentioned, and this only with difficulty, for in the wet weather the current is strong, while during the dry season the water in the river runs very low.

The river only navigable below Punalûr.

290. Three of the four branches of the river can be used for floating timber. The most southern is thus employed between Pinamanâr and the village of Kulattuppura, and at the time of my visit some timber contractors were conveying venteak planks by this route as far as the village. Below that point the river is impracticable on account of the cataract at Minmutti, so the wood has to be taken across country to one of the streams of the Itthikkara river, or by cart to below the cataract above mentioned.

Floating of timber on the Kulatthuppura river.

291. The Chenthrôni river can be used for floating from the top of the valley, there being no obstruction of any importance down to its junction with the Kulatthuppura branch. Beyond that, the timber can only be sent down in flood time, when the water is sufficiently high to carry the logs over the Ottakkal cataract. Nothing but light timber can be thus conveyed, for the rocks would inevitably burst off the bamboo floats attached to the heavier logs to make them swim.

The Chenthrôni river.

292. The Chālakkara runs level for a great part of its length, and can be conveniently used for floating timber till within a short distance of its junction with the Kalleda, where, for a few hundred yards, it rushes down over a cataract. A cutting has therefore been made by which timber can be hauled up and dragged across the land for about 200 yards, to a point whence the logs can be slid into the river, and so avoid the cataract. The timber brought down by this route is chiefly teak.

The Chālakkara river.

293. The 293 square miles of forest land, included within the basin of the Kalleda may be divided thus :—(1) Heavy forest 180 square miles. (2) Secondary forest, coffee estates, hill-cultivation, paddy fields &c. 40 square miles (3) Grass land intermixed with large trees, 60 square miles (4) Useless land 13 square miles.

The area of the Kalleda basin divided out according to classes of forest

294. Before the commencement of the coffee enterprise the 4 valleys which I have described must have been covered by one unbroken sheet of forest, stretching from the Peramakōtta station in the South up to the Nedumpára station, and beyond it in the North, but, within the last 20 years, this continuous expanse of noble trees has been broken by clearings in all directions. As it is, the forest lies chiefly in one large block which extends from the East nearly as far as the main road from Trivandrum to Shencottah. To the West of this road there are also large patches of forest, notably one that stretches from Kulatthuppura North-West to the Kalleda river. The rest of the country, West of the main road, is occupied by grass forest, which also occurs in patches within the large block of moist forest.

Extensive block of heavy forest in spite of the clearings made for coffee cultivation

295. The Kulatthuppura valley the most Southerly of the 4 to which I have described, measures about 12 miles long by 6 broad, and faces towards the West. It is bounded on the South by the Perumakōtta ridge which runs at an elevation of about 3000 ft., and on the North by the Churutta ridge, which leaves the main range at Alvā kuricchi peak, and runs along at an elevation of over 3000 ft. for about ten miles. At the Eastern end of the valley is an elevated grass plateau about 4,000 ft. high, which is called by Europeans the "Patenas". Here the land is covered with beautifully short grass, intermixed with patches of forest, and would be admirably suited for a sanitarium, if it was not so far from Trivandrum.

Description of the Kulatthuppura valley.

296. The highest hill in this valley is the Alvākuricchi peak which rises to about 4,500 ft., or 200 ft. higher than the most elevated point on the "Patenas".

The highest hill in it

297. The valley slopes steeply down on all sides, but opens out into a more or less level area at its bottom, with hills or ridges rising through it at intervals. The most elevated of these is Kumbada mala kunna (1,800 ft.), a ridge which lies to the West of the Pinamanār, and South of the Kulatthuppura river.

The hills situated in the lower part of it.

298. The chief trees growing within this area are the kongu and ānjili, both of which must have been very plentiful at one time, but the demand for boats has been so great, that all the ānjili trees of any size have been felled, and the kongu too in all accessible places. Farther inland kongu is abundant, and seems to thrive particularly well in this valley at an elevation of about 1,000 ft. Through the abandoned coffee land myriads of small plants are growing up. Teak is not found except on the Thēk mala ridge, a spur which runs South from Churutta mala. Blackwood is not common except on the slopes to the East of the village of Kulatthuppura and below Rockwood estate, where there are some nice trees. Ven-teak also occurs in quantity in the lower parts of the valley, and is largely sawn into planks for conveyance to Quilon.

The chief trees growing at the lower elevations in this valley.

299. On the hill slopes the following useful trees are found in abundance and are much employed on the estates.

The trees found on the slopes of the hills and in the moist forest.

Pūthankōlli (*Percilonenron Indicum*), iluppa (*Dichopsis elliptica*), punna (*Calophyllum tomentosiqn*), redwood (*Gluta Travancorica*), nãngu (*Mesua ferrea*), white cedar (*Dysoxylum Malabaricum*), enna (*Dipterorarpus turbinatus*), black kongu (*Hopea racophlea*), besides the real kongu, ãnjili and wild jack.

300. The soil in this valley is nowhere particularly good, and in some places it is exceedingly poor, but the climate is most forcing, especially at the lower elevations, where the trees attain an enormous height and size, and the undergrowth consists of the "eetta" reed (*Beesha Travancorica*), a sign of poor but free soil.

Soil in this valley.

301. Some parts of this valley, and chiefly those more to the East, suffer much from the land wind which blows with great violence in the months of December, January and February, but at a distance from the main range it dies down to a steady breeze.

High winds.

302. The Kullatthuppura valley was never occupied, except in a few places by the Hillmen, before land was sold for coffee cultivation. Out of the 5,000 and odd acres auctioned for this purpose, about 3,500 were opened, and all have been abandoned, but some 700 or 800 acres in 4 or 5 estates, which are planted with tea. These properties give employment to perhaps 800 maistries and coolies.

Coffee cultivation and population still resident on the hills.

303. In the lower land there are 3 colonies of Hillmen, who number 300 persons divided as follows :—

The hillmen.

Madatthura Kānies	...	...	70
Kallãda Kānies	...	...	70
Churuttamala Kānies	...	...	160

$\left.\right\}$ 300.

304. At the time of Lieut. Ward's visit, there was another colony of these people living on the ridge which is the prolongation of the spur on which the Perumakõtta station is situated. They called themselves Pulikõda Kānies, but about 15 or 20 years ago, I am told that they were nearly all killed out, probably by small-pox, and the few survivors joined the colony at Madatthurakkãni. Thus the population of the Kulatthuppura valley is about 1,100.

The Pulikõda Kānies.

305. As regards roads, there is a good cart road which leaves the Trivandrum-Shencottah road, a mile North of Madatthurakkãni, and running for about 12 miles to the Eastward crosses the Pinamanãr by a fine wooden bridge at 600 ft. About 12 miles before reaching this point it passes a saddle at an elevation of 800 ft., which is called Anakoppam, as elephant pits were dug here formerly, but they were not in use at the time of my visit. From the Pinamanãr one branch of the cart road runs for 2 or 3 miles to the South, while another continues due East as far as Chīnakkala, some 5 or 6 miles distant. From both these points bridle paths run up to the estates. There are no paths going over to Tinnevelly except 2 smugglers' tracks, one crossing the "Patenas" and descending to Pãpanãsam, the other following the Churutta ridge as far as Alvã-kuriechi peak, and dropping down to Chettikulam.

Roads and paths in the Kulatthuppura valley.

306. Outside the Kulathuppura valley a large stretch of undulating land extends Westward from the main road to Shencottah, as far as the forest line, and Northward from Madathurakkãni to the Kalleda river. This country is covered partly with grass forest, and partly with moist forest, the latter predominating in the Northern portion of it, and the former in the Southern.

Character of the land outside the valley.

307. Throughout this area teak is almost entirely absent, occurring only occasionally in the neighbourhood of the Kalleda river. Blackwood is not found in quantity, but is seen here and there. In the grass-land there is good thēmbāva and vēnga, and in the open forest large meili and ven teak, while in the moist forest kongu and ãnjili occur. About Kulatthuppura itself the forest is of large size, and is remarkable for the abundance of the enna (Dipterocarpus turbinatus), which grows to an immense height, with a straight clean stem, and thrives only on a clayey soil. Other peculiar trees are the rare nutmeg (*yristica r-ag-ifica*), chinna thuvara (*Diospyros microphylla*), and venkotta (*Lophopetalum wightianum*), all trees of very great size.

Chief trees found in this area.

308. A large unbroken expanse of forest extends from a little West of Kullatthuppura Northwards to the Kalloda river, and is called Elankoyittha káda.

Block of forest near the Kalloda river.

309. There are no high hills in this low country tract, the highest being Thávankóda, and Kódakal, and the curious Eramalúr rock, all of which rise to about 800 ft.

No high hills.

310. The only inhabited places here are Madatthurakkáni (600 ft.) where there are a travellers' rest-house and a chatthram, watched by a few Peons, and surrounded by an elephant trench, as the elephants are very troublesome : Kulatthuppura, a large village containing a rest-house, and a population of 200 or 300 persons, chiefly Mahomedans : (there is a celebrated pagoda here on the Eastern side of the river) : and a number of small settlements, each consisting of a few paddy fields, with the houses of the owners close by. These are Kadavankóda, Pulliyárúr, and Valiya-éla (Vulleayalau), making a total population of perhaps 400 persons.

Population resident in this tract.

311. The Chenthróni valley lies to the North of the Kulatthuppura valley, and is separated from it by the Churutta ridge already described. On the East it is bounded by the main range, extending from Alvákuricchi peak Northwards, and on the North by a spur running out to the Northwest, and ending at the Karutha-urutti river, the whole area being 12 miles long by 4 broad.

Area of the Chenthróni valley.

312. At one time all this valley was under forest, and that too not very long ago. At present it is cut up by clearings for coffee or tea cultivation. Some 3,500 acres of land were sold for the former product, of which about 2,500 were opened, and of these there are 500 or 600 now under tea, employing about 600 persons. There are no Hillmen resident here, though occasionally the Churuttamala Kánies cross over, and make clearings at the entrance of the valley.

Cultivation and resident population.

313. The trees growing in this tract are identical with those found in the Kulatthuppura valley, and the soil, climate, and rainfall are much the same and need not therefore be particularly described. Ebony is met with towards the East, and meili, and the white cedar *(Dysoxylum malabaricum)*, and red cedar *(Cedrela tuna)* are perhaps more common : teak and blackwood are absent.

Trees found here

314. Between the Umiyár (Manimuttár) and the Eastern watershed of the Chenthróni river, which is now the boundary between Travancore and Tinnevelly, there is a block of land covered with heavy forest, and aggregating about 6 square miles, which was at one time claimed by the British Government, but when the boundary was adjusted and surveyed some 12 years ago, this area was handed over to Travancore. The lowest point on the boundary on the Eastern edge of this block shows an elevation of about 2,500 ft., and at this place called the Alvákuricchi gap it has been proposed to locate a timber depôt for the sale of wood to persons in Tinnevelly.

Block of land at one time under dispute, but finally handed over to Travancore.

315. From the main road running from Trivandrum to Shencottah a cart road starts at a point about 2 miles North of Kulatthuppura, where the river is crossed by a ferry, and proceeds for some 12 or 15 miles to the Eastward : at the time of my visit it ended at a piece of flat land, called Eettapadappu on the Southern bank of the river, a branch of 3 or 4 miles being carried farther up the valley. Since then I believe it has been, or is being, extended. An old track used to run along this line, and before the Shencottah cart road was cut, a considerable quantity of produce was carried on pack bullocks up to the Umiyár (Manimuttár), and by the Alvákuricchi gap to the village of the same name in Tinnevelly, but this track has been long disused. Another path, which is still kept open, leaves the bandy road at Eettapadappu, and crossing the Chenthróni river rises up a steep ascent to the Karingaloda gap or Shínar pass (1,940 ft.) and descends to the Kannupilla mettu on the Eastern side (750 ft.), down a long and narrow valley. I believe it is intended to make a cart road through this pass in order to work out

Roads and paths.

the timber in the Chenthrōni valley, but it will be an expensive and difficult undertaking.

316. North of the Chenthrōni valley is that of Arienkāva, which is drained by the Karutha-urutti river. It is about 8 miles long and 3 or 4 wide, and has been largely cleared for cultivation. Some 40 years ago or more, a grant of 10 square miles of forest land for the purposes of cultivation was made to a Mr. Huxham, who had some property at Quilon, and at Patthanāpuram and its neighbourhood, and part of this grant comes within the basin of the Karutha-urutti river. About 1,000 acres of land were sold for coffee cultivation, and some was occupied by H H. the late Maha Rajah for the same purpose. This property, which lies close under the beautiful waterfall of Pālaruvi, to the South of the cart road, and not far from Ariankāva, has been handed over to the Forest Department for management. Of the other estates the greater part was opened, but now only some 200 acres of the land sold, and 600 or 700 of the ten square miles is still kept up. This land is mostly under tea, but there are upwards of 100 acres of very flourishing Liberian coffee belonging to Mr. Knight. At Ariankāva there is a celebrated temple of considerable antiquity still kept in good repair. At Karutha-urutti there is a chattram. The population resident within this valley may be set down at 800.

The Ariankāva valley. Its area and resident population.

317. The soil in this place is good and the forest of very large size, but the climate, rainfall and species of trees are the same as in the valleys last described, and need not be referred to farther here Teak and blackwood both grow here, and a small patch of some hundreds of sandalwood has been found on the ridge dividing the Shencottah from the Kottārakara Taluq, but, as the tree is not indigenous here, these trees must either have been planted or be the descendants of planted trees.

Soil, climate, rainfall and timber

318. The highest hill in this valley is Nāga mala, a conical grassy peak rising to a height of probably 3,500 ft., and called by the hillmen Thiruvāthara mala; the rest of the semicircle surrounding it reaches an elevation of about 2,500 ft. I was prevented from ascending this hill by bad weather, but I climbed to the top of Nedumpāra, (2,930 ft.) which is much lower.

Chief hills and peaks here.

319. From Karutha-urutti the Trivandrum bandy road follows the Northern bank of the river, and reaches its highest point at the Ariankāva pass (1,210 ft.), from whence it descends to the village of Puliyara, 600 ft. lower, where there is a timber depôt.

The main cart road from Trivandrum

320. From Karutha-urutti a bandy road has been cut to the North for 2 or 3 miles, as far as the Isfield tea estate. Another cart road connects the Pālaruvi estate with the main road, and a bridle path continues from the junction of these two, Northward up to the Lochnagar and Arundel estates, and this has, I am told, been partly cut out for the passage of carts to get timber out of these forests, but at the time of my visit the work of conversion had not been commenced.

Other cart roads

321. The old road from Shencottah to Kottārakara did not pass by Karutha-urutti, but, keeping along to the North of the Korakkunna ridge, passed through a saddle at an elevation of about 600 ft., and from thence continued straight down the Chālakkara valley to Patthanāpuram.

The old road through the Ariankāva pass.

322. The main road after leaving Kulatthuppura, runs North along the river side for about 7 miles, till the Karutha-urutti joins the Kulatthuppura river. Here the stream is crossed by a ferry, and on the North bank the road rises for 3 or 4 miles till it reaches the Karutha-urutti rocks. A mile before this point is gained there is a fine bungalow situated on the road side, and called Camp gorge. Built originally for the Engineer's Department, it is now regularly used as a resthouse. Near it a wire suspension bridge is stretched across the river, and a foot path is carried from this point to some estates in the Chenthrōni valley. Behind this bungalow a high ridge, Korakkunnu mala, runs North-west at an elevation of 2,000 ft., and parallel to the river, while in front of it a conical hill called Thēn mala (1,100 ft.) rises close to the road.

The present road. Camp gorge.

323. About two miles North of the ferry on the Kalleda river, a cart road

The cart road from Camp gorge to Punalūr. leaves the main road, and turning Westward, is carried parallel to the river for about 13 miles to the small town of Punalūr.

324. The land on both sides of this road was offered on very favourable terms

Land on both sides of this road given to applicants. 8 or 10 years ago to any one who liked to take it up, with a view to making the road more civilised, and to facilitate travelling. A good extent of land was thus occupied but it was speedily abandoned again, and now the road is as much deserted as ever it was.

325. The forests here consist of very large trees of the ordinary kinds occur-

Character of the timber here. ring in moist jungle; and farther to the West where grass is found, teak and blackwood are met with. Here for the first time is seen the irūl (*Xylia dolabriformis*), a tree absent from South Travancore, but common in all the grass-forests from here up to the Northern boundary. Here also occurs vekkāli (*Anogeissus latifolius*), a useful tree very abundant in the free dry soil of South Travancore, but absent from the moister clays of Nedumangāda, and its neighbourhood.

326. Continuing along this road towards the West, a chattram and a few huts

Villages upon this road. are seen at Ottakkal, and 5 miles farther the road passes through the small hamlet of Edamanna, the two places containing perhaps 50 inhabitants. A mile to the West of the last mentioned place the road leaves the forest area, and enters a country more thickly inhabited, and carefully cultivated.

327. The Chālakkara river is the fourth and last of those which combine to

Sources of the Chālakkara river. form the Kalleda. It has 3 main branches, one, the most Southerly, rising close to the Kōrakkunnu ridge, and called the Chālakkara; the Ambanāda river rising near the Mullu mala huts; and the Karuppan thōda whose source is near the ancient village of Kōmarangudi. All three of these streams run very level, and can be used for floating timber for a long way up them.

328. The valley drained by this river is about 10 miles long by 3 wide, and

Area and boundaries of the valley. is enclosed on the South by the Kōrakkunnu ridge, and on the North by a well marked ridge running out from the Nāga mala peak in a North-westerly direction, on which are Keikōnnu mala (Kycoonut Mullay), Elappakōda mala (Yellapacode hill), and Kadacchira para (Irraychera para).

329. The greater part of this area is covered with grass-forest containing good

Timber trees found here. teak, blackwood, venga, thēmbāvu, irūl and vekkāli. The first mentioned has been largely felled, and good trees are getting scarce. On the river banks kongu of good dimension is seen, and this especially on the upper portions of the Ambanāda river. Anjili is scarce. The upper part of the valley is covered with moist forest or secondary jungle. Here the talipot palm (*Corypha umbraculifera*) is found all through it, but I cannot think that it is indigenous here, as I have never found it elsewhere. It is more probable that the seed was scattered about in the jungle by Mr. Huxham, that he might get thatching material, as grass is scarce.

330. The old road to which I have alluded enters this valley at its upper end

Description of the old road. coming from Anankāvu: it descends to the old estate of Koravan thāvalam at 350 ft., part of the 10 square miles, where there is a bungalow and nutmeg, tea, cacao and sappan trees 40 years old and then runs along almost level through secondary jungle where there was cultivation a generation ago. After continuing for about five miles, the road, which is very much overgrown with the thorny sensitive plant (Mimosa pudica), reaches Māmbaratthura (Māmblatooray) where there are the remains of a small pagoda.

12

331. At the time of Lieut. Ward's visit there was *accommodation for travellers here, but all the buildings have long ago been levelled to the ground by elephants, with the exception of a small shrine of massive stone about five ft. square. From here the road passes on to Chálakkara a distance of about 5 miles, where it leaves the forest area. At this place there are the remains of an old shrine, and some nutmeg and other trees planted by Mr. Huxham.

Mímbaratthura and Chálakkara.

332. Kómarangudi (Covarangudy), at an elevation of 700 ft., near the source of the Karuppan thóda was in ruins at the time of Lieut. Ward's visit, and is not inhabited now. It is quite in the heart of the jungle, and all that can be seen now is a well, the remains of a shrine, and sites levelled for buildings. It may never have been permanently occupied, and possibly consisted of only a shrine and chattram on the path from Patthanápuram to Acchankóvil.

Ruins at Kómarangudi.

333. The soil of the Chálakkara valley is good and free, the climate and rainfall are the same as in the other valleys drained by the Kalleda river.

Soil climate and rainfall.

334. In this valley the only resident population consists of two colonies of Hill Pandárans and the coolies on the tea estate. The former live at the upper end of the Ambanáda river, and cultivate the hill slopes a little, but depend largely for food on wild roots and fruit. They are exceedingly timid, and when I arrived at their kudi all of them run away into the jungle except one boy who had fever, and was too weak to escape. They number about 40 persons. Besides these, 40 to 60 maistries and coolies are employed on the Koravan Thávalam tea estate, so that the total population is about 100. Adding this figure to those already given for the other valleys on this river we obtain a total of 3,050 or say 3,000 persons.

The resident population.

335. The forests of the Kulatthuppura basin are under the nominal management of 3 different officers. As far as I can make out, those portions which are to the West of the main road, and South of the Ayyúr-Kulatthuppura road are under the Aminadar of Nedumangáda, to whom I have already alluded. North of this, as far as the dividing ridge between the basin of the Chálakkara and Acchankóvil rivers, and Eastward up to the Karutha-urutti river, is under the Aminadar of Patthanápuram, while the remaining portion is in the charge of the Superintendent at Puliyara.

Management of these forests.

336. Although a great deal of timber felling is carried on in these forests, the only parts looked after are those under the Aminadar of Patthanápuram. This officer has only a small staff of peons, and a moderate extent of forest to supervise, and his work is very well done. He captures yearly in pits about 4 elephants, and he or his peons are well acquainted with all portions of his charge. The same cannot be said of the other officers controlling sections of these forests: from one year's end to another, the only Forest officers who come here are the peons attached to the Kadakkal Vijárippukáran's office, who collect wax and cardamoms, or occasionally some one sent by the Superintendent at Puliyara, to obtain some information. The timber merchants, in consequence, do just what they like, and fell where and how they please. I am speaking of the state of things at the time of my visit in 1887. Since then I understand that an officer has been appointed and stationed at Kulatthuppura.

None of them supervised except those under the Aminadar of Patthanápuram

337. The moist forests in the Ariankávu and Chenthróni valleys, and those in the neighbourhood of Kulatthuppura are worked for kongu for the Puliyara depôt, and every year the contractors have to go farther and farther to procure the timber. Teak (now very scarce), blackwood, and vênga are cut on the hills near Ottakkal for the same depôt.

Timber operations to supply the Puliyara depôt.

338. The timber merchants and contractors of Quilon cut ânjili planks, and a

<table>
<tr><td>

Timber operations for the supply of Quilon.

</td><td>

little thēmbāva, and mullu vēnga (*Bridelia retusa*) fiom the Ottakkol forest, teak from the Chālakkara valley, white cedar from the Chenthrōni valley, venteak planks, and ānjili, and

</td></tr>
</table>

little thēmbāva, and mullu vēnga (*Bridelia retusa*) fiom the Ottakkol forest, teak from the Chālakkara valley, white cedar from the Chenthrōni valley, venteak planks, and ānjili, and kongu boats from the Kulatthuppura valley and blackwood, kongu and vēnga, from the forests to the West of Kulatthuppura.

339. Kongu for the Trivandrum depôt is felled at Motha, a short distance to the West of Madathurakkāni. Thus all these depôts are drawing upon the same forests, and timber will soon become scarce there.

Also for Trivandrum.

340. I have recommended that the land enclosed within certain boundaries in this basin be formed into a reserve, extending over 120 square miles, and this has been done. Within this area there are many coffee estates, and the clearings of about 160 Hillmen, but as I shall describe the reserves in detail later on, I will not say any thing more about the matter now.

A reserve has been proclaimed here.

(10) The Thāmravurnni river.

341. The area of forest land within the Shencottah Taluq, and to the East of the main range, is drained by the Shencottah and Churanda rivers, which are tributaries of the Thāmravarnni, the chief river in Tinnevelly. Within the forest area the above mentioned streams are small, and on the hill slopes they contain very little water, which soon runs off after the monsoon has ended. On the flat land they are used for purposes of irrigation, but the quantity of water obtainable is small, and it has been more than once suggested to divert some of the streams in Travancore to the Eastern side in this Taluq, and there is no doubt that the water would be well utilised if the scheme were carried out, but whether or not the returns would give a good interest on the outlay I cannot say.

Sources of the river and its character.

342. The Shencottah forest area may be divided thus · (1) Heavy moist forest, 4 square miles. (2) Secondary forest and coffee estates, 4 square miles. (3) Grass land with trees, 5 square miles, and (4) Useless land, 7 square miles.

Division of these forests according to class.

343. The moist forest is of a very poor class, the trees being small and of little value, except in some of the upper valleys ; the same may be said of the grass land forest, the dry climate and high winds being against luxuriance of vegetation In fact the whole of this area might be omitted from the forests, but that it is valuable for charcoal, large quantities of which are daily burnt and brought to market.

Character of the forest poor.

344. The chief tree found here is teak which does not grow to a large size but is very abundant on the rocky slopes of the upper hills. It is of the class called kōl teak, and is hard, and heavy, and well suited for small work, or where large timber is not a desideratum. Blackwood, vekkāli (*Anogeissus latifolius*), meili (*Vitex altissima*), and vāga (*Albizzia procera*), are abundant. Sandalwood when planted, for it is not indigenous, thrives well. The flora of this part of the country is very peculiar, and includes many rare trees and plants : this is owing to the special character of the climate, which combines the heat and drought of Tinnevelly, with the moisture and cool air of the Western Ghâts. The jungle at the foot of the hills, especially near Puliyara, is very scrubby, all the large trees having been cut down by villagers, but its condition is very much better than that of the Mahēnthragiri slopes.

Chief trees. Peculiarity of the flora due to the climate.

345. The forests of the Shencottah Taluq lie in a long line on the Eastern face of the main range. Except up the Karingalōda valley they are of no breadth. The mountain back bone varies extremely in height in this vicinity. At the Southern extremity of these forests it is about 3,000 ft. high. It drops to 1,940 ft. at the Karingalōda pass or Shānar ghat ; rising again on the other side it runs at about 2,500 ft. through Thēnmala, and Vannātthi mala (Vunnady mullay) and at the Ariankāvu pass, where the road crosses it, the elevation is only 1,200 ft., being 500 ft. above Puliyara, and the plain of Tinnevelly ; rising again, and winding very much, it reaches a height of

Elevations of the chief hills.

nearly 4,000 ft., the Râmakkal rock to the North of Puliyara being 3,600 ft. From here it drops again to the Acchankôvil gap at 1,470 ft., and then rises again up a grassy wind blown ridge to Thuval mala (Coonumcul square rock) 3,000 ft.

Soil and slope. 346. Throughout this area the land slopes steeply, but the soil is very good.

347. The rainfall is light, and probably does not exceed 60 inches. When **Rainfall and climate.** heavy rain is falling in Travancore, during the South-west monsoon, a high wind, often carrying with it a fine drizzle, prevails throughout this Taluq, and the sky is obscured by clouds. By far the greater part of the rain falls during the North-east monsoon, from October to December. In April and May the country here is very feverish, and unhealthy, and the heat is intense. This continues until the cool winds, blowing over from the Western side, have reduced the temperature.

348. Within the forest area five or six estates have been opened, aggregating **Population within this river basin.** 500 or 600 acres. Of these all but two have been abandoned, and these are only partially kept up. One belongs to Thiraviam pillay and lies North-west of Puliyara, the other was opened by a former Tahsildar of Shencottah, and has still a little good coffee on it. It is situated under the Râmakkal rock, due North of Puliyara. Farther North, a few plantain gardens and houses are seen near the Acchankôvil pass. The total population living within the forests of Shencottah is certainly not more than 50.

349. Outside the forest area, and on the cart road to Palamcottah is the village **The Puliyara Depôt.** of Puliyara, with a population of about 1,000 persons. Here there is a large timber depôt, from which timber felled in Travancore, and brought here by cart, is sold to the value of over a lac of rupees per annum to merchants in Tinnevelly. The timber offered for sale here is almost entirely kongu and vênga with some teak. For bandy wheels, and poles other woods are used, chiefly two species of vâga *(Albizzia procera and A. odoratissima.)* This depôt was originally under the Revenue authorities, but it was transferred to the Forest Department in the year 1871, at which time the sales of timber were between 9,000 and 10,000 Rs. per annum.

350. The forest area above mentioned is under the Superintendent of Puliyara, **Management of these forests.** who has an Office and a considerable staff of peons and watchmen. At the time of my visit he had also under his charge a watch station at Kannupilla mettu, at the foot of the Karingalôda pass, and another at Mêkkara below the Acchankôvil pass. I understand that since then depôts have been opened at these places.

351. The timber for all these depôts is brought from the Western side, but **Charcoal made in large quantities and smuggled into Tinnevelly.** these forests yield a large amount of charcoal, which is a never ending source of trouble between us and the Forest Department in Tinnevelly. The rate we charge for charcoal to Travancore subjects is very low, while in Tinnevelly it is high, owing to the small extent of the Tinnevelly forests, and to the difficulty of making them pay. To prevent our coming into collision with the British Forest authorities, by interfering with the sale of their produce, it was decided to prohibit the export of charcoal from Travancore, but when I was at Puliyara I was assured that very large quantities of this article were daily smuggled out of the country, much of it not even paying our light charge. This is not to be wondered at, considering the facilities for smuggling offered by the great extent of our boundary, which it is not easy to guard, and also the great difference between the prices obtainable in Tinnevelly, and in Travancore. However vigilant the Superintedent of Puliyara and his subordinates are this smuggling will continue, until we make up our minds to guard the forests themselves, and have a regular staff patrolling the boundary, and until we raise the price of charcoal in Travancore to the British rates. This may seem hard on the inhabitants of Travancore territory, but they have other advantages over their neighbours in Tinnevelly, such as a shorter carriage of their fuel, and when we find that by maintaining this low rate we are actually depleting our forests and conniving at smuggling, it is time to make some change, and that a very radical one, in the system at present in force.

352. I have not suggested the reservation of any land in this Taluq, the total area being small, but it would not be difficult to propose the closing of certain forests, and the working of them in rotation. But, until I know that the Government would approve of increasing the rates on charcoal, I shall make no suggestion, as we should otherwise not be realising any income to cover the cost of supervision.

No reserve suggested.

353. There are no cardamoms found in these forests, but those collected by contract in the Kulattbuppura valley and elsewhere are delivered at Puliyara. Wax and honey are not abundant, but are brought to the Depôt from the valley of Acchankōvil.

Cardamoms, wax and honey.

354. These forests are well supplied with roads. The cart road from Trivandrum to Shencottah and Tinnevelly passes through them. A good bridle path runs from Kannupilla mettu up to the Karinga-lōda pass, and foot paths diverge from the villages in different directions, so that to examine this forest area is not a work of any great difficulty.

Roads and paths.

(11) The Acchankōvil or Kolakkadavu river.

355. The river rises on the Western slopes of the Thuval mala (Coonumcal square rock), and Rāmakkal peaks ; its streams decending rapidly to the bottom of the valley, begin to flow West at a low elevation. As it proceeds, this river receives many accessions from the waters of small streams flowing down from the hills on the North and South. At almost the same point the Pulikkandan ār (Pulleycoudenaur) joins it on the South side, and the Aravikkara thōda on the North. Two miles below this point, the river 40 or 50 yards wide and 2 or 3 feet deep, passes the village of Acchankōvil (250 ft.),and, a mile farther on, it is joined by the Sanguppilāva thōda on the South bank. Still flowing over a sandy bed, with occasional dark pools, it is joined 3 miles beyond this by the Chittār (Chittar), a large stream which drains the valley below Nāga mala. Unfortunately a fall occurs on this river at about half a mile, from its junction with the Acchankōvil river, and its channel cannot therefore be used for floating timber.

Sources and course of this river.

356. About 3 miles below the Chittār the river is joined on the right side by the Kalār (Kullaur), a stream much larger, and with a stronger current than the Acchankōvil river possesses. The Kalār takes its rise on the Western slopes of the main range, between Thuval mala (Coonumcul square rock), and Karinkavala (Currincowly peak), and on the neighbouring hills. At a point due North of Acchankōvil, it is a strong river 70 yards wide, and 3 ft. deep (elevation 490 ft.) From here it runs North-west for 5 or 6 miles till it is joined by the Arimba thōda, a large branch which unites with it on the North. Its course is then South-west, winding excessively, and its waters rush over shallows, and eddy between large boulders until they finally join the Acchankōvil river a mile above Thora. The rocky character of its bed, which has given the river its name, and the fact that it descends nearly 300 ft., while the Acchankōvil river has fallen only 50 ft., make the floating of timber much more difficult than in that river, nevertheless timber operations are sometimes carried on here, and teak has been floated by this route from the hills above the Arimba thōda.

Junction with the Kalār. Description of the Kalār.

357. In Ward and Conner's map of Travancore, the Kalār river is depicted as flowing into the Rāni river. This is a mistake. The Kalār, which takes its rise in the neighbourhood of the Karinkavala peak, runs as I have described it, and is quite distinct from the Kalār which rises on the slopes of Karadi mudi (Godaracull peak), and joins the Rāni river at Kumarampērūr (Comarapairoor.)

Course of the Kalār as given in the map of Travancore is incorrect.

358. A mile below the junction of its two main branches, the Acchankōvil river receives the waters of the Thora river (Thorayaur), a large stream which descends from the North-western slopes of the Nāga mala hill, and flows due North.

Course of the river as far as Thora.

359. At this point (200 ft.) called Thora, the river is 80 yards wide and 2 or 3 ft. deep during most of the wet weather, and it is navigable up to here. Continuing in an almost direct line Westward, and

Course below Thora.

13

running òver a clear sandy or gravelly bottom, the river receives one after the other the Chēmpāl thōda (Shamepall thode), Kacchira thōda, Karruppan thōda (Purpiu thode), Parayan thōda, Pānan thōda, Kokka thōda (Cokay thode), Nadavattha muri thōda, and the Kūnan thōda (Cooman thode) on its right bank, and the Pālakkura thōda (Palacooey thode), Mūnār pāra thōda (Munuar parae thode), and the Kadi ār (Curry aur) before it leaves the Forest area 4 miles above Kōnniyūr (Koneyur.)

360. This river is used for floating timber from almost its very source, and
This river very well adapted for floating timber.
certainly from some distance above Acchankōvil, and it has been utilised for this purpose from time immemorial. * In his diary Lieut. Ward mentions that large quantities of teak were ready at Acchankōvil for floating, on the occasion of his visit.

361. The Kolakkadavu river drains an area of 189 square miles within the
Area divided according to classes of forest
forest limits. This may be divided among the different kinds of forest as follows :— (1) Heavy forest lying in the Kalār, in the valleys below Nāgamala, to the West of Chōmpāla ridge, and along the Acchankōvil river, 70 square miles. (2) Secondary growth, teak plantations, coffee estates, hill cultivation &c. 40 square miles. (3) Grass forest with trees, occupying the ridges, and the greater part of the land in the West of the area, 59 square miles. (4) Useless parts 20 square miles.

362. The teak tree grows very well in this valley, and attains an immense size.
Teak and blackwood attain a very large size
I measured one on the hills above Acchankōvil, which reached 22 ft. in circumference at 6 ft. from the ground. I saw other fine trees on the Alappāda ridge, to the North of Acchankōvil, but these were all growing in places very difficult of access, the whole of the teak fit to fell, and obtainable in convenient localities having been cut down long ago. Blackwood is also found of good size, and in fair abundance ; but the grass fires do these trees immense damage, and it is no uncommon thing to see trees scorched for 20 ft. high, and with the bark killed by the fire.

363. The irūl (Xylia dolabriformis) is abundant in some
Abundance of irūl and its distribution in the grass forest.
parts, as may be seen from the following distribution of trees in grass land, to the North of the river.

Constitution of Forest near Ponganpāra, elevation 1250 ft.

(Irūl) Xylia dolabriformis ...	...	...	...	46	Good cover.
Terminalia paniculata	...	...	...	20	Only a few trees
Dillenia pantagyna ...	...	...	...	14	cut out.
(Teak) Tectona grandis	...	...	...	15	(100). Height
(Ven teak) Lagerstrœmia lanceolata		...	...	4	80 ft. 30—40
Other kinds ...	...	..	...	6	trees to the acre.

364. Of the other trees ānjili is very rare, as it is the first to go when mer-
Scarcity of ānjili.
chants commence to work a forest. Anjili boats are exceedingly valuable, and it pays to drag them by elephants a distance which would be quite prohibitive in the case of other timbers.

365. White cedar is not common : it is the second tree to go, and there is very
White cedar and ven teak also becoming scarce.
little left in these forests. The third tree to be felled is ven teak, as the planks are easily floated down to the sea coast, where they find a ready sale for export to Arabia. There is still some ven teak on these hills, but it is being largely felled.

366. Kongu boats are in demand, but not the timber, owing to its weight,
The heavier woods not felled with the exception of kongu for boats, and in places easy of access
and the consequent difficulty of transporting it. Some 6 or 8 years ago the bamboos in these forests seeded and died down, and it is not easy to get enough to float down heavy timber. Consequently kōngu, vēnga and thēmbāvu are not now felled, so much as they used to be. Near Kōnniyūr the two latter trees are often cut and sawn up, as the pieces can be taken by cart. to a point on the river where large boats can come to fetch them. The vēnga trees are often seen with incisions on the bark for the extraction of kino.

_____

* Diary November 1, 1818.

367. Besides the trees mentioned, manjakkadamba (*Adina cordifolia*) is
common and grows to an enormous size, 5 and 6 ft. in diameter,
Other useful trees so also are meili (*Vitex altissima*), pūyam (*Schleichera trijuga*),
and vekkāli (*Anogeissus latifolius*), all useful trees.

368. The Acchankōvil valley is bounded on the East by the main range
which at one point where the path crosses it, sinks to an
Boundaries of the elevation of only 1,470 ft. North of this, Thuval mala
Acchankōvil valley. (Coonunical square rock) is over 3,000 ft., and Chēr mala
(Shair male), and Karinkavala (Curtin cowly) are probably 1,000 ft. higher. The
Rāmakkal hills to the South of the gap rise to nearly 4,000 ft.

369. From the main range the Nāga mala spur runs West-ward, rising in
Nāga mala itself to about 3,500 ft. : from here the ridge, divid-
Elevations of hills on ing this valley from that of the Kalleda river, rapidly descends,
the South. and runs at an elevation of only 1,000 to 1,200 ft.

370. West-ward from Thuval mala a well defined spur called the Alappāda
mala runs in almost a straight line, and divides the valley of
The Alappāda ridge. the Kalār from that of the Acchankōvil river. At its upper
end this spur is covered with grass, and maintains an elevation of more than 3,000 ft.
for about 5 miles. It then begins to descend, and is only 1,400 ft. at a point due
North of Acchankōvil, and 600 ft. near where it terminates above the Kalār river
These lower portions are covered with forest of different kinds, and contain good
teak, but in many places the Southern slopes of this spur are so steep as to be
almost precipitous.

371. From the Karinkavala peak a ridge runs South-West, winding very
much, till it reaches Chōmpālakkara (Shamepallaca:ay), a
The hills on its North- high grass ridge 2,940 ft. high situated almost due North of
ern boundary. Thora, with a precipitous descent on the Southern side, but an
easy slope on the North. It can be easily ascended from the West. Native tradition states that when wild elephants feel their final dissolution approaching they
travel to this hill, and make a point of dying here, and that a quantity of ivory may
often be picked up at this spot, but though I visited the hill, and examined it, I
found no confirmation of the legend in the shape of bones, or other remains of these
animals. West of Chōmpālakkara, and 3 miles from the highest point the ridge is
only 1,300 ft. high, at Kūttampārn, this elevation it maintains for 7 or 8 miles
passing Pongampāra 1,250, and Pulikka mala 1,150 ft., it then drops to a saddle
(400 ft.) which is on the large elephant track running from the keddah to the teak
plantations ; from this point the hills dividing the valley of the Acchankōvil river
from that of the Kalar rise to about 1,000 ft. Near the river many hills attain a
height of 600 or 800 ft., but the most important of them the Pāppankuri (Paupencolay), a bare rock near the Kokka (thōda, reaches an elevation of 1,300 ft.
Kōtta mala (Cotua Mulla), at the upper end of the valley, and below Nāga mala,
a conspicuous rock, rises to about 1,000 ft.

372. The valley of the Kolakkadavu river is much broken up by steep rocky
ridges : between them the soil is fertile and free, a condition
Soil, rainfall and climate. most favourable to the growth of the teak tree. The rainfall
is about 150 inches, and the climate similar to that of the
adjoining river basins. The wind does not blow severely here, except on the more
exposed and higher ridges.

373. About a century ago the whole of this valley, which at that time
belonged to the Rajah *[*] of Punthalam, was under cultivation,
The Acchankōvil valley as shown by the records in possession of the family. Native
once thickly populated. tradition states that wild elephants, who are at the present
time exceedingly abundant here, drove the people out, and the only place where
there is still any resident population, excluding the coffee estates, is Acchankōvil.

---

* According to the Memoir of Travancore Vol. IV p. p. 99 the administration of the Punthalam State
was assumed by the Travancore Government in 1812, and the Rajah was reduced to the condition of a pen-
sioner.

374. At the time of Lieut. Ward's visit this village boasted a large pagoda and a chowkey, and contained a population living in about 30 houses, and a havildar's guard, stationed here to prevent smuggling. When I visited the place in 1887, the number of houses was only 15 and the pagoda, whose copper roof * Lieut. Ward mentions, had been burned down in the year 1035 (1860). There still remains a high stone parapet approached by steps, and within the walled enclosure there is a well, and a tall copper flag staff, with an arm carrying a couple of bells and crowned by a copper horse, but the deity is located in a temporary building at the side. The temple is maintained by a yearly allowance from Government of 6,000 parahs of paddy, and rites are still performed there, while the population of 60 or 70 persons seems to consist almost entirely of Brahmins and Nāyars connected with it. The havildar's guard is no longer maintained, nor are the different chowkeys, here and at various points along this route, kept up, for, since the cutting of the Arinukāvu cart road, all the traffic has been diverted thither. A little below the Acchankōvil gap and on the Eastern side of it, there used to be a chowkey with a guard, and there is still a strong gateway built of stone in the narrowest part of the pass. Similarly, in numerous localities there are signs of former habitation. § Lieut. Ward mentions that about 30 years before his visit there was a large madam at Vākkānam (Wacanam tavalam), but that he saw no signs of it; also that at Parakkolam (Pana colum), to the South of the river, 3 families were still residing, and that there was a pagoda. a spring, and the ruins of a small palace surrounded by a deep ditch there. At Kunthōdi kōvil, farther down the river, there are remains, as also at Kakkāri kudi on the ridge to the North, and doubtless at many other places that I did not visit. The presence of these ruins supports the local tradition, and makes it probable that at one time the population hereabout was more scattered than it is at present, though it does not follow that it was more numerous.

The village of Acchan kōvil, and signs of habitation in other places.

375. About 1,000 acres of land, in the neighbourhood of Nāga mala were sold for coffee cultivation, and of these about 200 are still kept up, giving employment to over 200 maistries and coolies. In addition to these people there are often several hundreds living for a time in this valley. These are timber merchants, and their followers who form numerous camps during the wet season, and villagers from the low country, who come up and sow paddy on land they clear, and then return to their homes, coming back again for about a month to watch and reap their crops.

Coffee estates and timber camps.

376. The Hillmen who reside in the valley of the Kolakkadavu are Pulayans, and number about 30 persons. They do not cultivate much land, though, I am told they do have small clearings. They live principally on the fruit of the eentha (*Cycas circinalis*), and on roots. Occasionally they manage to purchase paddy with the money paid them for the wax and honey they deliver at Puliyara. They seldom remain for more than ten days in one place.

The Hillmen.

377. Thus the resident population of this valley does not exceed 300 souls.

Total resident population.

378. The forests in the basin of this river are under the charge of the Assistant Conservator of Forests who lives at Kōnniyūr, and who has under him a numerous staff. At that place there are two nice bungalows, one occupied by the Conservator on the occasion of his visits, and the other called the Assistant Conservator's bungalow, though never used by that officer. There is also a Cutcherry, and elephant cage for training wild elephants. On the river, about 4 miles from Kōnniyūr there are 700–800 acres of teak plantations, besides two small patches at Aruvipālam, and Kalannyūr, but these I will describe later on.

Management of the forests.

379. These forests have, in consequence of the great suitability of the river for floating timber, been worked for a long time: at present, owing to the want of bamboos, the heavier timbers are not exported by water, nor to the same extent as formerly, but teak and the lighter timbers are largely felled.

These forests have always been worked.

---

* Diary October 30, 1818.
§ November 6, 1818.

Annual yield of timber. 380. Mr. Thomas estimates the number of logs taken out of this valley every year at 4,000 teak, and 1,000 of other woods.

381. Mature teak of the first or second class (above 15 inches in diameter) is getting very scarce, and difficult to procure, but there is still in this valley a good supply of trees of small size which require careful attention. Much of the teak timber brought down every year is immature, and would grow much larger if left for 20 or 30 years more, and it is a great pity to sacrifice it merely for the sake of showing a large present revenue.

A great deal of immature teak felled every year.

382. I recommended that the whole of this valley should be retained as a reserve, but the Southern boundary proposed was slightly altered, and the river itself is the Southern limit now chosen, while the Western is the path running from the teak plantation to the keddah.

This valley recommended for reservation, and a reserve proclaimed.

383. The reservation of this valuable forest is an excellent thing. Hitherto it has been customary for the villagers to come up in large numbers, and make clearings for paddy on the hill sides. Of course all the best land was taken for this purpose, and after giving one crop of grain it was abandoned. This will all be prevented now.

This will check the hill cultivation.

384. On the hills to the North-east of the teak plantations, and at an elevation of 1,200 ft. there are several cardamom gardens, where the Kanni variety of cardamom is grown. Some of these belong to Government, and some to private owners. At the time of my visit a small bungalow was being erected in their neighbourhood, for the supervision of these gardens, and the storage of the spice.

Cardamoms.

385. The wax and honey and other minor forest produce collected on these hills is delivered partly at Kōnniyūr, and partly at Puliyara.

Wax and honey &c.

386. The valley of the Acchankōvil river is traversed by a foot path running from Kōnniyūr along the North bank as far as Thora, a distance of 17 miles. At this p'ace there is a ferry boat maintained by Government, and the path crosses to the Southern bank, and runs along it for 6½ miles to Acchankōvil. From this point to the frontier is 7 miles, making a total of 30½. For the whole of this distance the path runs level till within a mile of the pass, when it ascends rapidly, and falls with equal rapidity on the other side to the small hamlet of Mēkkara.

The path from Kōnniyūr to the Eastern boundary.

387. The numerous streams which cross the track, and flow into the river, make the journey at times very difficult, for they rise with great rapidity, and the water in them, owing to the high level in the river, is backed up, and does not fall. If this path were much used it would be necessary to erect numerous bridges.

The path often rendered impassable by the streams.

388. A proposal has been made to carry a bandy road from the Puliyara depôt to the top of the pass, and down to Acchankōvil in order to get timber for the depôt, but the cost of such a work, owing to the rocky nature of the ground, would be very heavy, and besides that, the timber is required in Travancore. Later on, when we are able to estimate the annual yield of these forests with some attempt at accuracy, it may be advisable to cut such a bandy road, for the removal of our surplus to Puliyara, but for the present I deprecate the felling of timber in quantity, till we know how much the forests can be expected to yield.

A cart road from Puliyara to the top of the Acchankōvil pass not recommended.

389. The Kalār valley is very inaccessible, and a good path should be cut along the side of the river to enable it to be more easily examined and worked. Some good teak, which is only waiting to be felled and brought down, is to be found on one of the branches of this river. In certain places where the channel of the Kalār is blocked by rocks, which render the passage of timber difficult, the impediments should be removed by blasting.

Improvements required in the approaches of the Kalār.

14

390.   Though the Acchankōvil river is so well adopted for the floating of timber,
it is not at all easy to get logs, which are at some distance from
it, down to the water, and so to utilise its resources to the full
extent.   This is because no systematic attempt has been made
to open out timber ways up each side valley, which would serve
the whole of the valley.   At present every timber merchant clears his own track to
the trees he wishes to " exploit ", and this makes the cost of removal greater, while
very much injuring the forest by the numerous openings which traverse it in all
directions.

Improved means of exit required in the side valleys of the Acchankōvil river.

(12)   *The Rāni river.*

391.   The Rāni is formed by the junction of 3 large rivers,
the Kalār, the Kakkāda ār and the Valiya ār.   In the last named
are united two small rivers, the Pamba ār, and the Arutha ār.

Chief tributaries of the Rāni river

392.   The most Southerly of the three, the Kalār, takes its rise in the valley
North of Chōmpālakkara, and on the hills at its Eastern extremity between Karinkavala (Curriu Cowly peak), and Varukka
pāra (Wacaparei Mogle), two streams, the Chellakkal ār, and the Kāni ār take their
rise, and, after uniting, flow Westward past Vilanga pāra (Shelly cull.)

Sources of the Kalār.

393.   At this point the combined stream is very rocky and full of rapids (700 ft.),
and descending quickly it shows an elevation of 500 ft. where
it is joined by the Mani ār (Munneanaur), which falls into it on
the right hand.   A mile below this point, where the Pākkal ār
(Wacullaur) joins it, the elevation is 410 ft., and above this
point the river is far too steep for the floating of timber.   The Pākkal ār takes its
rise somewhere in the vicinity of Chiranga mala (Sherungum mullay), and running
due West falls steeply until within a mile and a half of the main river, here it is
joined by the Ettālungal thōda, and the combined streams unite with the Kalār, as
already stated.   Continuing in a more Westerly direction, the Kalār, now a broad
river 80 yards wide, receives the waters of numerous small streams, the Asāri thōda,
the Mūkkan thōda, and others, and after flowing for about 6 miles it is joined by the
Thēk thōda (200 ft), which runs down from Nānāttappāra, and Panniyara (Púnny-
yerra).   Turning to the South the Kalār passes the keddah after 4 miles, and the
forest bungalow at an elevation of 180 ft., a mile below it.   It is here 100 yards wide
and hardly fordable.   Proceeding North-west it is joined by the Kūttukal thōda
(Cotacal thode), the Karingāda thōda (Curringud thode), and the Chēlaviriccha thoda,
all on the right hand, and a mile below Kumaram pērūr (Comaru pairoor) it unites
with the Rāni river, at an elevation of about 30 ft.

Course of the river up to its junction with the Rāni river

394.   The sources of the Kakkāda river are between the lofty peaks of Chiranga
mala (Sherangammullay), and Pulicchi mala (Pollatchy mullay).
Turning North-west the streams, which have united to form
this river, pass close under the Valanya catti (Neely mullay)
ridge.   It proceeds West for about 6 miles, and above this
point is called the Manian ār (Muneean aur), and here it is
joined on the South by a small river formed by the combination of the Thīkkal ār,
the Vetti ār, and the Mūli ār, and on the North by several small streams.   Turning
Northward so as to circumvent a ridge, and passing close beneath Thēvara mala
(Tayvara mullay), it receives at this point the Palaya ār (Palawa aur), whose waters
are in their turn supplied by the Kaduvā thoda (Cuddavay thode), and turning South
it is from this point called the Kakkāda ār.   This river, after running for a short distance, is joined on the South by the Chīkka thōda (Sheeca thode), and, changing its
course, it proceeds North, and then North-west, and is here increased by the waters
of the Chittār, a small river whose branches rise near Vampuliala pāra (Bumbalaly
parae), and are called the Pulivalli thōda, and the Ambalam thōda, and by the Mani
ār.   These rivers fall into it on the South.   The Kakkāda river, winding very much
between high hills, receives to the East of Vadana Kunnu (Wadany Coon), the
Mūnnār pāra thōda, a stream which drains a fertile valley on its North bank, and a
mile above Peranāda it is joined by the Kāri thōda, and. at the first mentioned place,
it unites its waters with the Valiya ār at an elevation of 40 ft. to form the Rāni
river.

Sources of the Kakkāda river, and its course to its junction with the Valiya ār.

395. The Kakkāda river is navigable for small boats for about 5 miles above Peranāda, though on account of its strong current, even these ascend with difficulty. Logs can be floated down from the neighbourhood of Thēvara mala, and teak was in former times felled near Nellikkal (Nelly kul), and transported by the Kaduvā thōda, and the Palaya river, but these Government works have long ago been stopped, though timber contractors still fell in the hill sides overhanging the river lower down. At its point of junction with the Valiya ār the Kakkāda river is about 50 yards wide.

The Kakkāda ār used for floating timber.

396. The Valiya ār, as already stated, is composed of the Pamba ār, and the Arutha ār. The former rises on the high hills to the North of Pulicchi mala (Pollutchy mullay), and runs through heavy forest, at an elevation of over 3,000 ft., for several miles, before falling into the valley below. It then drops very rapidly, between very high cliffs, to an elevation of about 800 ft., and turns North-west, and at 12 or 15 miles from its source, it is joined at Anavattam (Anawattam) by the Kalār. This river rises slightly to the North of the other in dense, matted jungle, and runs West at an elevation of over 3,000 ft., being here 20 or 30 yards wide and knee deep. It then falls perpendicularly over a cliff, some 2 miles East of Koretti mudi (Cornyty moody), into the valley below, and flowing Westward is joined by a large stream which drains the valley West of Shabari mala (Chowry mullay), and, a few miles below that point, unites with the Pamba ār. At this point the river is probably about 300 ft. above sea level. Three miles below the junction is a spot called the Pambakkadava (Pumbay Caduvoo), and, 7 miles below that again, the Pumba and Arutha rivers combine.

The Southern branch of the Valia ār, the Pamba ār, and its sources and course.

397. The Arutha river does not take its rise so far to the East as the three rivers last mentioned, but it is composed of several considerable branches, which drain valleys full of heavy jungle and it is therefore a large stream. The most Easterly branch in the valley between Chiragirunna mala (Cherigooandan mayle), and Chuttambalam mala (Chitumbalam mogle); and the Northern branches the Pāmban ār, Arutha and Panni ār rise on the Peermerd plateau, and fall over its edge into the valley below, and joining their waters with the Eastern branch, run South-west, and the Arutha river thus formed continues till it unites with the Pamba ār and becomes the Valiya ār. The latter river proceeds West for 6 miles without any obstructions, and then falls over the Peranthēn aruvi (90 ft.). At this point the river is crossed by a reef of rocks, and, except in very high flood, rushes through a gap in it, and falls about 20 ft. Below this fall there are two more barriers of rocks 3 to 6 feet high, but not sufficiently abrupt to prevent navigation.

Sources and course of the Arutha river.

398. Small boats can ply up this river as far as the above waterfall, but timber can be floated down the Pamba ār from above Anavattam, and on the Arutha from a good distance above its junction with the Pamba river.

Timber can be floated down the Valiya ār.

399. The Valiya ār is joined by the Kakkāda ār at Peranāda (40 ft.), and 3 miles lower down by the Kalār, at which point, Kumaram Pērūr (30 ft.), the combined rivers now called the Rāni river leave the forest area, as a powerful stream nearly 200 yards wide. The Rani river runs, with a winding course, in a Westerly direction for nearly 30 miles, when it is joined by the Mani mala river, and 6 or 8 miles farther on, at Vīyapuram, the Kolakkadava river unites with it, the whole of them then pouring their waters to the North-west, and eventually into the Vembanāda lake.

Course of the Rāni river below Peranāda.

400. The area of forest land drained by the Rāni river amounts to 487 square miles, an extent which shows it to be the most important river in the State, the Periyār only excepted. This may be divided as follows (1) Heavy forest, 367 square miles. (2) Secondary forest, hill cultivation, coffee estates &c., 80 square miles. (3) Grass land with large trees, 20 square miles. (4) Useless land, 20 square miles.

The area of the Rāni basin classed according to the forest on it.

401. It will thus be seen that by far the largest portion of this basin is covered with moist forest. About half of this is worthless, or I should say, we can make no use of it at present, for it is at a high elevation, and is difficult of access from any quarter. In addition,

Large extent of moist forest, much of it worthless.

this forest lies at such a height, that it contains none of the saleable trees such as kambagam, white cedar, ānjili, and so on. Probably the most useful trees found here are the red cedar and the kurangādi (Acrocarpus fraxinifolius), but the latter has no market value.

402. The heavy forest growing at the lower elevations, on the slopes of this higher plateau, as well as in the low country at the foot of the hills, contains ānjili, white cedar, ven teak, kambagam, chittagong wood (*Chickrassia tabularis*), pūn-spar (*Calophyllum tomentosum*), nāngu, black dammer (*Canarium strictum*), and others, but timber merchants have been at work here for a long time, and from the more accessible places, the first 4 trees mentioned have been removed. In the neighbourhood of Nellikkal I found numerous trees of ebony, which wood does not seem to have been telled in this part of the country.

The moist forest at the lower elevations much more valuable.

403. This moist forest, broken only here and there by patches of cultivation, or by ridges covered with rocks or grass, extends in one block from the Eastern most hills to within 6 or 8 miles of the forest line, thus leaving a strip of about that width between the moist forest to the East, and the cultivated country to the West. The line of forest runs from Pāppankuri (Paupencolay), a rocky hill above the Kolakkadava river to the keddah, and thence proceeds almost due North to Nāladipāra (Nellady paurae), and continuing in the same direction, strikes the Valiya ār two miles East of the Peranthēn waterfall.

Large block of forest almost unbroken.

404. In the strip outside this line, there is no moist forest, except in very small patches, the original forest having been felled in former times, and the land cultivated, and to the same spot the low country people return at intervals of 10 or 12 years, cultivating other parts until the soil has recovered some of its fertility.

Strip of partly-cultivated land outside the block.

405. On the ridges, dividing these partly cultivated valleys, grass and large trees grow, and here are found teak, venga, blackwood, thēmbāvu, nūl, and such trees. The first mentioned tree grows here very well, and on all these ridges it is abundant everywhere, the soil being free and well suited to it.

Distribution of the more common trees

406. But the quantity of teak of large size still remaining is small, for these forests have been worked for this timber for a very long time. Even at the time of Lieut. Ward's visit the felling of teak had extended as far as Nellikkal, and Nānāttappāra. What still remains is teak of small size, which should not be felled till it has reached a girth of at least 10 borrels, equivalent to a diameter of 17 inches in the middle of the log, or, say, 20 at its base.

Quantity of large teak is small, and what is left should be carefully husbanded.

407. The Kalār river is confined between the Chōmpālakkara and the Chennāttakara (Mosrinya curray) ridges, each of which rises to a height of about 3,000 ft. the latter being some 100 ft. the more elevated of the two. Both of these ridge are covered with short grass on their highest points, and stand up conspicuous objects from a distance. As they extend to the West they break up into numerous spurs, which slope down to 600 or 800 ft., and are covered with forest and cultivation.

Ridges to the North and South of this river.

408. At the upper end of the valley the higher peaks rise to more than 4,000 ft., and are clothed with vegetation almost to their summits. The slope of the land in this valley is easier than in the adjoining valley of the Acchankōvil river, but the soil is poorer. After once being cultivated, the land becomes covered with the eettareed (*Beesha Travancorica*), a sure sign of thin, poor soil, whereas land near Acchankōvil, similarly treated, grows up in soft stemmed bushes of different kinds, and eventually, if it is not cleared again, trees appear once more.

Soil in the Kalār valley much poorer than in the adjoining Acchankōvil valley

405. The Kakkáda river runs between the Chennáttakara ridge, and another
grass ridge of considerable height, which descends at Nellikkal
to 1,100 ft., and then rises again to a height of 1,500 or 1800 ft.,
and is covered with forest. West of Nellikkal it is called the
Thévara mala murippa. Half way along it is Mélppáru (Meil paraethavalam) 1,720
ft., and on its Western extremity are Náladi pára, and Rájan pára, the former a
small rock by a stream, and the latter a flat rock of 2 or 3 acres in extent: the elevation of each of them is about 1,500 ft. After this the ridge runs down steeply to
Peranáda. In the centre of the valley is the Valanga catti (Neely mullay) ridge,
covered with grass, and detached from it is the Thévara mala, a conical hill rising to
a height of 1,800–2,000 ft., and covered with forest.

Water-shed of the Kakkáda river.

410. The Pamba ár drains the valley between Thévara mala murippa and
Kanakkan mala (Conaco mole). Its sources must be at an elevation of nearly 5,000 ft., on the high mountains which form
the back bone of the Peninsula. Koretti mudi (Corayty moody),
a grassy peak on the plateau usually called " Paradise, " is 3,750 ft. ; Tholappa thávalam (Tolapay tavalam) 3,100 ft. ; Nullathannipára 3,400 ft. ; Mánammutti mala
3,970 ft. ; and the average elevation of the Padinyátta pára plateau is about 3,600 ft.
Chuttambalam mala (Chitumblum mole) is 3,540 ft. Kanakkan mala (Conaco mole)
3,200 ft. The two last mentioned points are on a grassy ridge called " Hannyngton's
plateau " by Europeans, and running out from the Padinyáttapára plateau towards
the West. Being somewhat inaccessible, this place is much frequented by elephants
and bison, though its extent is not large.

Highest hills in the basin of the Pamba ár.

411. The Arutha river is confined between the Padinyátta pára plateau and
the Mayiládi ridge (Myladdy mole) on the South, and the
Peermerd plateau on the North. The average elevation of the
latter is 3,500 ft., but at some points it runs up much higher.
Amratha méda (Vembaulay mudi), a high grass hill above the
sources of the Arutha river is 4,500 ft. Perambukótta (Peruinby cotay), a peak on
the grassy spur running South from Peermerd, and Chemmani kótta (Chemmuny
cotay) are 3,050 ft. Perumpára mala (Perimparae mole), a conical hill in the valley
below, is about 2,500 ft. high, Kombukutthi is 1,240 ft., and the Gópura mala, a
pointed hill covered with forest on the North, with a precipitous cliff on its Southern
face is about 1,800 ft.

The chief hills and peaks on the basin of the Arutha.

412. The rainfall on the Ráni river is about 150 to 200 inches. During the
South-west monsoon the water remains at one high level, but
after the commencement of the North-east, the river is liable
to very high floods, showing that its sources lie far to the Eastward, and are well under the influence of that monsoon.

Rainfall.

413. A good deal of land in the valleys of the Kakkáda and Arutha river
is steep, but the soil is good and better than in the Kálar
valley.

Soil.

414. There seems to be very little wind in any of the
forests drained by the Ráni.

Absence of wind.

415. Within this basin the chief places of interest are Nánáttappára (1,770 ft.),
a rock West of the Chennáttakara ridge, where cardamoms are
dried. There are several gardens in this neighbourhood, and
the produce of them is annually collected, and handed over to
the Forest Department. During the off season no one lives here, but during the
cardamom crop time there is a large camp. Vilangapára (Shelly cull) on the upper
part of the Kálar river at 1,000 ft., is a similar station with no resident population.
A great part of the land in the valley close by has been cultivated, and is now covered with a dense growth of reeds.

Chief places of interest. Nánáttappára.

416. Pongampára a similar rock at an elevation of 1,350 ft. serves the cardamom gardens situated on the ridge dividing the Acchankóvil
and Kálar valleys.

Pongampára.

15

417. Upon the Kalār, about 4 miles North of the teak plantation bungalow,
and 8 or 9 from Kōnniyūr, a keddah for capturing elephants
was constructed about 15 years ago. After being used for 9 or
10 years it was abandoned, and elephants are not caught there now : a small bungalow and a training cage were built on the same river, and both are still standing.

The keddah.

418. The Panniyara fields (Punniyerra) marked in the map, were, at the time
of Lieut. Ward's visit, occupied by people from the low country,
who had sown paddy there, but the spot seems never to have
been permanently occupied any more than dozens of other places which are cultivated
in a similar manner.

Panniyara fields.

419. The only villages within the basin of the Kalār are Kumarampērūr, and
Mādōnam, situated at the junction of this river with the Rāni.
The former contains a population of about 30 persons, and the
latter 20. At Kumarampērūr there were in Lieut. Ward's
time 25 houses, now there are only 6, and the land once occupied is now abandoned to
the elephants who destroyed the buildings, and drove the inhabitants from their
homes.

Villages and their population.

420. Proceeding Northwards to the valley of the Kakkāda, the most noteworthy
place is Nellikkal situated on the dividing ridge between this
river, and the Pamba ār, at an elevation of about 1,100 ft.
* When Lieut. Ward visited the place, he found a pagoda, and
a Pandāran in charge of it. This man told him that about 300 years before there
was a population of 300 Nāyars there, as well as a number of Syrian Christians.
What reason induced them to leave the place cannot be ascertained, but probably the
elephants drove them out. One tradition states that the people were destroyed by
flying cockroaches or dragons, who attacked them in pairs, and while one removed
the roof of a house, the other seized the inhabitants, and ate them up. But whatever the cause, Nellikkal is now entirely deserted, the pagoda has tumbled down,
the Pandāran has died, and found no successor, and all that remain at the present time
are numerous ruins, and a small natural tank close by the path.

Places of interest on the Kakkāda river Nellikkal.

421. Thavalappāra (Towlapparae), and Vampuli-ala-pāra were timber camps
when Lieut Ward visited them. No one lives there now.
Lēkha mala (Lauga male) on the Thēvara mala murippa, is not
inhabited, but is noteworthy as marking the boundary between
the area supervised by the Forest Department and by the cardamom Department, in regard to the collection of the cardamoms.
South of this point the spice is collected by the Forest Department, and North of it
by the subordinates of the cardamom Superintendent.

*Lēkhamala, the boun-
dary between the Forest
Department and Carda-
mom Department.*

422. Peranāda and Kakkāda situated at the junction of the Kakkada river
with the Rāni, the former place on the North, and the latter
on the South bank of the Kakkāda river, contain a good
number of houses, but both of them were at one time larger
than they are now, many of the inhabitants having died of fever or having been
killed by wild animals. Peranāda is situated on the river bank, and has a high and
steep ridge rising behind it. Its population may be about 200 souls and consists of
Nāyars, Syrians and Chogans. All the compounds are protected by stone walls to
keep out elephants, but in spite of this I noticed several enclosures which had been
broken down, and from which the owners had been expelled. Kakkāda lies upon
more level ground, and contains a population of about 100 persons. All these people
engage largely in hill cultivation, upon which they entirely depend for their food.
They have no irrigated paddy fields, and they do not hold much communication with
the country to the West, for they grow sufficient coconuts, and arecanuts to supply
their wants. Paddy is the medium of exchange here, and money is of scarcely any use.

*The villages of Pera-
nāda and Kakkāda.*

423. The other small villages situated within the forest area, and in the basin
of the Rāni river, are so similar that one description will do for
them all. They are built on high ground, and the houses are
placed in compounds protected by revetments, the passages or

*Other small villages all
very similar in character.*

, vi..age pu.. running at a much lower level, often 10 ft. below that of the compounds. Each village has, or had at one time, a pagoda (its ruins testifying to the fact), and a tank, and every one of them shows that its population was at one time much larger.

424. The inhabitants are chiefly Nāyars, or Mahomedans, or Chogans : Syrians prefer to congregate in the towns, or larger villages. The *Their names.* names of these villages are Karivellam (Curricolum), containing 15 houses, and, say, 75 people, Chēttakkal (Shatacul) 20 houses and 100 inhabitants, Kunam (Coonatoo) 15 and 60, and Nūrvakkāda (Noorvacaud), where there is said to be only one Nāyar family I did not visit the place myself. Yeddamurroogoo (sic) is abandoned.

425. In the interior, the famous pagoda of Shabari mala (Chowry mala) merits attention. Lieut. Ward describes 2 temples, and a small *The Shabari mala (Chowry mala) pagoda.* kottāram as in existence in the beginning of the century, and states that on the hill above there was a tank, and a small chattram, for the convenience of travellers coming from Tinnevelly and Madura. At the present time, only the chief building remains : the elephants have knocked down the others, and the chattram up the hill. The pagoda is on a level spot, upon the top of the Nīla mala ridge, at an elevation of 1,700 ft. ; here a revetment of enormous slabs of dressed stone has been erected, measuring 15 ft. high, and 50 yards long and broad. The area enclosed within these walls has been filled up, and 2 or 3 small buildings with copper roofs have been erected upon the terrace, and a row of coconut trees planted round its edge. During the greater part of the year this pagoda is deserted, except by a couple of watchers, and sometimes by a recluse, who may have selected this spot for his retirement. Certain priests are nominated to perform ceremonies there once a month, and if they carry out their instructions regularly they must have hard work, for they have to travel a very long and difficult way from the low country. The annual festival commences at the beginning of the month of Magaram (Jan.-Feb.), and lasts for a week, and very large numbers of pilgrims from different parts of Travancore collect here at that time.

426. Besides the population described as residing in the villages at the foot of the hills, there are several colonies of Hillmen located here. *Colonies of Hillmen. The Malayadiyār.* The most Southerly is a tribe of Malayadiyār, numbering 20 persons, who live close to Nāuāttappāra. They used to clear land about the valley of the Kakkāda, and were there in Lieut. Ward's time, but they lately moved up to their present position in order to assist in collecting cardamoms, and working down timber. They are Kuravar by caste, and seem a feeble race. They cultivate little, but depend for food largely on roots, and on the fish they kill by poisoning the rivers. To the Forest Department, and the owners of the cardamom gardens, they are very useful, as they know the forests well, and are handy in collecting produce, or running up huts.

427. There are said to be 2 colonies of Vālens living near the confluence of the Arutha and Pamba rivers. I did not come across them in the *Vālens.* course of my travels, but some years ago Mr. Dighton, while on a shooting trip, visited their " kudies ". They number about 50 persons.

428. In the basin of the Arutha river there are, to my knowledge, 5 villages of Arayans at Valavakkuri, Kālaketti, Kōthayadi, Kombukutthi *Arayans.* (Kumbucooty), and Kirakka (Kellekkay). These people are much more settled than other Hillmen, their houses are substantially built, and surrounded by groves of jacks, arecanuts, and palms, and are always located in the same places. It is the cultivation only that is shifted from one spot to another. Their number is perhaps 100 or 120, not more, for their colonies are not large.

429. The Hill Pandarans a perfectly wild race, said to be broken up into 3 gangs numbering 40 or 50 in all, live in rocks, and caves, or *The Hill-Pandarans.* build themselves temporary huts in the innermost recesses of the Arutha and Pamba forests. They do not clear land at all, but subsist on roots, and fruit, or fish.

430. Upon the Peermerd plateau and close to the sources of the Artha river
there is a cluster of coffee estates, a Residency, Magistrate's
cutcherry, Hospital, and station of a D. P. W. officer. The
number of acres sold for coffee cultivation was about 4,000, of
which more than half was opened. At present about 900 acres
are kept up, tea having almost everywhere replaced the coffee, and the population on
the estates and attached to the Sircar offices may amount to 1,000.

Population on that part of the Peermerd plateau included within the basin of the Arutha.

431. Adding up all these figures we get a total of 1,838
persons resident within that portion of the basin of the Rāni
river included in the forest area, or in round numbers 2,000.

Total population of the basin of the Rāni.

432. At the beginning of the century the forests of the Rāni river must have
been more extensive than they are at present, and though the
small villages in the interior boasted a larger population than
they do now, yet the land lying between them was probably
covered with large forest trees, and not with secondary forest,
and scrub. There was then but little demand for any timber
except teak, and perhaps ānjili. White cedar would not be required at all, as the
coconut oil trade was then in its infancy. Similarly those timbers now largely felled
for export ven teak, kambagam, vēnga, irūl, and thēmbāvu would not then be cut.

Forests of the Rāni very extensive at the beginning of the century. No demand for timber then.

433. It is only within the last 20 or 30 years that there has been a demand
for any of these trees, and even now, in the interior, the heavier
woods have been very little cut, because there is a great scarcity of bamboos on this river, and therefore it is impossible to float
them, and there are no cart roads. Anjili has been very large-
ly felled from all accessible places, and it is very rare to meet with a tree of any size.
The same may be said of white cedar in a lesser degree. The other trees must remain untouched until some means of working them down is devised, either a road, or
a slide, or something of that sort.

The demand has much increased in the last 20 or 30 years.

434. At the time of Lieut. Ward's visit, teak was regularly felled every year
on this river. The Deputy Conservator Mr. Walcot had a
bungalow at Rāni, and during the monsoon used to carry on
extensive operations in various parts of the forests. Timber
felling was going on at Nellikkal and Nānāttappāra in 1818,
and the forests about Thavalappāra had been worked a short time previously. These
places are not the Easternmost limit of the teak forests, and it is not therefore any
matter for surprise that the quantity of large teak on this river, is small, for even
though 70 years is a long time, ever since then, and especially during the last few
years teak has been felled, and, though the number of trees still standing is large,
they consist almost entirely of 3rd class timber.

Teak has been felled here for a very long time, and large trees are scarce.

435. The forests of the Rāni are partly under the Hill Aminadar of Kōnniyūr,
and partly under the Aminadar of Kānjirappalli, who at the
time of my visit was living at Rāni. These officers go as far as
the strip of semicultivated land, and sometimes prosecute the
hill cultivators for felling, or girdling trees growing within their
paddy-clearings, but they seldom enter the forests proper, and make no attempt to
supervise the felling.

Management of these forests Supervision of the interior imperfect.

436. I recommended that the interior forests of this river should be included
in a reserve, and this has been done, although the boundaries
I suggested have been altered, and the area reduced. The
reserve now proclaimed extends over 300 square miles, partly
on the Acchankōvil river, and partly on the Rāni.

A reserve proclaimed on this river

437. The forest area included in the basin of this river is very badly supplied
with roads, in fact it has none. A trace for a bandy road has
been made from Kōnniyūr to Rāni, but this is outside the
forest line, and the road itself has not been completed. From
Rāni to Manimala it is proposed to cut a bandy road, but of this not even the trace
has been finished. This too is outside the forest line.

Cart roads. None at present.

438. As I have already shown, it is possible to ascend the branches of this
river in small canoes for some distance, but, after they cease to
be navigable, the only way to travel is by shockingly bad foot-
paths, sometimes running up hill, sometimes descending, and often overgrown with
thorny bushes, or abandoned for years together. The most important of these foot-
paths is the one that leads from Peranāda up to Rājan pāra, and then proceeds along
the Thōvara mala ridge till it descends to Nellikkal. Beyond this there is no regular path, though at the time of the Shabari mala festival a good path is cleared to
the pagoda, and beyond this an elephant track runs up the hill to the Padinyāttappāra
plateau. "the Plateau" as it is called, and on to the forest to the East.

Foot-paths.

439. Many other such foot-paths exist, and run in different directions; they
are kept in good order, or entirely abandoned, according as
there is cultivation in the vicinity of them or not. This absence
of roads partly accounts for the manner in which these forests
have escaped destruction : as the rivers have been the only
means of removing timber, and bamboos being scarce, boats
and light timber alone have been hitherto exported. It will be part of the duty of
the Forest Department to open out roads through these forests.

These foot-paths quite obliterated if there is no cultivation in their neigh-bourhood. Roads much required.

(13) The Manimala river.

440. The largest branch of the Manimala river takes its rise close under the
Mothavara (Kanadi hill), and drains the valley to the West of
Amratha mēda (Vembanlaqmudi). It is called the Palaya ār.
After flowing for about 6 miles, it is joined at Kūttukal by the
Nyārampulla ār, which drains the valley South of Amratha mēda, and by the
Perinkolam thōda, which flows into it from the North-west. The combined streams
running South are joined at Mundakkayam (190 ft.) by a small river, which is said
to be the Southern boundary of the land claimed by the Chengannūr chief, and
whose course is North-west : a mile lower down, the Manimala river is joined by
the Ulakandan thōda, on the right bank, and after receiving many smaller streams,
and following a very winding course, it is met 5 miles lower still by the Pēra thōda
and Gōpura thōda.

Sources and course of the Manimala river.

441. Up to this point boats can come, but timber can be floated from above
Kūttukal. The river now turns Westward, and proceeds with
a tortuous course, and after 2 miles it is joined by the Kāri
thōda which flows into it on the South. Four miles below this
Chennāyi pādi (Shennapaudy) is passed, and a mile farther on
the Manimala unites with the Kānnyirappalli, and turns almost due South.

River navigable as far up as its junction with the Gōpura thōda.

442. The latter branch takes its rise on the hills near Thambalakkāda and
Pālappura, and is joined by the Pūthappāndi thōda, a mile below Kānnyirappalli. This branch is navigable up to a mile above
its junction with the main river, at which point a reef of rocks,
5 or 6 ft. high crosses it, but timber can be floated down from higher up, and especially by the Pūthappāndi thōda.

The Kānnyirappalli branch of the river.

443. A mile below its junction with the Kānnyirappalli branch, the Manimala
river leaves the forest area, which it has been skirting, and
running South-west eventually joins the Rāni, and pours with
it into the Vembanāda lake. This river runs very low in the
dry weather, but during the rains it rises with great rapidity, and its current is very
strong.

Course of the river outside the forest area.

444. The area drained by the Manimala river, and included within the forest
line, is 168 square miles, which may be divided as regards the
character of its forest as follows:—(1) Heavy moist forest, 16
square miles. This is chiefly in the Kūttukal valley: and to
the South of Peruvanthānam. (2) Secondary forest, culti-
vation &c., 110 square miles. (3) Grass land, with large trees, 30 square miles. And
(4) Useless land, 12 square miles.

Division of the area according to classes of forest.

16

445. It will thus be seen that there is not much original forest in the basin of this river; by far the larger portion of the land is covered with secondary forest, the result of hill cultivation, which has extended, and done immense damage here, owing to the proximity of the Peermerd cart road, which enables the cultivators to cart their produce to a distance, and to obtain a better price for it.

Large extent of secondary forest

446. Of the trees growing here teak is to be found most abundantly South and East of the river, and in the vicinity of Erumēl and Alappāra it is excellent, though small, for the contractors have been, and are still at work in these parts. North and West of the river, and in the Kūttukal valley, teak does not occur, except near Edakkōnam.

Teak abundant to the East of the river but is almost absent from the West of it.

447. The reason of this is that South of the river the soil is much more open and full of stones, which teak likes, whereas to the North of it a deep red loam is found, which is retentive of moisture, and admirably adopted to the growth of pepper, and the ānjili, blackwood, and venteak trees, but not of teak. The only exception is near Edakkōnam, where some rocky hills rise to the height of 500 or 600 ft. above the surrounding country, and the soil around them, made free by the admixture of stones and sand from the hills, becomes suitable for the teak tree.

Reasons for this.

448. Anjili as a wild tree is very scarce in this valley, and white cedar may be said to be absent, it is seldom met with.

Scarcity of Anjili and white cedar.

449. The excellent meili (*Vitex altissima*) is practically unknown, but, where the villagers have not felled the trees for their clearings venteak, vēnga, thēmbāvu, irūl or irumulla, and vāga (*Albizzia procera*) are abundant.

Other trees.

450. The most important hills and peaks in this basin are Karuvālikkāda (Curvalecad hill), on the South-west corner of it, which rises to a height of about 800 ft. Manitthūkka nēruvamala (Munitook neruvoomullay), a forest clad hill rises to about the same height. Gōpura mala (Goparal mullay) is 1800 ft. high, and is at the Southern extremity of the Puliyan-chēri ridge. It is a conspicuous land mark, and is precipitous on its Southern face. Kombukutthi 1,240 ft., Chemmanikōtta 3,050 ft., Perumpāra mala 2,500 ft. and Perambukōtta 3,050 ft. I have already described.

Chief hills and peaks.

451. On the Chemmanikōtta ridge is found a line of thin stones 3 ft. high, some of them set upright in the earth, and others fallen over. Such stones are common all over the Peermerd plateau, and the usual explanation of their origin is that they were set up as boundary marks. Another suggestion is that they are the remains of habitations once occupied by a diminutive race of aborigines, but this seems improbable.

Curious stones found all over the Peermerd plateau.

452. Close to this point and on the same ridge is a heap of rocks called Edutthuvecchānkal, which tradition states were brought from Madura by a race of Titans, who meant to carry them to Peravanthānam, for the purpose of erecting a pagoda, but being disturbed by the approach of dawn, for these giants could only work in the dark, they threw down their loads, and withdrawing to their homes, never resumed their work. A flight of stone steps, part of the old road from Madura to the West, is found close to this spot, and from here a ridge runs almost due West to Peruvanthānam and beyond it, along the side of which the Peermerd ghāt road has been made.

Other remains

453. This forms the Southern side of the valley of the Nyāranpulla (Naran pulla) river. From Edutthuvecchānkal the line of the ghats runs North at a level of 3,300 ft. or thereabouts, and rises to 4,500 ft. at Amratha mēda (Vembaulay mudi). Dropping again on the other side to about its former level, the watershed of this river runs round in a semicircle, and reaches a height of about 3,500 ft. at

Elevations of different hills overlooking this valley.

Mothavara (Kunudi hill), a cone of rock and grass visible from an immense distance. From this point the ridge drops to 1,200 ft., and from thence spurs run out in various directions, sinking finally into the lower land. The isolated hills in the valley, such as Kuruppalli (Cooroopally hill), Kūvappalli (Coovapully hill), and the hill near Edakkōnam, and Pālappura rise to about 700 ft.

454. Except in the immediate vicinity of the hills the slope of the land is easy, *Soil exceedingly rich* and well adopted for cu tivation. The soil South of the river is perhaps not quite so fertile, but North of it, and near Mundakkayam, and. Kānnyirappalli it is deep and rich, the valleys of the Manimala and Pālāyi rivers containing probably the finest soils in the low country of Travancore.

455. The rainfall and climate are much the same as in the rest of Travancore, *Rainfall and climate.* though, owing to the breadth of the mountain plateau East of Mundakkayam, the North-east monsoon is felt here less than further South, and the drought is therefore of longer duration.

456. There are no coffee estates in this valley except 3 or *All the coffee estates abandoned.* 4 on the Peermerd ghat road, and the Mothambāyi estate to the West of Perumpāra mala, all of which have been abandoned.

457. The whole of the Kūttukal valley and a great part *A great part of the land is alienated from Government.* of the small valley South of Peravanthānam belong to the Chengannūr chief, and some of the land near Edakkōnam is owned by a Jenmy.

458. The Manimala basin is thickly populated, the most important villages in it being Kannyirappalli, a straggling place with a population, *Places of interest. Kānnyirppalli.* including its suburbs, of about 1,000 persons, Mahomedans, Roman Catholics, Pulayans and others. It has a Roman Catholic church, a mosque and a resthouse. At the time of Lieut. Ward's visit it was falling into decay, as the traffic along this road was decreasing, but since the opening of the coffee estates on Peermerd, and the construction of the cart road, a great impetus has been given to trade, and the place has revived.

459. Mundakkayam is a village of 20 or 30 houses with a population of 100 persons, of whom all but 4 families of Mahomedans are Chris- *Mundakkayam and Edakkōnam.* tians. There is a Protestant church here and a rest house. Edakkōnam, half way between these two places, but not on the bandy road, contains 50–60 houses, and a population of 200, Nāyars, Mahomedans, Chogans, and Arayans.

460. Alappāra is a very old place, situated at an elevation of 200 ft., and East of Manimala. The people here who number about 100, living *Alappāra.* in 20 houses, are Nāyars and Mahomedans. They are very anxious to get a cart road here, for they are at present cut off from the rest of the world, but the place itself is situated at the bottom of such a steep and narrow valley that, when the road from Rāni to Manimala is completed, it will not come within 2 or 3 miles of them. There is a pagoda here, and the remains of terracing on the hill sides.

461. Erumēl has 20–30 houses, and a population of 150. At the time of Lieut. Ward's visit there wore six families, half of them Nāyars, and *Erumēl and Pālappura.* half Mahomedans, but the latter have driven the Nāyars out, and there are only one family of Nāyars, and one of Chogans there now. This village is on one of the paths to Shabari mala, and boasts a mosque and a pagoda, both of them kept in repair. Pālappura (710 ft. on the ridge between the Manimala and Pālāyi rivers, has 2 or 3 houses inhabited by Nāyars, numbering about 12 souls.

462. At the " small but populous " village of Peravanthānam, as Lieut. Ward calls it, there is a pagoda and mosque, and a population of about *Peravanthānam Chennāyippādi and Karikkāttur.* 100 Nāyars and Mahomedans. Chennāyippādi (Shenna paudy) contains 30 or 40 houses, and a population of about 150, mostly Nāyars. At Karikkāttūr there are a few houses, and 20 or 30 persons.

463. The Arayans muster very strong in this valley, and in the neighbourhood of Kūttukal number upwards of 300, besides 200 Chogans and others from the low country. Besides these there are colonies at Kalangara mala, Chiramala, Kūttanpāra, Athappan, Puliyanchēri, Karumala, Thelli thōda, and other places, their total number amounting to quite 500. These, added to the population resident in the villages, makes a total of about 2,500.

The Arayans very numerous. Total population.

464. Between Mundakkayam and Gōpura mala, there are the remains of an extensive village, with the ruins of a pagoda and a tank, at a place called Pacchilakkānam, but there is no one living there now.

Old ruins at Pacchilakānam.

465. The forests in this basin are partly under the control of the Kānnyirappalli Aminadar, and are in the central Division, and partly under the Thodupura Aminadar, and are in the Northern Division, the road from Cottayam to Peermerd being the boundary.

Management of this area.

466. Teak is the chief tree felled here, and has been much cut about Edakkōnam, Alappāra, and Erumēl, as well as along the river bank just outside the forest line. Blackwood has also been felled in small quantities. Kambagam, ānjili, vēnga, and other woods are cut and sawn up, and the scantlings conveyed into Cottayam, where they find a ready sale. I have reason to believe that there has been a great deal of smuggling in this part of the country, and as it lies on the borders of two Divisions, and under the supervision of two Aminadars, it has received less attention than it deserved. A great deal of ānjili, and other such woods is cut down from the chērikkal lands in the neighbourhood of Kannyirappalli, Vārūr, and other large places, and is used for house building, and we know nothing about it.

The chief trees found in this valley.

467. None of this valley will be reserved, both because there is not much extensive forest here, and because large portions of it belong to private owners, rendering it difficult to find in one block any great extent of land belonging to Government, but arrangements should be made for a better supervision of this part of the country, with the view of putting down the illicit practices mentioned above.

No reserves recommended, but better supervision required.

468. One or two cardamom gardens are found to the West of the Amratha mēda, and at the head of the Kūttukal valley. Though the land itself is, for timber purposes, under the Forest Department, the collection of the spice growing in these gardens is superintended by the Varukkappāra Vijāripukāran acting under Mr. Maltby's orders. Very little wax is obtained here.

A few cardamom gardens in this area.

469. The valley of the Manimala river is well supplied with roads. A cart road runs from Manimala to Kānnyirappalli, and on to Erāttapētta, and forms the Western boundary of the forest area. Another cart road passes through Kānnyirappalli, and Mundakkayam, and then beginning to ascend the hills goes to Peravanthānam and Peermerd, and, after crossing the plateau, descends to the Kumbam valley in Madura. A good bridle path leaves Mundakkayam, and runs for 3 or 4 miles to Kūttukal, and from thence various footpaths diverge in different directions. Numerous tracks lead to the Arayans' clearings in other parts of the valley, and it is not difficult to travel through it, though a guide is acquired. One of the largest tracks, half foot-path, half elephant-path, starts from Mani mala, goes to Erumēl, and is continued to the Shabarimala pagoda, crossing the Arutha river at Valavakuri.

Roads and paths.

(14). *The Pālāyi River.*

470. The main branch of the Pālāyi river rises on the Peermerd plateau at an elevation of 3,500 ft., a little above the Nallathannipāra, and, running North-west and West, descends rapidly over very rough ground till, after 7 or 8 miles, it joins the Kavana ār (120 ft.) which drains the valley North of Erāttapētta, and takes its rise on the slopes of Mēlakāvu. The combined stream, here about 60 yards wide, and 3 or 4

Sources and course of the Pālāyi river.

feet deep in the wet weather, runs fairly level, but with a strong current, and after a course of 2 miles due South, it is joined by the Kudamurutti (Codamoorty) river, which rises very near the other, and follows a more direct line, passing the palace of the Pūnnyātta chief on its right bank, at the place of that name. The Kavana ār is navigable as far as 4 miles above Eerāttapētta, but, on account of reefs of rocks, boats cannot ascend higher than that, nor can they proceed above Eerāttapētta itself, on the Kudamurutti branch of the Pālāyi river, but timber can be and is floated down both of these streams from considerably higher up.

471. This river leaves the forest boundary at Eerāttapētta, and is here more than 100 yards wide, and 5 or 6 feet deep in the monsoon. It is admirably adapted for the floating of timber, for no rocks or rapids bar its progress, all the way down to its mouth. Its course is South East, and leaving Kondūr on the left, and Lālam on the right hand, it eventually finds its way into the Vembenāda lake by two channels, which separate at Pārampura one passing Cottayam, and the other running farther to the North, and debouching near Thannirmūkkam.

Course of the river below Eerāttapētta well suited for floating timber.

472. A small river, the Chittār, which takes its rise at Vengatānam, and Chēnāda (Shaynaud), joins the Pālāyi river on its Southern bank, a mile to the west of Eerāttapētta, and therefore outside the forest area. It is 20 yards wide and 4 ft. deep, and is used for floating timber.

The Chittār.

473. The area of forest land drained by the Pālāyi river is 102 square miles, the greater part of which is, or has been, under cultivation. Dividing it into the different classes of forest, we find of (1) original moist forest scattered in small blocks, of which the largest is near Kayyūr (Kyeur), 5 square miles, (2) secondary forest, 52 square miles, (3) grass forest with large trees, 30 square miles, and (4) useless land, rock, and short grass, 15 square miles.

The area divided according to classes of forest.

474. The teak tree is very numerous on all the rocky hills which abound within the area indicated. This valley seems particularly well suited to the growth of this tree, which is seen marking all the hill sides with its white blossoms. Probably no where in Travancore is teak more common than here, and owing to the absence of wind the stems grow straight, and without a flaw, but very few of the trees are of any size, for the timber merchants have been at work here for years, and are still carrying on their operations. On the flatter land teak is not found.

Abundance of teak, but all of small size.

475. Of other trees wild ānjili and white cedar are quite absent. Blackwood is occasionally found, but is far from common. Vēnga, venteak, and thēmbāvu grow on the hills not occupied by cultivation, and vāga (Albizzia procera), which springs up very rapidly after the land has been abandoned, is a common tree. Speaking generally, the valley is too thickly populated, and too much cultivated, to be able to boast much timber. The wood most commonly felled is teak, which is in great demand at Cottayam.

Other species of trees found here Too much cultivation for the existence of forest.

476. The upper end of the Pālāyi valley is bounded by a semi circle of lofty hills. Mothavara on the South, 2,500 ft. has been already referred to; next to it are the Kudamurutti cliffs, about 3,500 ft., a short distance north of them is Kallālmēda, 4,000 ft., a rocky ridge which runs out into the low country, and separates the two branches of the river. North of this another spur juts out towards the North-west, and on it are Ponmudi (Pun mudy), Melappulla mēda (mailly pulliy mode), Illikkal (roykykall), Erumāppura mala, Valiya mala, Mēlakāvu mala, and Kodayatthūr (Kodiatour), all rocky peaks, rising to nearly 4,000 ft. Of these the last named hill is the highest, but the Illikkal is the most remarkable as it consists of a number of rocks clustered together on a flat terrace, and looks almost like an artificial fortress, built of enormous stones.

Elevations of the chief hills and peaks in the valley, to the East and North.

477. To the West of Motha vara the ridge above the Arayan village of Manna is 1,200 ft. high, and from this height it sinks by Vengatānam to the plain. In the centre of the valley, and North of Kudamurutti river are Kāttillapāra 650 ft., and Māvadi mala 1000 ft.

Elevations of the hills in the centre of the valley.

Vetti mala, Adakkam, Thallanáda (Thulla naad), Ayyampára (Iampara), and Mankombu are about 1,000 ft. high, and Thèvara mala, a detached hill near Kayyûr 600 ft. To the west of Mêlakávu mala, the ridge dividing the valley of the Páláyi from that of the Mûvâttupura river falls rapidly to about 800 ft., the elevation of Nîla mala (Neelur mala), and Chemmala, forest clad hills upon it. Where the road crosses the ridge at Nelláppára (Karinji pass) the elevation is about 450 ft.

478. The soil of this valley is exceedingly rich, and, in consequence, the whole
Great fertility of the soil. area is thickly populated. On the upper parts the presence of stones in it makes the earth free, and here teak thrives, but in the plain below the red loam is dark and tenacious, and the pepper vine, the area and the ánjili tree thrive amazingly in it. A great part of the land is slightly undulating; it is only close under the higher hills that the slopes are steep.

479. The rainfall and climate are the same as in the adjoining valleys, and
Rainfall and climate. there is no wind. Altogether, this is a very favoured spot, and the people living here ought to be, if they are not, better off than anywhere else in the country.

480. As in the case of the valley of the Manimala river the Chengannûr chief
The greater part of the valley claimed by the Punnyátta chief. puts forward a title to a large portion of the land in it, so the greater part of the area of the Páláyi valley is claimed by the Pûnnyâtta chief. This petty sovereign owns, in addition to the land near his palace, a tract of considerable extent (230 square miles) in the North-east corner of Travancore, including the Anjináda valley, and the High Range which is leased to a company.

481. His rights do not seem to be very clearly defined. The Government
His rights, especially as regards timber, not properly defined. royal timbers growing on his land cannot, as I understand, be touched by him, and we cut teak and blackwood from his territory, without any objection being raised, but I am told that he is in the habit of cutting, and using sandalwood, which is found in the Anjináda valley, though it is a royal timber. On other woods grown in his territory he charges a royalty of his own, but as soon as the timber enters the Travancore territory it is liable to seigniorage according to our rates.

482. There are no coffee estates in this valley, but the population in it is large.
Population resident in this valley. The most important places are Eeráttapêtta, containing about 400 persons, of whom the Mahomedans live on the North bank of the Kudamurutti river, and between it and the Kavana. The Syrians are located on the Southern bank of the former river, and have here a large church. Pûnnyára is the residence of the chief, who lives in great seclusion with his numerous relations around him. I have no means of knowing the population of the place, as it is off the road, and at some distance from it, and I had no occasion to go there, but I put it down at 400.

483. At Chedanáda (Thedanaud), on the road between Eeráttapêtta and
Places of interest. Kánnyirappalli, there is a temple and a population of 80 or 100 Náyars and Chogans. Nîlûr (Neelur) at the North-west of the valley is inhabited by about 30 persons, mostly Náyars, Chênáda (Shaynaud) can boast of 5 or 6 houses, and about as many inhabitants as Nîlûr. At Erumáppura there is a colony of 5 or 6 Náyar families who live on the hills, and whose ancestors have lived there for many generations. They have a small pagoda of their own, and seem very much more independent than the Náyars of the plains. This makes a total of about 1,000.

484. In addition to the people already mentioned the Arayans reside in this
Arayans very numerous. valley in great numbers, and are very well off. The greater number of them are Christians, and at Mêlakávu (1,200 ft.), Erumáppura 1,100 ft. and Mankombu 1,000 ft. (close under the Illikkal rock, and to the West of it), where the Rev. A Painter has a bungalow, there are churches. The congregations attached to these probably number over 2,000 souls.

485. There is a great contrast between those Arayans who have been brought under the influence of civilization and Christianity, and those who are still wild men. Drunkenness has always been one of the vices of these people, and it still clings to those who have not been taught better, while the more civilized Arayans are almost all teetotallers. The latter live in substantially built houses erected on terraces cut on the hill sides, and plant jack trees, palms and coffee in their gardens, and many of them have settled down, and have registered land in their names. Besides the Arayaus there is a tribe of Hill men, called Ullādans. I only know of one colony in this valley, but there may be more; altogether there must be quite 3,000 Hill people in the valley of the Pāḍyi river, and 1,000 low country people.

Some of them have embraced Christianity, and are much more advanced than the rest. Total population.

486. The valley of the Pālāyi river is under the charge of the Aminadar of Thodupura, who has a very large area to supervise. Owing to inadequate supervision, and to the facility with which teak can be sold at Cottayam, there used to be a great deal of illicit felling and smuggling of this timber, for there was at one time no watch station on the river above Pārampura, and all the smugglers had to do was to run past that depôt during the night.

Very much smuggling on this river in former times.

487. A station has lately been erected at Pālāyi, which is a great check on this smuggling, and last year a "boom" was thrown across the river at Pārampura, to prevent the passing of timber at night. Unfortunately, this was made of arecanut trees, as no other woods could be obtained at the time, and after a couple of months these trees became water logged and sank, but they served their purpose for a time. No doubt it will be properly constructed this year.

The new station opened at Pālāyi, and the boom erected at Cottayam.

488. Teak is almost the only tree brought down this river; sometimes a few logs of blackwood, or of ānjili, vēnga, or other woods pass, but the total number is small. At intervals timber may be taken along the cart road, but the facilities for floating being so great, the occasions when the wood is carted and not floated, are very rare indeed.

Teak the chief timber taken down this river.

489. There will be no reserves in this valley, the amount of available land being so small, but an improvement can be made by stationing an Aminadar at Pālāyi, to have charge of this valley, and some of the valley of the Manimala river, or else by reducing the work of the Thodupura Aminadar so as to enable him to visit Eerāttapētta oftener than he can do now. The last mentioned place is a regular hot bed of thieves, indeed, the facilities for stealing timber are so great, that it is no wonder that so many people take up the profession.

No reserve recommended, but the supervising staff to be strengthened.

490. No cardamoms nor wax are obtained in the Pālāyi basin. If any are found they are taken to the Pūnnyātta chief who claims them, as well as any ivory discovered within his territory.

Cardamom and wax claimed by the chief.

491. When the bridges now being constructed in this valley are finished, and especially those along the road, from Pālāyi to Eerāttapētta, it will be very well off for roads. A cart road runs from Kānnyirappalli to Eerāttapētta, following the forest line. From Eerāttapētta a road has been cut for about 3 miles due East, and leaving Pūnnyāra on the right, it ends at a Syrian church. From this point a bullock track used to proceed up the valley, and ascended the Kudamurutti ridge, but it was much frequented by smugglers, and many years ago the Government partly destroyed it, and made it impassable for four-footed animals. Since then, and since the Peermerd road was opened, this old track has been quite abandoned. From Eerāttapētta, after crossing over to the North bank, a cart road runs to Pālāyi, and about a mile to the West of Eerāttapētta, another cart road starts from a place called Ambāra, and goes North-east, crossing the Kavana river, and circling round the Ayyampāra rock on the North; it finally ends, after proceeding for about 8 miles, at the foot of Mankombu. The Arayan colony of Mēlakāvu is also provided with a cart road, which is 6 or 7 miles

Roads and paths.

long, and leaves the Palayi to Thodupura road about 6 miles from the former place. Throughout the valley numerous foot-paths, generally very rough, run in all directions, connecting one · village or patch of cultivation with another, but it is unnecessary to describe them in detail.

(13) *The Mūcáttupura river.*

492. This river is formed by the junction of three smaller rivers—the Thodu-
Affluents of this river. pura, the Vadakkan, and the Kothamangalam, which take their rise on the Western slopes of the Peermerd plateau, and running in a Westerly and North-westerly direction unite at the small town of Mūvāttupura, which is the head of a Taluq.

493. Of these three the most Southerly is the Thodupura river. Its largest branch takes its rise in the moist forest at an elevation of
Sources and course of the Thodupura river. about 3,500 ft. near Ponmudi (Ponumudy hill), and flows North-west for 5 or 6 miles, in a valley between the lofty Kodayatthūr ridge on the South, and the Idāda mala which runs out from Alacchippāra (Purwawlly) on the North. Then it is joined by another stream rising near Olippūnni mala (Ulappunney mala). Two miles below this, the river now running at an elevation of 150 ft. passes the village of Arakkulam, and a mile lower down the village of Thittila (Keelthailey): about the same distance farther on the small hamlet of Kodayatthūr is reached, and here the river may be crossed by an insecure bridge of sticks, suspended from the branches of two trees. A mile lower still it is joined by the Varippura at a point called Kūttukayam.

494. The Varippura drains the valley North of the Arakkulam valley, and parallel to it, the strong Thumbippāra ridge dividing the two.
The Varippura. This river rises on the edge of the Peermerd plateau, and is here called the Kingini thōda. Falling rapidly to an elevation of about 300 ft., through dense forest it passes Thēvara pāra, and its name is changed to the Chēnnan ār: a few miles further on it is called the Varippura, and turning South-west it bisects the path from the hills to Thodupura, and joins the main river as already stated.

495. A mile below Kūttukayam the river is crossed by another very insecure bridge attached to the boughs of trees at a place called Kōlap-
Course of the Thodupura river below its junction with the Varippura. pra, and the same distance lower down it passes the hamlet of Chungapalli. Two miles farther on it is joined by the Mundappilākkal thōda, a stream which drains a narrow valley to the South of the Kodayatthūr peak, and a mile farther it leaves the forest area at Mrāla (Mirthala), a part of the country once famous for its teak, but never regularly inhabited any more than it is now.

496. Outside the forest area the Thodupura river continues flowing in a North-westerly direction, with a stream 40 yards wide and 6 or 8 ft.
Its course outside the forest area. deep, and 2 miles below that point it is joined at Olimattam (Ulmattham) by the Thekkambbāgam thōda, which flows into it on the right, after watering a small area of low lying, undulating land within the forest line. The town of Thodupura itself is reached a mile lower still, and from here to Mūvāttupura, where this river joins the other branches, it follows a winding course of about 14 miles.

497. The Thodupura river is very well adapted for floating timber, on account of the absence of bad rocks in it, and of its flowing at a more or
Suitable for floating timber and for navigation. less uniform level. As its sources are not found far back on the hills, it is liable to very sudden floods and ebbs, the effects of a single heavy shower becoming at once apparent in the rise of the water. Logs can be, and are, floated down from some distance above Arakkulam on the main branch, and from Thēvara pāra on the Varippura. Boats can ascend, with difficulty on account of the strong current, as far as Koddyatthūr, but only in the wet weather.

498. Moist forest clothes the upper part of the valley of the main river, and extends down to within a mile or two of Arakkulam. The
Extent of moist forest in this river basin. slopes of these hills are steep, and unsuited for cultivation, but the more level land on the banks of the river has been cleared,

and abandoned again, and is growing up in secondary jungle. The adjoining valley of the Varippura is covered with moist forest over about half its area, on its Northern and Eastern slopes, and it contains no resident population except Hillmen.

499. The rest of the land within the forest area drained by this river was at one time covered with heavy forest, but it is now occupied by clearings for hill-paddy, by compounds, and paddy fields, or by secondary growth springing up after abandonment. This part presents the appearance of undulating hills covered with thorny scrub, and destitute of any trees of large size or of value, though the higher hills, such as Onāra mala and Muthiya mala, rising out of the plain, are often covered with groves of teak.

The rest of the area covered with secondary forest or with cultivation.

500. Some years ago the Sircar, to prevent the extension of hill cultivation, put a mark up at a place called Mūlakkal near Velliāmattam, beyond which no one was to be allowed to clear land, but as no officer was deputed to see the rule carried out, and as the rule itself could not be enforced under any law, the restriction was ignored.

Attempt to confine the clearing of forest.

501. The Arakkulam and Velliāmattam valleys were in former days famous for the teak they contained, and even at the present time there is a considerable quantity of this timber still to be had there, but, owing to the heavy fellings which were continued without cessation up to quite modern times, what remains is of small size. On the Nyaralanthanda, the ridge overhanging Kudayatthūr, there are some nice trees of the 2nd class, and at the time of my visit last July, there were some hundreds of logs there which had been felled by a contractor, and were being brought down by the Department owing to his failure.

Abundance of teak here in former times. Nothing now left but small timber.

502. Some capital blackwood is found on the ridge in the centre of the valley, but it would require some skill to bring it to the river, as the sides of the ridge at this point are almost perpendicular.

Some good blackwood.

503. Of other trees kongu or thambagam is found on the river sides, but not in large numbers. Irul is abundant in the Velliāmattam valley springing up in abandoned clearings. Ven teak and white cedar may be found sparingly in the interior forests, and I heard from a merchant of the existence of ānjili in one place, but so difficult of access, that no one had ever attempted to get the trees out. Vēngu is met with here and there and perhaps the most useful second rate wood occurring in this river basin is vāga (Albizzia procera).

Other timber to be found here.

504. Taken as a whole the forests in the valley of the Thodupura river cannot be considered rich in timber. I could hear of no felling operations by timber merchants, except of a few thambagam trees for boats on the Chēnnan river. The Government elephants were engaged in working down teak at the time of my visit.

Forests of the Thodupura not rich.

505. This valley is confined by the lofty Kodayatthūr ridge on the South, and the Nediyattha, and Perambukkāda ridge on the North. The former I have already described, in writing of the valley of the Pālāyi: it runs along at an elevation of over 3,000 ft., and presents an exceedingly steep rocky face to the North, on the upper part of which there is not much vegetation. The Nediyattha ridge on the other hand, though steep, is covered with trees to its summit. Uppu kunnu, a bare rock with grass on its top, at the upper end of the valley, cannot be less than 2,500 ft., and Nediyattha (Midiettbu mullay hill sto) is perhaps 2,000 ft., while the elevation of Perambukkāda is only 1,000 ft. From here the spur sinks rapidly, and merges in the lower land near Alakkōda (Aulay code).

Chief hills of this valley.

506. On the East this valley is bounded by the Peermerd plateau, the elevation being about 2,500 ft. From the centre of this line a strong spur runs to the West, starting from Olippūnni mala (Ulappunney male), and continuing through Nādakāni (1,500 ft.) to Thumbippāra (1,200 ft.), Pūcchappāra (1,000 ft.) and Bhagavathi kōvil (Bhagavaddicoil). From here it sinks rapidly to the low land, and at Velliāmattam at its end is only 300 ft.

Its Eastern boundary.

507. In the centre of the valley Onāra mala, about a mile West of Yellungūr,

Other hills. and Muthiya mala are the chief hills. The former has a large Arayan colony located on it, and numerous teak trees growing there. The latter is a rocky hill South-west of Velliámattam, and is also covered with teak. Neither of them are of any height.

508. The slopes of the upper hills are steep, but the lower land is very

Soil, wind &c. gently undulating. Every where the soil is good, and is more free on the hill sides, but it is rich and deep in the plain. There is no wind in this valley.

509. The more important villages are Kodayatthūr, which contains about 100

Chief villages. houses, and say 500 persons, most of whom are Mahomedans: Arakkulam with a population of about 150, mostly Brahmins and Nāyars: Thittila 100 persons: Chungapalli 70: Mattam and Perumattam each with 50. In these 4 villages last named there is a mixed population of Nāyars, Mahomedans, and Chogans: Konthālappalli (Yellungur) is a Mahomedan village of about 50 persons: at Anjērikkal there are a few houses of Nāyars, the paddy lands there being the property of the Kidangūr Dhēvasom.

510. At Arakkulam there is a large temple dedicated to Ayyappan, and kept

Places of interest. in good order. A certain allowance is made to support this institution. The village of Velliámattam was of some extent at the time of Lieut. Ward's visit, and contained many houses, of which the ruins still remain. Indeed it was inhabited till within 9 or 10 years, when the last of the inhabitants was driven out by the elephants. This valley is much troubled by these animals, which come down the Thumbippāra ridge from the East, but the Arakkulam valley is quite free from them, as its sides are all very steep; consequently, the people prefer to reside at Kodayatthūr, and the neighbourhood, and to clear land in the Velliamattam valley, where they need not remain for long at a time.

511. There is a chowkey at Velliámattam, with a small staff in charge of it.

Velliámattam chowkey Before the Peermerd to Madura road was cut, a great deal of traffic passed this way, and cattle were brought to graze even as far as this, all the way from Kambam, consequently the returns from the chowkey were considerable, but the whole of the traffic has been diverted, and the receipts no longer pay the charges of the establishment.

512. Some of the land of the Velliámattam valley is said to belong to the

A Jenmy. Kirakkumbhāgam Nambuthiripāda, who has rights of seigniorage on timber felled within its area, but I do not know if this is true.

513. Besides the resident population mentioned above, there are very many

Arayans and Urālies. Arayans living here. One colony resides on the Perambukkāda hills, another at Onāra mala, another on the Thumbippāra ridge, and 5 or 6 on the slopes below Kodayatthūr peak. There are also a few Urālies. Their total number may reach 500. Adding these figures to those already given we get a total of 1,500 in all.

514. North of the Perambukkāda ridge is the valley of the Vadakkan ār. The

The sources and course of the Vadakkan ār. sources of this river are near the peak called Allachay pāra in the map, though this is not its proper name. After running for 6 or 7 miles, the Kámbār as it is called here, is joined at Irukūtta by the Thōni ār, a strong river running from the East, and 2 miles lower down by the Alakkāda thōda a small but torrential stream which takes its rise between Nediyattha and Peringaccheri (Peringachagry). These combined streams, running at an elevation of about 250 ft., and falling rapidly, are crossed near Vēlūr by the path leading from Udumbannūr to the cardamom gardens near Nagarampāra. Two miles lower down the Muthuvarakkuri thōda joins this stream at Kacchira mūla; and 3 miles on again, a large stream running down from the Manakkāda valley, and the cardamom gardens beyond Thoppi mala falls into it. At this point the Vadakkan ār is crossed by a most precarious bridge, which consists of not more than 3 or 4 sticks as thick as the wrist tied to each other and to two trees on the river sides. Without the assistance of a hand rail it would be impassable. The river which has

10. ...ort .west is in the
been running ...le forest ...ere deflected to the West, and after winding much for
six miles it leave...e the dry e...; a stream 60 yards wide and 8 or 9 ft. deep in the
monsoon, 2 miles a...-r 15 miles. ...ion of the Shangarapilla or Kodikkulam thôda with
it. From this junctic.rest ext...ttapura is above 10 miles.

515. This river can ...avig.ted as far up as the point where the bridge
crosses it but not higher, and indeed, as there are no villages
Suitable for navigation   above this point, boats have no occasion to ascend beyond it.
and floating.   Timber can bo floated down from Irukûtta when the river is in
flood, and almost at any time, except in the very dry weather, from a point 3 or 4
miles lower down.

516. The upper part of this valley is covered with heavy forest, which extends
down as far as the Nediyattha hill, and a line running from this
Moist forest here.   point to where the path crosses the river near Vê.ûr. North of
that place, the North Eastern or right bank of the river is covered with forest, except
immediately by the side of the water, where clearings have been made for paddy.

517. This interior forest in the neighbourhood of Irukûtta is called Pûmattam,
and every year elephant pits are dug here, as the place is much
Clearings and carda- infested by these animals. The continuous forest on the upper
mom gardens in it.   slopes is broken by the clearings of Arayans who live at Chôla-
kkâda, Peringachêri, and Nediyattha, and in the same locality there are several
cardamom gardens belonging to the Sirkar, and to ryots of Thodupura or the neighbourhood. Cardamom gardens are also found near Nagarampâra, and Kïrippilâva
kânam.

518. The rest of the country included in the forest area of this valley is covered with shifting cultivation, or with the scrubby jungle that
Secondary forest in this   grows up after the fields have been abandoned. In some parts
valley.   there are stretches of grass land with large trees growing
through them, and containing moderately good timber.

519. Curiously enough, considering its abundance in the next valley, the teak
is almost entirely absent from the basin of the Vadakkan âr,
Curious absence of teak   being found only on the slopes about Chelvam, and in detached
trees.   groves, which are evidently artificial, and are usually seen in
the vicinity of old shrines.

520. This anomaly is evidently due to the more gentle slope of the land, and
to the greater quantity of moisture found in the clayey soil of
Due to the more clayey   this valley. In places outside the forest area and within the
soil.   basin of the Vadakkan âr teak occurs in suitable localities where
there is a good drainage through the soil, but with these parts we are not at present
concerned.

521. Of other trees blackwood is said to be found, but I cannot say I saw
much of it. Anjili, white cedar, and ven-teak are met with very
Trees in the moist for- sparingly in the moist forests of Pûmattam, and by the Kombau-
est.   chêri ascent, but they have all been in demand on account of
their floating properties, and those that remain are of small size.

522. In the grass land there is a fair amount of irûl and thêmbâvu and some
vênga, all of which are heavy woods, or they would have disap-
Other trees in the grass peared long ago. All but the small trees of ven-teak have
forests.   been cut.

523. The Vadakkan âr valley lies between the Nediyattha or Perumbakkâda
ridge on the South, and the Kottappâra ridge on the North.
Chief hills here.   The former I have described, but the latter runs in a straight
line from Nagarampâra to the Northwest, though it is broken at one point by the
stream that rises East of Thoppi mala. This ridge stands up very boldly, and its
upper part consists almost entirely of rock, without any vegetation on it. The
Padikkabam peak (Kydhapara hill) rises to about 1,500 ft. Close above Manappuram
the elevation of the ridge is 800 ft., and, continuing pretty level, it finally ends in a
pointed conical hill called indifferently Thïyyâttum mudi or Sâsthâmmattam.

524. On the East the valley is bounded by the ¼ a mile West ...y and the slopes are excessively steep, sinking hills. The f...e over 2,000 ft. right to the plains below. He and numerous ani and Nagaram-

Other hills here.

pāra hills stand up against the sky, the latter (marke ... iāmattam .ie hill station) being by far the higher of the two, and rising well over 3,00 ... .ie very high grass hill called Pālkkulam mēda (Cunnacaud or Peremale hills) i ... ..0 ft. high, and from its slopes the Kirippilāva stream flows. Thoppi mala is about 2,000 ft. high, and Mundan mala, a rocky hill close under the Kydhapāra peak and to the South of it, though not high, is a conspicuous object.

525. The chief places of importance in this valley are, Edamuruga (Yedamur-rahu), with 25 houses and, say, 100 persons, Vellanthānam, Karuganāda, and Pānnūr each with about 40 persons. Chelvam is a large place on the hill slopes, and contains perhaps 80 people. Vēlūr (Vaillur) has long been abandoned, indeed it was uninhabited at the time of Lieut. Ward's visit, though he mentions having noticed the signs of old iron smelting so that it could not have been long deserted then. At Manappuram (Munnapuram), and Nyārakkāda (Needracaud) there are 2 or 3 houses each, and extensive paddy fields, employing perhaps 30 people, making a total of 330 persons living within the forest area.

Places of importance and the population.

526. The Hill men consist of Arayans and Urālies, the former live at Ulliruppa, Chēlakkāda, Peringacchēri, and Chelvam, and number about 120 persons, while the Urālies confine themselves to the Manakkāda valley and the neighbouring hill slopes. They number about 50, making with the low country people, in all some 500 heads.

Arayans and Urālies.

527. The village of Udambannūr is just excluded from the forest area; it contains more than 100 houses, and supports a large population which depend chiefly depends for its food on hill cultivation. Nearly all the inhabitants are Mahomedans, but there are some Chogans.

Udambannūr.

528. North of the Vadakkan ār lies the Mullaringāda valley enclosed between two long ridges, running parallel to each other. One of these is the Kottappāra ridge already mentioned, and the other or Northern the Bānan mala (Bama male) or Thusi mala ridge. The breadth of the valley is not more than two, or two and a half miles, and its length about 10 miles.

The Mullaringāda valley.

529. The Mullaringāda river rises in the upper slopes of the valley and running West North-west, passes out of the forest area after a course of about 10 miles, the line cutting across the entrance. Outside the valley several streams join the river, and two of these the Nēthālampāra thōda, and the Kūri thōda, the last of which falls into the river at Kōthamangalam, drain land included within the forest area, though it is of small extent.

The Mullaringāda river.

530. From the forest line at the entrance of the Mullaringāda valley to Kōtha-mangalam is by river a distance of about 12 miles, and on to its junction with the other branches of the Muvāttupura river 6 miles.

Its course outside the forest line.

531. This river is navigable as far up as 10 or 12 miles above Kōthamangalam. The Kūri thōda, though winding very much, can also be used for the same purpose. Timber can be floated down from the upper end of the valley, the river running fast, but evenly : as far as I could ascertain, it is not obstructed by rocks. The Kūri thōda also is used for the floating of timber, and it is a very common practice for timber merchants to drag boats overland from the Periyār to this stream in order to avoid the rocks in the Periyār, and the duty that would have to be paid at the Varāppura chowkey.

Suitable for navigation and floating of timber.

532. In a previous para I pointed out the absence of teak in the valley of the Vadakkan ār. In the valley under consideration teak is again found and grows to a large size, owing to the free character of the soil on the slopes. The tree is not found in the flat below, but is seen abundantly clustering up the hill sides.

Teak is found here.

533. ...ys on the river sides, and the trees of the moist forest may
ere be found in some abundance, but I had not time to make a
lengthy examination of them, as I had to return the same day
to Thodupur...he 14 or 15 miles, which I had travelled in the morning.

Other trees.

534. Th moist forest extends over the upper part of this valley, but, about
half way down it, patches of secondary forest occur, showing
where land was cleared for paddy a few years back. At the
entrance of the valley all the land has been thus cleared.

Moist forest and secondary forest.

535. There is no resident population here, nor are there
any places of interest,

No population.

536. This valley, owing to its secluded character, and to the abundance of food
to be had, is a favourite haunt of elephants and pits are annually dug here, with the result that one or two of these animals
are caught every year. It is much to be desired that low coun-
try paddy cultivators should not be allowed to clear land in the valley and frighten
away the elephants, besides destroying the forest itself.

Much frequented by elephants.

537. The Mûvâttupura river, formed of the junction of these three tributaries,
flows past that town with a stream 100 yards wide and 10 ft.
deep in the monsoon, and continuing for about 8 miles in a Westerly direction, it is joined on the North by a large stream, the
Kadamattam thôda, which unites with it near the village of that name. Then turning South, and winding very much, it passes Râmangalam, Piravam and Vettikkâttu
mukku which is about 28 miles from Mûvâttupura. At this point the river forks,
one branch running in the direction of Cochin, and the other to Thannir mukkam.
To prevent smuggling a depôt is fixed at Vettikkâttu mukku, and watch stations at
Mûvâttupura and Piravam.

Further course of the Mevâttupura river.

538. The area of forest land drained by this river is 175 square miles. Of this
about 50 square miles are heavy forest, 70 are covered with present cultivation or abandoned clearings, 30 square miles are grass
with trees through it, and the other 25 square miles are useless.

Different classes of forest.

539. The population in this area amounts to 2,000 persons of which three quarters are credited to the basin of the Thodupura, and one quarter
to that of the Vadakkan ir, the Mullaringâda valley supporting
none.

Population.

540. Throughout the area of these 3 rivers the soil may be considered as decidedly good. On the more level land it is stiff and not suited to
the growth of the teak tree, but it produces excellent crops of
paddy. The hill slopes, except those far to the East on the ghâts proper, are admirably adopted for teak, and this tree has been felled in this neighbourhood for a long
time past.

The soil is good.

541. Wind is absent, and the rainfall is about 150 inches, and well distributed,
the drought seldom lasting more than 2 or 3 months. Owing
to its distance from the sea, this tract of country is unhealthy
during April and May, and fever is prevalent, but it does not assume a deadly type.

Rainfall and climate.

542. The whole of this land is under the control of the Aminadar of Thodupura,
who has a very large tract of country to supervise, including the
valley of the Pâlyi, part of the Manimala valley, and a considerable extent of open land to the West of Mûvâttupura. The
consequence is that he can hardly do justice to the whole.

Supervision here is insufficient.

543. Teak has been felled within the area for many years. Besides the places
already mentioned in the valleys of the Thodupura and Mullaringâda rivers, Mrâla (Mrthala) was formerly well supplied with
this useful wood. Writing in his diary in M. E. 1042 (1867)
Mr. J. S. Vernede the Assistant Conservator at Malayâtthûr,
mentions with admiration the existence of very fine teak of the first class in this

Teak has been felled here for a long time, and there is not much left.

10

neighbourhood, and also of large numbers of young plants which ... carefully preserve. When I visited Mrāla last year I found very few ... sts, and
not one large tree remaining : they had all been felled.

544. Many of the teak contractors have failed, and their logs are lying all
about the country, in places of which we have no knowledge.

Quantity of teak timber left about by the contractors. In the Mullaringāda valley there are said to be 400 large logs
which we shall have to bring down ourselves, as the contractor's
elephant has died.

545. Blackwood is often met with and can be obtained of good size in certain
localities, but the supply is limited. Kongu or kambagam, is
Blackwood and kamba-gam. felled for boats, but is becoming decidedly scarce. This valuable tree is used almost exclusively for this purpose in North
Travāncore, and indeed the only reason why any of it still remains is, that it has been
employed in this way only in recent years. Formerly nothing could be done with
the wood, in the absence of roads, as it was too heavy to float.

546. The light woods ānjili, white cedar, and ven-teak, have mostly been felled
as already stated. Sometimes boats made of the first named
Other woods. wood are brought down this river, but they are felled either beyond Kōthamangalam on the Periyār, or on the same river, but to the East of Naga-
ram pāra. A good quantity of the heavier woods irūl, thēmbāvu, and vēnga &c.
can be obtained here, but in the absence of bamboos they are difficult to float.

547. I have recommended a reserve on this river basin, which I have selected
so as to include all the moist forest still remaining, and some of
A reserve recommend-ed. the land in the Velliāmattam valley which grows teak so well.
The line runs down the Thumbippāra ridge, and takes in the
level land to near Onāra mala, the whole of the Pū-mattam forest, and the steep slopes
of jungle on the ghāts, and most especially, the whole of the Mullaringāda valley,
which, from its isolated position, the existence, of some good forest there, and its excellent soil, while at the same time the land is for the most part too steep for
permanent cultivation, is well suited for reservation. This reserve has not yet been
proclaimed.

548. The forest land in the basin of the Mūvāttupura river is destitute of cart
roads, but foot paths traverse it in different directions. The
Path from Thodupura to Kambam. largest of these leaves Thodupura and runs South-east to Velliamattam, a distance of about 9 miles, from here it begins to as-
cend the Thumbippāra ridge, and in parts it is very rough. Nādakāni hill is reached
at a distance of 7 miles from Velliāmattam, and from here the path continues Eastward inclining to the South, and crossing the Peermerd plateau descends to Kambam. Before the Peermerd cart road was cut this path was much used by merchants,
but it is now almost abandoned ; the only persons who travel along it are visitors to
the pagoda on the Periyār, Sircar servants who go to collect cardamoms or wax, and
Hillmen. Every year, before the cardamom season commences, it is cleared.

549. Another path runs from Thodupura to Udambannūr, passing the village
of Karimannūr, the distance being about 10 miles over fairly
Path from Thodupura to Thēvāram. level country. From here it runs 4 miles to Vēlūr where the
Vadakkanār is crossed : the path then continues level for about
a mile, and then begins to ascend very steeply, the name of this place being Kombanchēri. In spite of the steepness of the slope this path used to be followed by
merchants with pack cattle when there was traffic with the villages in Madura. After marching for 4 hours from Udambannūr, Nagarampāra is reached at an elevation
of 2,550 ft. Here a camp is built every year for the collection and storage of cardamoms from the gardens around. Beyond this, the path runs to Thēvāram by Udam-
ban chōla, but it is very uneven.

550. Other paths traverse this basin in various directions, but they need not
be particularly described as many of them are abandoned for a
Smaller paths. time, and others are opened, as the cultivation is shifted from
place to place.

(16) *The Periyār.*

.51. The Periyār enters the sea at Cochin and Paravūr, some miles North of
the Mūvāt'upura river, but its sources lie far to the South of
Thodupura, and are in the same latitude as some of the streams
of the Rāni. Unlike many other rivers in Travancore it is not composed of several
affluents, of more or less equal length, any one of which might be the main stream.
On the contrary, the Periyār rises alone in the Shivagiri forests, and though it is
joined in its long course of 142 miles by many considerable rivers, there is never any
doubt as to which is the river itself.

The Puriyar.

552. The water shed which separates the sources of this river from those of
the Rāni is a ridge running from the Padinnyātta pāra plateau
South-east to the lofty hill South of the Shivagiri peak.
Among the numerous high rocks and hills collected together in this place, it is not
easy to say where the Periyār begins and where the Rāni, more especially as all this
country is covered with forests of dense jungle, which no one is known to have
penetrated; but the source of the Periyār may be located within a small area, and
with sufficient precision to answer all practical purposes.

Its sources.

553. When the Periyār emerges for the first time from the dense forest, it is a
river of clear cold water 30 yards wide and 2 feet deep, even in
the driest weather. At this point, close to the Mlāppāra
cardamom station (Sangany Tavalam), the elevation is about
3,000 ft., and the direction slightly West of North. About half a mile from
Mlāppāra the river is joined by the Lakshmippāra thōla which drains the heavy
forest 4 miles to the East, and enters the river on its right bank.

Its course below Mlā-ppāra

554. From here it continues running North-west for about 4 miles, winding
between stony grass hills, and flowing over a rocky bed; it
then receives the waters of the Varukkappāra thōda (2,950 ft.),
a considerable stream which rises in the moist forest near Aragadi thāvalam (Arr-
cuddy Tavalam). About 2 miles South of Varukkappāra thāvalam rises the high
grass hill of Kathira mudi, and at an equal distance to the North-east of it Varayāttum mudi (Vauragaut hill), both of which are over 5,000 ft.

Its further course

555. The Periyār now turns to the West, and runs in this direction for 2 or 3
miles, the banks being steep and stony and covered with grass.
Passing round the end of the Kumarikulam ridge it flows due
North for 4 miles, till it reaches the Mutlayār thāvalam, and is joined by the river of
the same name. The elevation is here about 2,880 ft., and the thāvalam, where huts
are built every year for the convenience of the Cardamom Department, is on the side
of the river and in a thicket of bamboos. To the West of the river rises the Kūttan
Mēda (Colun mole), an undulating grass plateau which lies at an elevation of about
4,000 ft.

Near Mullayār.

556. The Mullayār rises on the slopes of the Kōtta mala, a peak which reaches
an elevation of 6,400 ft., and is one of the most Easterly points
of Travancore. From Mlāppāra up to Kōtta mala and from
here on to Mēthāram mettu which is on the boundary, due North of the Mullayār
thāvalam, the Western slopes of the range of hills forming the water-shed of this
river, are covered with dense forest for an average breadth of 4 miles from their
summits. It is not therefore surprising that the Mullayār always has a constant
supply of water in it, although its length is not above 15 miles; for it is fed by
numerous streams rising in the forest. The largest of these joins it close to the
thāvalam, and is called the Erātta Mullayār. It rises near the Mēthāram Mettu,
and flows due South for about 6 miles, through a narrow gorge with rocky grass
hills rising about the forest on each side, past Kattappa Nāykkan thāvalam (Conda-
panaik Tavalam).

The Mullayār.

557. The Periyār, augmented by the waters of the Mullayār, turns due West
and runs for about 10 miles in that direction, its bed being
sandy rather than rocky, and the fall in it not more than at the
rate of 10 ft. per mile. On its Southern bank the land rises quickly to an undulating
plateau cut by deep ravines, called Arukuriccha mēda, which lies at an elevation of
600–800 ft. above the river and is covered with grass on the summit, the hollows

Below Mullayār.

and the river banks being clothed with forest. The Northern bank, as a rule, lies much lower and the land is often swampy.

558. About 7 miles below Mullayár thávalam the hills rise to a considerable height on either side of the river and approach each other more closely, so as to form a sort of gorge, between Véttappára (Vaētaypara hill) and Máradi mala (Mavady mullay). It is at this point that a dam is being thrown across the Periyár by the Madras Government to a height of 160 ft. and a width of 1,200 ft. to form a lake, which will help to irrigate the land in the valley of the Veiga in Madura.

The Periyâr Project.

559. The result of the construction of the dam will be to cause the Periyár to back up for a considerable distance, fully as high up the river as the Varukkappára thávalam. All the low lying land on the North bank of the river will also be submerged, the water will extend up all the side valleys, and will reach to within a mile of Kumili (Gudalur thávalam). From here it is intended to tunnel a channel through the hill side over a mile long, by which the water will be conveyed into one of the streams that go to feed the Veiga.

Its objects.

560. It was estimated that the lake thus formed would cover an area of 8,000 acres, and, accordingly, this extent of land was leased to the British Government by Travancore in 1886 for a yearly rental of 40,000 Rs. with a promise that if more land was required it should be leased on the same terms and at the same rate of 5 Rs. an acre. It was also agreed that the representatives of the British Government should be allowed to use, free of charge, any timber growing on the land assigned to them for works in connection with the Project.

Terms of the agreement between the British Go-vernment and Travancore.

561. The forest that will be thus submerged cannot be considered first rate. In the neighbourhood of Thēkkadi where the tunnel will be made, there is a good quantity of teak, but the elevation is close on the height (3,000 ft.) above which teak will not grow, and the trees though numerous have short boles, and frequently show signs of having suffered from fires. The class of teak that is to be got here is only suitable for small work, furniture, and so on. Besides this, teak is only found over a portion of the land that has been leased, much of which is swampy, and has no trees at all growing on it.

Character of the forests on the area leased Teak.

562. Of other trees found on this area I may instance the ánjili which attains a large size, and is much used by the Engineers for planking, blackwood, vēnga, ven-teak, and vekkáli, the former two of good size but not very abundant, the latter numerous but small. Red cedar and the kurangádi (*Acrocarpus fraxinifolius*) are also found here, but whether or not they are in demand in Madura I do not know. At the Kumili depot, teak and vēnga are the only trees saleable. The gallnut tree (Terminalia chebula) grows in this neigh-bourhood in abundance, and in the moist forest the black dammer (Canarium strictum).

Other trees found here.

563. When the work was commenced in 1887 it was anticipated that it would be completed in about 5 or 6 years, but though every allowance was made for delay and interruptions, progress has been so retarded by sudden floods sweeping away temporary dams, that it will probably take some years longer than was at first anticipated.

Progress of the work

564. It has often been argued that when the dam is completed and the water of the Periyár begins to flow into Madura, the diversion will affect the water supply of the cultivated lands bordering on the river lower down, and by decreasing the depth of water in the Periyár will make travelling in the dry weather impossible, or at least very much more difficult than it is at present.

Probable effect of the diversion of the water

565. By examining the figures relating to the catchment area of this river we may be able to draw some conclusions on the subject. As will be shown lower down, the total area of the basin of the Periyár above Malayátthúr, where it crosses the forest line, is 1,432 square miles, of which 500 consist of moist forest, 200 of secondary forest, 100 of

Catchment area of the Periyár above the dam.

grass forest, and 632 of useless land. The area of the land lying above the dam is approximately 217 square miles, of which the four classes occupy 105, 15, 10, and 87 square miles respectively.

566. Supposing that the rainfall over the whole river basin was equally distributed, we should find that of the total quantity falling on the 1,432 square miles, about one seventh would be lost to Travancore, but we know that the rainfall on the Eastern side of the Peermerd plateau is not so heavy as on the Western, probably not more than one half as much, and therefore the total rainfall which would be diverted to Madura would be from one eighth to one tenth of the whole.

One eigth or one tenth of the rainfall diverted.

567. In the wet weather this would in some respects be a positive advantage, as the floods in the river would not rise nearly as high as they do now. The only persons who would have cause to complain are those who have low lying land on the river banks, which is annually submerged and fertilised by the deposits of mud brought down by the river. It is difficult to say what extent of land would thus be adversely affected. Much of the low lying land is annually and deeply submerged, and the lowering of the flood level in the river would not therefore make any difference to it. It is only the more elevated lands, that even now occasionally get no benefit from the floods, that would suffer. On the whole, I think we may hope that during the monsoon the diversion of the water to Madura will not make much difference to Travancore.

During the rains the quantity of water diverted will make but little difference.

568. The case is somewhat altered in the dry weather. The grass hills which form so large a portion of the Peermerd plateau, retain but little water in the hot season. What water there is in the river flows out of the moist forest or secondary forests, where the ground, covered with leaves, acts like a sponge and parts with its moisture slowly.

The case different in the dry weather.

569. These two classes occupy about 120 square miles above the dam, or about one sixth of the whole 700 included within this river basin. Not only so, but the forests on the upper part of the Periyār are watered mostly by rain which falls during the North-East monsoon, and showers may be expected there at any time during the dry weather, while the rains that descend on the lower and Western portions of this river basin, though more copious at the time, cease early in December, and no more falls till the end of April.

The catchment area above the dam includes a good deal of forest.

570. Consequently, by cutting off one of the chief sources of supply, the level of the water will be considerably lowered in the dry weather. At present, the river is very shallow above Aluvā from December to April, and with perhaps one fifth or one fourth less water in it, the difficulty of travelling will be much increased. However, the traffic above Aluvā is quite insignificant.

There will be one fourth or one fifth less water in the dry weather.

571. Three miles below the dam the Periyār bends round to the North-west, and at this point a chowkey, called the Māvadi chowkey, has been erected on the left bank to prevent the smuggling of tobacco and salt from the Project works (where these articles are allowed to be imported at the British rates) into Travancore. Here the Parannavari thōda (Paradawary thode) which drains the Padinnyātta pāra plateau falls into the river. From this point the river runs 5 miles with a very winding course, and is constantly augmented by small streams till it reaches the ferry on the Peermerd cart road, the elevation of which is 2,700 ft. Here there are the remains of a large wooden bridge, which was broken by a high flood in 1882 soon after it was completed.

Course of the Periyār below the dam.

572. The old crossing on the Periyār used to be a mile below the present ferry: at this point there were formerly a pagoda and a chowkey which have both been abandoned.

The old crossing.

573. The river continues to flow in a general direction North-west, but with numerous bends in its course, for some miles, when it passes through a narrow gorge formed by the Aramana ridge (Arimaunay) on the South and the Uppukulam mala (Pacolum hill) on the North. The elevation of the river is here about 2,500 ft., its descent being at the rate of about 30 ft. a mile. Its bed is full of rocks over which the water rushes in numerous small cataracts, preventing the floating of timber.

Course of the Periyār below the crossing.

574. The character of the country on either side of the river remains the same, and consists of open hills covered with grass of moderate length on their summits, but over 6 ft. high in the more sheltered parts. In the hollows and lower land where the soil is exceedingly rich are found 4 or 5 coffee estates, while in other places are seen the clearings of Hill men, or land once cultivated by them and covered with thorny scrub. Down these hollows streams, often of considerable size, flow, such as the Cheyitthān ār (Shatanar), and the river that flows down from the "Mary Ann" estate.

Character of the surrounding country.

575. Below the Chenkara estate and the gorge referred to above, the river runs for 3 miles to the West, until it is joined by the Perinthura which drains a large extent of undulating grass land. It then turns due North, and a mile and a half lower receives the waters of the Chenkal ār, which follows a valley parallel to the Perinthura and takes its rise 8 miles to the West.

Below Chenkara.

576. Three and a half miles lower the crossing at Thodupura Periyār is reached at an elevation of 2,300 ft., the river having descended 200 ft. in 8 miles. Here, on the right bank of the river is a pagoda falling into ruins, but still visited by pilgrims. One or two attendants are stationed here, and a yearly allowance is made by the Sirkar for the performance of ceremonies and the payment of the officiating priests. The Periyār is famous for its fish, the mahseer and Carnatic carp being especially common; at this place there are numbers of these creatures, which are quite tame from being daily fed with rice from the temple.

Thodupura Periyār.

577. At this point the Periyār, swollen by the accession of numerous streams, is, even in the driest weather, impassable except by a raft, so, as this is on the road from Peermerd to the Cardamom Hills, a ferry raft and ferry man were always kept here. I believe that since the time of my visit this has been discontinued, and a new road has been opened from the Cardamom Hills to Kumili which obviates the necessity of crossing the Periyār, except at the ferry on the cart road.

The Periyār crossing.

578. Below Thodupura Periyār the river turns almost due West for a mile and a half and then recovering itself, flows due North again for 5 miles till it is close under a very long and lofty ridge of rock and grass called Kalyānappāra by the Mannans, but marked Cunnaroy mulla in the map. Here it is joined by the Kattappanayār (Cuttapenaur) which unites with it on the right bank.

Course of the Periyār.

579. The Kattappanayār rises on the Northern slopes of the Cheyitthān mala ridge which forms a continuous line between Amratha mēda and the hills above Rāmakkal. After turning for 3 miles due North it is deflected to the West by the Kalyānappāra ridge, and runs close under it for more than 6 miles before it joins the Periyār. The Thōni ār (Conyee aur) a small stream, flows into it about half way along its course.

Small streams.

580. The Periyār now turns nearly due West, and runs under the lee of this curious ridge for about 6 miles. The Kalyānappāra is part of a ridge which extends across the Peermerd plateau from, and at right angles to, the Cheyitthān mala ridge to Nagarampāra. It is broken at intervals here and there, but the general line can be followed from one point to the other. The Kalyānappāra itself is about 10 miles long and maintains a pretty uniform level, but it is over 4,000 ft. in the centre, and sinks gradually to near 3,000 ft. at either end.

The Kalyānappāra.

581. The Periyâr after running along in a North-westerly direction as described, suddenly turns North-east and flows thus for a mile. At this *The Idukka gorge.'* point it is confined in a narrow gorge of no great length but only 50 yards across, formed by the Idukka (Idokky) cliff on the South, and the Nelli chôla rock on the North. Turning again to the North-west the Periyâr is joined by the Cheruthôni or Chittâr, a large stream which flows into it from the South-west.

582. The Chittâr takes its rise in the Mêthanam valley which lies to the West of Vanjurampadi (Pearmode hill. S), and Channanakkâna mak- *The Chittâr.* ham (Chundana kao mugham) at an elevation of about 2,300 ft. Collecting all the streams that flow down from the hill slopes it turns North-east, and after a course of 7 or 8 miles winding between high grass hills, and narrow valleys clothed with secondary forest, it joins the main river at Chittâr kûtta.

583. Three miles below this point and due North of i·, the Madatthin kadava crossing is reached (Keelperryaur), where the river forms an *The Madatthin kadava.* island in its centre, and where the path from Nagarampâra to Udambanchôla meets it. The elevation at this point is 1,650 ft., the river having descended 650 ft. in the last 17 miles. The Periyâr here is over 150 yards wide and is always too deep to be forded even in the driest weather, but at the crossing there is a still pool, in which there is very little current.

584. For the last 9 or 10 miles above the crossing the banks on either side of the river have been clothed with more luxuriant vegetation, the *Character of the vege-* valleys containing heavy timber and even the hills and higher *tation.* ground being covered with large trees mingled with grass, in- stead of with grass alone. This kind of jungle continues all the way down the river, from this point the character of the timber improving as the elevation decreases.

585. Half a mile below the crossing the Periyâr is joined by the Perinchêri and Panthappilâkkal streams at almost the same point. The *Small streams.* former takes its rise on the Kâmâkshi mêda (Nachuurbulla hill) and flows North-west, and the latter on the rocky slopes of Pâlkulam mêda (Pere- male hills) and runs South-east.

586. A mile lower still, the Uppu thôda joins the river at Karumânkayam. This stream drains the valley North of Karuka mêda (Inar hill), *Karumânkayam.* and its course is parallel to that of the Perinchêri thôda. Here the river spreads out into a large pool of deep water.

587. The Periyâr now turns West for a mile and then North again for an equal distance when it is joined by the Mûkkanchêri thôda *The Mûkkanchêri thôda.* (Mookancherry tode) which rises near Perumânkayam thâvalam, and draining a densely wooded valley has a considerable volume. Its length is about 8 miles.

588. A mile and a half below this point the Periyâr whose course has been nearly due North for some time is joined by the Pirinyânkûtta *Course of the Periyâr.* âr, and a mile further on by the Môthirappura âr.

589. The elevation at its junction with the last mentioned river is about 800 ft., and if that of Karumânkayam is taken at 1,600 ft. which *Great fall in the river* would be about its height if we allow 30 ft. a mile for a mile and a half, measuring from the crossing, we find that there is a fall of 800 ft. in 4½ miles between Karumânkayam and the Môthirappura.

590. In his diary of the 25th November 1817, Lieut. Ward describes a fall in the river 30 ft. high just below Karumânkayam, where the *The Kukkarani pâra.* water rushes through a narrow gorge only 40 paces across. This is probably on a line between the Anakkunna (Andycoon) and Kallippâra (Cully- paurae) ridges. The Mannâns told me of a fall which they call Kokkaranippâra, where the river is said to tumble over a cliff 100 ft. high. This they described to me as just to the South of the junction of the Pirinyânkûtta âr, and it is probably on a line with the Thêra mala (Tairmulla) ridge and the ridge to the West of the river. Whether this is the same fall as that described by Lieut. Ward, or is quite a

different one, I had not the opportunity of ascertaining, but, considering the rapid descent of the river in these 4½ miles, it is very probable that they are distinct.

591. Above Karumānkayam the Periyār, as I have shown, has a pretty even *Average fall in the river.* fall of between 10 to 40 ft. a mile from the point where it *Too rough for floating* emerges from the jungle at Mlāppāra to Karumānkayam itself. Throughout this length the bed of the river is full of rocks which sometimes lie quite flat, and sometimes jut up so as to form small cataracts, and though a portable boat can be used here and there in the still pools, the river cannot be said to be navigable. Neither can timber be floated easily, although in very high floods it may be possible to send logs down it so as to reach a point below.

592. But below Karumānkayam every one I asked told me that it was utterly *No timber can be floated* impossible to float any timber, a conclusion that is borne out by *down the river above its* the difference in elevation of the river between that place and *junction with the Mo-* the Mōthirappura; and this is further supported by the fact, *thirappura.* that boats cut in the forest to the East of Karumānkayam have to be laboriously dragged by elephants all the way up to Nagarampāra and down the steep Kombanchēri descent, before they can be transported by water. This sets at rest for ever the suggestion that the timber growing on the sides of the Periyār from Peermerd Northwards should be floated down the river to Malayāttūr. I have no hesitation in saying that it would be absolutely impossible to use the Periyār for the transport of timber, above its junction with the Mōthirappura river.

593. The Pirinyānkūtta ār which unites with the Periyār 3½ miles below Karu-*Sources of the Pirinyān-* mānkayam and drains the country to the East of the junction *kutta river.* is a large river. It rises on the hills near Vandanvatta (3750 ft.), and on the slopes of Chiragirunna (Chengoarandan) hill, and flows due North. At Rājākkulam, close to the Eastern boundary of Travancore, the water of the streams was collected into one spot to form a tank, with the object of diverting some of it into the Madura District, but whether the object was ever carried out or not I cannot say; in any case the quantity of water collected here was so small that it would have done very little good to the fields below.

594. North of this, the streams that drain this elevated and open plateau unite *The course of the* to form the Cocchār, a small river which winds among the bare *Kalār.* grass hills, and contains but little water during the dry weather. After proceeding for about 4 miles to the Northward it is joined by the Karinkal ār (Kurrunkalaur) which rises in the forest near Puliyan mala (Pooli mulla), and even in January has plenty of water in it. Here the combined streams incline to the North-west, and, descending into a better wooded country, are joined by the Sannāsi ār (Sunnachi aur) which also rises near Puliyan mala. The course of the river from this point is first North, and then West, and it is here called the Kalār.

595. Where the path from Kalkkūnthal to Kumili crosses it, that is, in a line *Its further course.* between Thōvāla mala (Towaulah mulay) and Chettiyān pāra (Chettyar rock), the Kalār flows down a very rocky bed. In the dry weather it is crossed with ease, but when rain has fallen it is often exceedingly dangerous to a traveller, as a failure to cross at the right spot results in his being carried over the rocks below. Four miles to the West of this point the Kalār joins the Vanjikkadava thōda (Wunjacuddavoo thode), a stream almost as large as itself.

596. The Vanjikkadava thōda drains the valley between Thōvāla mala on the *The Vanjikkadava* East and Kalyānappāra on the West, a country thinly peopled, *thōda.* (the only inhabitants being one camp of Mannans who sometimes move elsewhere), very rough, and the home of many herds of elephants.

597. The two streams, now combined into a large river, flow for 4 miles North *The Pirinyānkūtta* along a deep valley and are joined at Pirinyānkūtta by the *junction.* Thuval ār (Thooval aur), whose streams rise in the cardamom gardens and moist jungles that lie between Kalkkūnthal and Udambanchōla. Here too another stream unites with them, which waters the Pottan kāda, marked in the map as "high rugged table land overgrown with an impenetrable forest."

598. The elevation of Pirinyänkütta is 1,700 ft., and from here the river runs
The elevation. Westward for 8 miles between two very lofty ridges, rising on either hand fully 1,000 ft. above its bed, and clothed with forest from base to summit. It then joins the Periyär which flows on to the North, and meets the Möthirappura river coming due South, as already described.

599. On the banks of the river where the Kalär and Thuval meet, which is
Teak trees at the Piri- called Pirinyän kütta, there is a grove of very nice teak trees,
nyän kütta. and here the path from Nagarampära to Udambanchöla crosses the water. I made enquiries of the Hillmen as to whether it would be possible or not to float this timber down the river, and they assured me that it would be quite impossible on account of the cataracts in the river bed. This also seems most probable, when we consider that the fall of the water down to its junction with the Periyär must be over 100 ft. a mile.

600. The Möthirappura river is one of the largest tributaries of the Periyär.
Sources of the Möthira- Its source is found at an elevation of 7,000 ft. near Chitta vara
rpura river. on the Eastern ridge of the High Range plateau. From this point the stream, gathering volume as it goes, passing in its course Kundala, and Pälak-kadava where it is crossed by a bridge. Continuing in a South-westerly direction it is joined near the Anamudi estate, after a course of 14 or 15 miles, by a large stream which rises close to Anamudi itself. The Möthirappura river now turns due South, and receives the waters of two powerful streams which flow into it from the West.

601. The map at this place is incorrect as it shows a river rising near Aut
Map incorrect. mulla and running to the West, whereas the drainage is to the
East, and into the stream running from Anamudi.

602. The part of the river where these several streams meet is called the
The Münnär. Münnär. Its elevation here is 4,700 ft.. and for a short distance it winds through a beautiful meadow of short grass, the water being always 2 or 3 ft. deep, clear and cold, while the breadth of the river is about 30 yards.

603. The Möthirappura river at this point passes, during the monsoon, through
Disappearance of the a rift in the rocky ridge which extends from Parvathi peak
Münnär. (Nautanchayee) to Chokkan mala, and falls over a precipice into the valley below. In the dry weather the water disappears down a passage through the ridge, and it is then possible to cross from one bank of the river to the other without wetting the feet. The water which thus disappears emerges some way down the hill.

604. At the Möthirappura thävalam 3½ miles below the top of the fall, the
Rapid descent of the elevation of the river is only 2,500 ft, showing that it has
river. descended 2,200 ft. in this short distance, indeed about 2,000 ft. of this fall is at one spot, but owing to the precipitous character of the banks it is impossible to take exact observations. At the last mentioned thävalam the river is 50 yards wide and 3 ft. deep, running over a sandy bed. The banks are covered with forest, or grass, and game of every description abounds here.

605. A mile below the thävalam the Chinnär, a stream which rises on the
The Chinnär. slopes of Muraccha mala (Moolchindy hill), and drains the valley below Chokkan mala, joins the river, and two miles further to the South, the Panni är meets and unites with it.

606. The sources of the Panni är lie in the moist forest that clothes the hill
The Panni är. forming the Eastern boundary of Travancore, about 12 miles in a direct line due East of its point of junction with the Möthirappura. The streams which drain this undulating country, covered partly with grass, and partly with forest, and lying at an elevation of about 3,000 ft., unite to form the Panni är. Its course is almost due West, though it winds considerably.

607. The Chemman ár which joins the Panni ár about 6 miles above its junction with the Móthirappura is a considerable stream, and drains the valley North of Udambanchóla, which is famous for the numbers of elephants that congregate there.

The Chemman ár.

608. The Móthirappura river, now largely augmented by so many additions, turns due West after uniting with the Panni ár, and continues in that direction for 5 miles, when it receives the waters of the Kalár, and alters its course to the South.

The Móthirappura.

609. The Kalár is a river of considerable depth though it is not broad. It drains a large extent of forest land, and during the monsoon can only be crossed on trees, as the water is 7 or 8 ft. deep.

The Kalár.

610. At its point of junction with the Kalár the bed of the Móthirappura is full of immense boulders, and if the water is not very high, sticks can be laid across from rock to rock, by which a passage can be effected, but when I was there last December we found the river in flood, and had to build a raft in order to cross.

Junction of these rivers.

611. The Móthirappura after flowing for 3 miles due South and descending with great rapidity, meets the Periyár at an elevation of 800 ft. The height of the river where the Kalár unites with it and which is called Thirukallan párakkútta is 1,340 ft., so that the fall is 540 ft. or at the rate of 180 ft. a mile between the two places.

Course of the Móthirappura.

612. It is hardly necessary to point out that this is too steep for the floating of timber, indeed, the river in some places rushes over cataracts that extend for a long way. This is unfortunate, as there are some (not a great many) nice teak trees on the banks of the river, which could easily have been sent down it, had the condition of the river admitted of water transport.

Impracticable for floating timber.

613. The Periyár after receiving the waters of the Mothiroppura, turns at right angles to its previous direction and flows West North-west. In the course of 8 miles the fall is between 400 and 500 ft. Sometimes the river pours over rocks, or eddies between them, and sometimes it flows tranquilly along with hardly any current at all.

Further course of the Periyár.

614. After covering a distance of 8 miles the Periyár pours under a large rock which has fallen from the hill side. At some former time a landslip must have occurred, which deposited a pile of stones in the middle of the river, thus forming an island. From this island to the North bank a boulder, falling after the formation of the island, stretches across and makes a natural bridge called the Uttharam or "barrier". During the dry weather, when the volume of water is small the whole of it flows under the rock above mentioned, but in the monsoon a good portion of the river passes round the island, and pours down between it and the South bank.

The Uttharam.

615. This freak of nature has been exaggerated by people who have never been to the spot, into a disappearance of the river underground for miles. Before I visited the place I was told that the water passed into a chasm, and did not emerge again for a long distance. As a matter of fact the appearance and position of the rock are nothing wonderful, and similar natural bridges may be seen on all our rivers though on a smaller scale. Why so much has been said about this particular rock I do not know, but perhaps the difficulty of reaching the spot has caused people to exaggerate its curious appearance, and to magnify the labour of getting there.

So called disappearance of the river.

616. The scenery at this place is most magnificent. For the last 9 miles before reaching the Uttharam the river runs down a gorge between immense cliffs rising to a height of 1,000 or 1,500 ft. above its bed. The Northern bank of the river is quite precipitous, but on the Southern bank the hills though very steep, are sometimes clothed with forest.

Fine scenery.

617. Close above the Uttharam and to the South of it, marking its position,
The Minirangiyān kal. an immense bastion of grey rock rises perpendicularly to an elevation of quite 3,000 ft. The peculiar shape and lofty height of this curious rock, called Minirangiyān kal, and marked "Station" in the map, draws attention to it from a great distance off, and makes it a notable landmark. On its Northern side it is precipitous, but it slopes to the South and may be accessible from that quarter.

618. Below the Uttharam the Periyār pours over cataracts, and eddies among
The Chirppa. numerous pointed rocks for 2 or 3 miles and descends in that distance probably about 200 ft. At one spot the path which usually runs along the river bank is taken up a sloping rock on the Southern side, called the Chedayan pāra, for the river here rushes between two walls of rock not 50 yards apart, and cannot be approached. This place is called the Chirppa (a plug or stoppage) and the combined name of Uttharan-chirppa is given to these 2 miles of dreadfully broken water in which no boat could live for an instant, nor log avoid being splintered into fragments.

619. The question has often been raised whether it might be possible to make
Impossible to make the river passable for timber at a cost that would pay. the Periyār fit for the floating of timber above the Uttharam, by blasting some of the rocks that bar its progress, or by cutting a timber road for a short distance along one bank at this point, and I made an expedition to the spot in August last year for the express purpose of forming a definite conclusion. I decided that it would be quite useless to attempt any work of the kind. The rocks to be removed do not lie at one spot only, but extend along the bed of the river for upwards of 2 miles; at the same time the descent is very rapid, so that timber would be carried down at a great velocity and if it came into contact with any rock would be broken to pieces in a moment. Nor would it be in any way possible to construct a road along the bank for the conveyance of timber, for, at the Chedayanpāra, the sides of the river are perpendicular walls of rock at least 200 ft. high. If the forests above this point were very extensive, and contained large quantities of valuable wood, it might be worth while to construct a timber slide to bring the logs down, which might cost one or two lacs of rupees. But, as I have shown, the Periyār and its tributaries become quite useless for the passage of timber above its junction with the Mōthirappura, so that these works would only serve the forests up to that point. These forests contain but little timber. I made careful enquiries on this subject, and was assured that the quantity of teak or any other valuable tree growing here is very small. The Northern bank of the river is quite precipitous, and, beyond a few trees of kambagam and teak, nothing could be expected from it. On the South bank also the cliffs approach the river closely, and where there was formerly any forest the Urālies have cleared it for their cultivation, so that here too the quantity of available timber is very limited.

620. Mr. J. S. Vernede, a former Assistant Conservator, visited the spot about
Mr. J. S. Vernede's opinion. 20 years ago, and arrived at the same conclusion that I have. It may therefore now be considered definitely settled that the Periyār cannot be used for floating logs above the Uttharan-chirppa, and that none of the timber growing on its banks can be utilised except what can be taken over land.

621. The first point on the river at which it becomes navigable, or suitable for
Karimanal. the floating of timber, is Karimanal, which is 10 miles below the junction of the Mōthirappura with the Periyār, and 5 miles above its confluence with the Dhēvi ār. Here the river is 300 yards wide with steep cliffs rising up to Kuthirakuttippāra on the North side, while on the South bank there is an extent of flat forest land, varying in breadth, between the river and the hills.

622. From Karimanal the Periyār flows for 4 miles without any impediment,
The Pinakkan chāl. but a mile above the Dhēvi ār a barrier of low rocks crosses it, to which is given the name of the Pinakkan chāl. This reef forms an obstruction at all times except when the river is very full, it is therefore an annoyance to the timber merchants, who have to collect their logs on the river bank above it, and to wait till the water is sufficiently high to float them over the reef.

At the time of my visit the river was in flood, and the rocks were covered, so that I could not see them properly, but I am told that if a passage was cleared through the reef by blasting, timber operations would be much facilitated.

623. The Dhěvi ār (Muduka ār), which joins the Periyār a mile below Pinak-
The Dhevi ár. kan chāl, takes its rise at a place called Mannān kandam near
Kallithāvalam (Culley tawullam), and a mile to the South of the curious Petti vara (Cure male). This is a rocky hill standing up above the surrounding country and carrying on its summit a large square rock, shaped like a box, from whence it takes its name. It is visible from a great distance off, even as far as the hills near Chāntha pāra.

624. Mannān kandam is a tract of swampy grass-land, lying at an elevation of
Mannān kandam. 1,700 ft. between two rocky ranges of hills Tradition states
that the whole of this land, which extends over several hundred acres, was at one time cultivated by the Hillmen, but that the elephants destroyed the fields and so put a stop to the cultivation. These animals are exceedingly common here, and trouble the Hillmen much.

625. Collecting the waters of this swamp, the Dhěvi ār drains slowly to the
Course of the Dhevi ár West, preserving a uniform level of about 1,600 ft. and running
between two parallel ridges; the lofty Thuruva mēda separating it from the Periyār on the South, and Iruppurayil thanda (Irruppurulley thundu) hemming it in on the North. Arrived near Kuthirakutti hill after a course of 8 miles; the river bends somewhat to the North and descending in a waterfall which is visible from Kōthamangalam, it turns to the South-west and unites with the Periyār at an elevation of about 200 ft.

626. The course of the Dhevi ār is wrongly marked in the large map (1 mile
Large map wrong. to the inch) where it is shown as joining the Periyār nearly
opposite the Mīnirangiyan kal station, but the small map (4 miles to the inch) is correct.

627. A mile below the Dhěvi ār the main river passes the site of the village of
Nēriyamangalam. Nēriyamangalam, which was at one time populous. At the
time of Lieut Ward's visit in 1817 Nēriyamangalam was not inhabited, but he noticed the ruins of a village, showing that when the traffic between Cochin and Madura passed this way, there was a resident population there. At present, one Mahomedan and his family constitute the only inhabitants.

628. Shortly below Nēriyamangalam the Pulikuttha thōda flows into it on the
The Púchāl. right and a couple of miles farther, the reef is crossed by a
barrier of rocks at Pūchāl. This is not so great an obstruction as the Pinnakkan chāl, but here too a passage should be blasted through it.

629. The banks of the river are covered on both sides with large forest as-
The river banks. cending the hills to their summits, occasionally broken on the
western side by patches of cultivation, but on the East untouched, save by the operations of the timber merchants. On the left the hills do not rise to any height, but on the right a lofty ridge follows the line of the river for several miles, and parts of it are called Pulikuttha mudi, Kokkalipāra, and Pōttha pāra, and the whole of it Chōlatthanda mudi (cholay thunda mudee.)

The Avalchāl. 630. Another barrier of rocks called the Avalchāl crosses
the river 2 miles below Pūchāl, and should have a passage blasted through it.

631. From this point the Periyār flows with an unimpeded stream for 6 miles,
Junction with the Idyara. when it unites with the Idiyara river, which falls into it from
the east. A mile above this junction the Thatthakkāda crossing is reached.

632. The Idiyara or Idamala river takes its rise on the slopes of the lofty
The Idyara. Paruttha mala (Payrat male,) on the boundary between Tra-
vancore and Coimbatore, at an elevation of over 7,000 feet. From here it descends with great rapidity through a steep gorge clothed with forest on either side, and reaches an elevation of less than 1,000 feet in a very short distance.

633. As already stated in the introductory chapter, the line of the Idiyara
river between Paruttha mala and Pannimāda kuttha is incor-
rectly marked in the map. Instead of bending to the South
and then turning West, its direction is almost straight from point to point, a distance
of some 15 miles.

The map incorrect.

634. Many streams, whose sources are on the mountain slopes and peaks to
the North-west of Anamudi, join the Idiyara on its left bank.
The most Southerly of these, the Nilakkallōda rises to the East
of that mountain, and close to Vāgavara at a height of 7,000 ft. Flowing to the
West it falls rapidly till it reaches a lower plateau covered with forest, through which
it winds before reaching the river.

The Nilakkallōda.

635. Another stream the Eettilayār, which is now the boundary between Travancore and Coimbatore, rising on the Eravi mala plateau, runs
due West, to the South, and under the lee, of a well marked spur
of the plateau, and also joins the Idiyara in the valley below.

The Eettilayār.

636. The main river, collecting all these streams from the forest above, runs
through a narrow gorge at an elevation of probably not more
than 400 ft., and foams over a cataract. To the North of it,
the immense cliffs of Pannimāda rise to a height of 3,800 ft. and to the South equally
lofty precipices, called Kōdanāda thanda (Codainadee thundu), stand like a wall.

The Panui māda.

637. Some 3 miles below this point the Irulakayam (where there was a timber
camp in former times) is reached : a mile below this the Kili
thōda, which takes its rise in Cochin territory, joins the river.
Two miles lower the Kaliman thōda falls in on the South bank, and a mile farther on,
the Anamādan thōda, which drains a large valley in Cochin territory, unites with the
Idiyara.

Course of the Idiyara.

638. At the Kilipparamba ēla half a mile down stream, there used to be a timber camp in former times when these forests were worked by
Government for teak, but there is nothing there now. The
river is here some 50 yards wide and its elevation is 300 ft., with no rocks to obstruct
it. Three miles lower the Vādamuri camp is passed, the river flowing gently over a
sandy bed, and so it continues running to the West between the Thēratthanda ridge
on the South and the high cliffs that mark the boundary with Cochin on the North.

The Kilipparambu āla.

639. Fifteen miles below Pannimāda kuttha the Idiyara is met at a place
called Perumuri by the Manimāda thōda, and alters its direction
abruptly to the South. In former times timber used to be float-
ed down all this distance, and the valley was then famous for its teak which was of unusually large size. The level character of the river bed also facilitated timber operations. As, below this point, it becomes impossible to float timber except in high
floods, the logs used to be collected at Perumuri until the water rose sufficiently high
to submerge the rocks at the māda, and the tree is still shown to which they were
attached, and which served as a water gauge to show when the river had risen sufficiently for floating operations.

The Perumuri.

640. The famous Idiyara māda is situated a mile below the Perumuri. Here
the river, which has been flowing sullenly along, begins to descend more rapidly, and in one place, for a distance of 300 yards,
its bed, here narrowed to no more than 40 yards, is filled with enormous builders 40
and 50 ft. in height. Between these the water twists and eddies, the total descent
being perhaps 30 ft. When the river is in full flood and the water is sufficiently
high to overtop the rocks, the sight must be magnificent, but this does not occur
every year, for the records show that it was sometimes 2 or 3 years before logs collected
at the Peru muri could be sent down. Any attempt to despatch them before the
river rose to the proper height resulted in their being hopelessly wedged between the
rocks in such a position that no power on earth could remove them. There are many
logs lodged in this way still to be seen, though timber operations in this valley ceased more than a score of years since.

The Idiyara māda.

22

641. To prevent this loss of timber, Mr. Kohlhoff, a former Conservator of Forests, blasted out a portion of one rock in the centre of the river, but to be of any use thousands of rupees would have to be spent in levelling all the large rocks through the whole of the máda.

Mr. Kohlhoff's blasting operations.

642. A more feasible plan for getting down the timber was suggested by Mr. Vernède the present *Conservator, who in the year 1869, proposed to Government to cut a timber track from the Peru muri down to Makaracchál below the máda. The total distance is only a couple of miles, and as far as I could see from a casual glance at the place there would be no difficulty in carrying out the work. Whether the proposal was ever approved or not I do not know, but as, shortly afterwards, Government ceased to work down teak on its own account, the money if ever sanctioned was never spent.

Mr. Vernède's proposal.

643. On the hill side above and East of Makaracchál there is a very curious pointed rock, poised on the ridge. It is called Kákkénippára, and marks the position of the above mentioned máda with exactness. It is a famous place for bees nests.

The Kákkónippára.

644. The Idiyara river is joined shortly below Makaracchál by three streams which drain the valley South of Thérathanda and run from East to West. They are the Kiri thóda (Kunnee thode), Kótta muri thóda (Kotuvunye thode), and the Elukavara thóda (Illahavura thode). About a mile South of the stream last mentioned, the large Kandanpára river unites with the Idiyara. The Kandanpára river has two branches, the Northern and smaller is the Parishakkuttha, the main stream is called the Kandanpára.

Tributaries of the Idiyara.

645. The Parishakkuttha takes its rise on the slopes of Churi mala (Shuleo mala) and the high grassy ridge to the East of it, and rapidly descends to a low elevation. The first point of any note is the Varunan kadava about 5 miles from its source, where there used to be a timber camp in former times and from which timber used to be floated. Five miles lower, the stream is joined by another called the Kurambar thóda on which also there used to be a timber camp. About a mile and a half below this, the fall which gives its name to the river occurs. The height over which the water tumbles is not more than 14 or 15 ft., but some jagged rocks stand up above the fall, and so make it necessary for timber to be taken round by land, and not shot over into the pool below. From this point to its junction with the Kandanpára river, the Parishakkuttha runs very level, and though not more than 30 yards wide, is often 3 feet deep even in the dry weather, though generally the flow of water is very little.

The Parishakkuttha.

646. The Kandanpára river takes its rise near the hill marked Putchay malo in the map: its upper portion is called the Pacchappullóda. From this spot, about 5,000 ft. high, this stream runs to the West down a deep valley partly wooded and partly covered with grass, and is joined by several others. On reaching a point due South of Pannippárakkúmba (Punneypauray cuppu) the stream falls into the valley below : a couple of miles lower down, it is joined by another stream which drains the valley North of Patthakatthi mala (Perrunputtey cul). Arrived at this lower valley it winds through it at a pretty uniform level for a dozen miles, receiving several streams which flow into it from both sides, the largest of which is the Mangalatthóda (Mankellay thode), which drains a large valley North of the Dhévi ár valley.

The Kandan pára.

647. Just below Kólapára mudi (Kollay pauray mudee), and due South of the letter "k" in that word, on the big map, the Kandanpára river is precipitated over a waterfall called the Pöndimádakuttha. It is here about 50 yards across, and the fall is between 100 and 200 ft. high, but below it again there are some minor falls and cataracts, so that the total descent must be over 200 ft. From this point to its junction with the Parishakkuttha is a distance of 4½ miles, and over this length the fall is not great though here and there a cataract occurs which would bar the passage of logs.

The Pöndimádakuttha.

---

648. The Kandanpâra river after receiving the waters of the Parishakkuttha, flows due West between low ranges of hills covered with grass and trees, or the " eetta " reed (*Beesha Travancorica*). During the dry weather it is tiresome travelling by boat here, as the river is crossed by numerous reefs of rocks, up or down which the boat has to be dragged ; but when the water rises these are all submerged, and then travelling or floating of timber are very simple matters. About a mile below the junction, the Vadâvana stream (Madavan thode) joins the river on the right bank, and 2½ miles lower the Kandanpâra pours its waters into the Idiyara river.

Junction of the Kandanpâra with the Idiyara.

649. The Idiyara after receiving the waters of the Kandanpâra flows due South for a mile, when, at Urulupâra, it turns sharply to the West, and at the same time the Kûttapâra thôda unites with it on the South bank. The Kûttapâra thôda drains an extensive valley of low lying well timbered land. It takes its rise on the high hills to the North of the Dhêvi âr, and after a course of 7 miles reaches Urulupâra. During the dry weather there is scarcely any water in this stream, and near its mouth its course is blocked by numerous rocks : this and the rough character of its bed, probably account for the fact that no timber was cut out of the jungles in its basin till a few years ago.

The Kûttapâra thôda

650. The Idiyara river after flowing for 5 miles due West joins the Periyâr as already stated. During its course over this distance it passes between banks well wooded on both sides. On the South two immense cliffs of rock, precipitous on their Northern faces, tower above the river, one of them called Pulimblâ and the other Nyârapalli. Between them and the river a narrow strip of forest extends, and here numerous elephant pits are situated, in which one or two of these animals are captured every year. On the North bank the land is more level and the forests more extensive.

Junction of the Idiyara with the Periyâr.

651. After receiving the waters of the Idiyara the Periyâr, now a grand river upwards of 400 yards broad, turns due West and after flowing for a mile and a half is joined by the Pûthatthânkatta or Prânthan thôda, which drains a considerable area of forest land on the South bank. A little lower down, the river passes between two prominent rocks called the Pûthatthânkatta, and is narrowed down to about 40 yards in width. Shortly below this its course is impeded by the Vattakkal, a large rock which rises in the very centre of the narrow channel, and proves very troublesome to timber merchants when floating their logs down the river. Money would be well expended in blasting away this obstruction.

Course of the Periyâr.

652. A little below this at a bend of the river, is the Mûnnûrâm pâra where logs are often carried by the force of the current and deposited, so that it is most difficult to get at them, for pot-holes abound, and prevent elephants from approaching the spot. Here the Chenkara thôda falls into the river on the South bank, and a little farther on the Thundattha thôda unites with it on the right hand.

The Mûnnûrâm pâra.

653. This large stream joins the Periyâr above the Ayyanchâl island, and not below it as shown in the map. It takes its rise on the high ridge to the Nort-west and waters a valley of forest 9 or 10 miles long before reaching the main river.

The Thundattha thôda.

654. The Periyâr continues winding to the West and North-west, sometimes spreading out into large reaches and sometimes running down narrow channels between rocks ; some 2 miles below the Ayyan châl island the lesser Pôra is reached. This is a small fall in the river which is troublesome for boats. Here, and a little above it are situated the big and little Panni rocks, which interfere much with the floating of timber.

Further course of the Periyâr.

655. In his report for the year 1044 M. E. (1869) the Conservator of Forests refers to these rocks and recommends an expenditure of 400 Rs. on blasting them, but up to the present time the work has not been carried out : whether it was ever sanctioned or not I do not know.

Blasting needed.

656. About a mile below the lesser Pōra the greater Pōra (Purryauraypur-
thuruthee) is reached, which is by far the worst obstruction to
the navigation of the river below Kərimanal. At this place
the Periyār spreads out into several channels, and pours over a shelf of rock 20 ft.
high. When the river is full it dashes over the rocks in a broad cataract
of foaming water which stretches from bank to bank; but when it is low the greaten
part of the river bed is exposed, and the water is confined to a few channels up which.
boats have to be dragged with much labour. According to the state of the river
timber is floated down one or other portion of the cataract; but the logs are often.
damaged or completely ruined by striking against rocks in their descent. It would
not be possible to effect any improvement to this part of the river bed by blasting,
but a side channel might be cut, or a "portage" might be constructed for the conveyance of the logs overland for a short distance.

The "Great Pōra."

657. A stream of some size joins the Periyār on the North at this point.
Three miles below the Pōra the Vembūram island (Womb-
thuruttha) is reached. It lies in the centre of the river, and
is about a quarter of a mile long, with an average breadth of 100 yards. This island.
has been under cultivation from time to time, and Lieut. Conner in his memoir of
Travancore states that a Brahmin "illam" was formerly located here. In more-
modern times it was taken up for coffee cultivation and abandoned; it is now cleared
and sown every five years by a wealthy Mahomedan who lives at Varāppura.

The Vembūram island.

658. At this point the Thekkanperin thōda joins the Periyār. It is a slug-
gish stream of no great breadth, but deep, and is used for floating timber. North of the river and rather lower down the-
Vadakkanperin thōda unites with it.

The Thekkanperin thōda.

659. This stream drains a very large valley of forest lying at a low elevation,
and surrounded on all sides save one, by high ridges. Its
source is near the Vādamuri, the pass on the way to the Kōttasshēri river, and this part is called the Pullāni thōda. A little
lower down, the Kuravankuri thōda which drains the valley to the West joins it;
and about a mile below this, the Muthukuri timber camp is pitched to command the
neighbouring forests of Eramukham (Yairmayillawurray). Lower still the Mauyap-
pāratthōda joins it on the Western side, and the Chēlatthōda on the East. This
valley of forest is most convenient either for conservancy, or for the planting of
teak, as it is quite sheltered from wind and is easily accessible from Malayāttūr.
Timber operations have been going on here for some time back; but, with the exception of irūl and such heavy woods, the number of trees of valuable sorts still remaining is not great. A year ago, before this forest was declared a reserve, people from
the low country, many of them from Cochin territory, cleared and burnt upwards of
2,000 acres of original forest, which were sown with paddy, and gave a very heavy
crop; in the land thus cleared and abandoned it is now proposed to sow or plant
teak.

The Vadakkanperin thōda.

660. The Vadakkanperin thōda runs perfectly level, and is largely used for
the conveyance of timber cut in the Cochin forests, and in our
own forests beyond the Vādamuri; for, owing to obstructions in
the Kōttasshēri river, the latter is not adapted for the floating of
timber.

*Useful for timber oper-
ations.*

661. The Periyār sweeps on below the Vadakkanperin thōda a magnificent river
400 yards broad, and passes on its left bank the Malayāttūr teak
plantations (to be hereafter described) which stretch along the
river for several miles, and aggregate nearly 150 acres, of which.
about one-third have failed and been abandoned.

*Malayāttūr teak plan-
tation.*

662. On the right bank the configuration of the country has not been properly
laid down by the surveyors; for, instead of the ridge extending
in a straight line from Kurishamudi East to the Vadakkanperin
thōda, it stops abruptly about a mile to the East of the Cochin boundary, allowing a
stream, the Vadakkanpalli thōda which rises North of Kurisha mudi, to enter the river.
A mile farther down, the Illitthōda, the boundary between Travancore and Cochin is.
reached, and here the Periyār leaves the forest line.

Mistakes in the map.

663. To the North and North-west of the tract of land belonging to Cochin
at Ma'ayāttūr two small valleys, the drainage of which is into
The Maunyapra thōda. the Mannyapra thōda, lie within the forest area. This thōda
debouches into the Periyār at Kottaniam where there is a watch station.

664. The Periyār, after leaving the forest area, passes Malayāttūr, and after a
winding course of 14 miles reaches Aluvā, where it is divided
Further course of the into two branches, one flowing to Paravūr and the other to
Periyār. Cochin.

665. Aluvā is a favourite hot-weather resort of the people of Cochin, who also
always get their drinking water from this place. On Peermerd
Aluvā. it is considered a most foolish thing to drink the water of the
Periyār during the hot weather months, as fever is almost sure to be the result : it
has often been asked, why the same water that is so much avoided on the hills
should be eagerly sought after in the low country.

666. The explanation seems to be this. On Peermerd the water of the
Periyār is full of decaying vegetation which runs into it from
Reason of the purity of the forest above Mlāppāra, and on Peermerd itself, whenever
the Aluvā water there is enough rain to carry down the dead leaves, but as the
river descends over cataracts and rapids the water gradually becomes aerated, until
it has lost much of its poisonous properties. Besides this, in the Periyār below its
junction with the Idiyara, and in the Idiyara too, the bed of the river is divided into
deep pools by barriers of rocks, which cross it at intervals In these pools the water
is almost stagnant and they therefore act as natural filters, the decaying vegetable
matter being deposited in them, while the pure water runs down the channel into
the pool below. This occurs over and over again, and finally, before reaching Aluvā,
the water is further purified by passing over some 14 miles of perfectly clean sand.

667. I have explained that the Periyār is navigable for small canoes as far as
Karimanal on the main river, Makaracchāl on the Idiyara
Periyār navigable. river, and the junction of the Parishakkuttha with it on the
Kaudan pāra river : but these are very small boats and will only contain 3 or 4 men
with their loads. Larger boats with luggage on board cannot ascend above
Malayāttur.

668. Timber can be floated down from Karimanal on the Periyār, from near
Pannimāda kuttha on the Idiyara, from the Varuman kadava
Suitable for floating on the Parishakkuttha · and from Pendimāda kuttha on the
timber. Kandanpāra river ; but owing to the rocks in the Periyār itself
which obstruct the passage, nothing but boats and light timber can be brought down
by this way, for, even if there were many bamboos on the banks of the river, which
there are not, the rocks referred to above would inevitably burst off anything tied to
the heavier logs to cause them to swim ; it thus happens that we have a good number of thēmbāvu, blackwood, and vēnga trees growing in the vicinity of this river,
which can only be brought down after the rocks have been removed.

669. These pools and cascades have all received different names which are
convenient for fixing the position of any point on the river.
Names of places on the Thus on the Idiyara river we have (1) Makaracchāl and the
Idiyara kayam (pool) below it, (2) Paravanchāl and kayam, (3) Varukkappāra, (4) Kannan chāl and kayam, (5) Putthan chāl, (6) Urulupāra, (7) Nyara-
ppalli chāl, (8) Māmutti kayam, (9) Pulimbla and the pool below it.

670. On the Periyār river there are (1) Pūthatthānkatta, (2) Vāttappāra-
kkayam and chāl, (3) Mēl Vadakkam kayam, (4) Cheriya Pōra,
F Names of places on the (5) Kīr Vadakkam kayam, (6) Valiya Pōra, (7) Perinkal chāl,
Periyār. (8) Pānēlikkayam, (9) Vembūram island, (10) Pānam kuri,
(11) Pōthatthān kayam, (12) Vadakkampalli kayam.

671. The total area of land drained by the Periyār within the forest line
amounts to no less than 1,432 square miles, which comprises
Area of the Periyār ⅔ths of all the land enclosed within the forest boundary line I
basin. have described.

23

672. This extensive tract may be divided as follows (1) Moist forest 500 square miles, (2) Secondary forest, cultivation &c. 209 square miles, (3) Park land forest, grass with trees, 100 square miles, (4) Useless land, grass, rock &c. 632 square miles.

Divided into classes.

673. The moist forest at the head of the Periyâr, comprising both those parts where the river rises and the Western slopes of the water-shed running between Mêthâram mettu on the North and Shivagiri (Shivagerry) hill on the South, extends over an area of about 100 square miles, if we include the patches of jungle in the neighbourhood of Thêkkadi, and between that place and the Periyâr.

Moist forest at the source of the Periyâr.

674. The next large block of moist forest to be met with, proceeding North, is the forest that clothes both sides of the Cheyitthân mala ridge, and extends Eastward to Vandan vattu, and Nort-east to Puliyan mala and its neighbourhood. Its area is fully 20 square miles, in spite of the damage done to it by the Puliyar hill-men of that part of the country, who, not content with re-opening old land, often make new clearings in the forest. A good deal of this land is occupied by cardamom gardens, some of which were in existence in 1817 when Lieut. Waid visited the spot

Forest on the Cheyitthân mala ridge.

675. In the neighbourhood of Kalkkûnthal and Udambanchôla, there are about 20 square miles of forest, including the cardamom gardens and land outside them. This block lies between the two places mentioned, above and to the North and East of Udambanchôla, as far as Ilayakâda (Yellayacaud garden).

Forest near Udambanchôla

676. Northward from here as far as Thondi mala (Tonedu mulla garden), the forest stretches along the Western slopes of the Travancore-- Madura boundary, and varies in breadth from 2 to 4 miles. From Thondi mala a tongue of forest runs out due West to Pottankâda near Moolchindy hill. The area of this land may be altogether 20 square miles.

Forest near Thondi mala.

677. East of the Môthirappura river and the Vanjikkadava thôda, the extent of moist forest, excluding those blocks that I have already described, is not great; perhaps 15 square miles represent the whole of it: but to the West the elevation is lower, the valleys are more sheltered and neither the Hill-men with their axes, nor the Madura cattle men with their grass fires have made much inroad on the forest. Above Karimanal on the Periyâr, that is to say to the East and South of it, and between it and the Môthirappura river, there may be 60 square miles of forest land, all of which is unfortunately inaccessible for the utilization of timber.

Forest in the Môthirappura.

678. Of the other 265 square miles of moist forest in this basin, about 30 square miles between Panninâda kuttha and the Eravi mala plateau, and, say, 35 square miles on the High Range, at an elevation too high to be accessible cannot be worked, but the other 200 square miles are available, and are included in the large Malayâtthûr reserve which has lately been declared.

Moist forest in other places.

679. Of secondary forest coffee estates, Hillmen's cultivation &c. are answerable for 200 square miles, and the land occupied in this way is so broken up into small blocks that it is not easy to describe their position.

Secondary forest &c.

680. In the neighbourhood of Mlâppâra are found several valleys which the Mannâns cultivate, shifting about from one place to another, and it is noticeable that they usually select steep land, probably in order that they may be out of the way of elephants.

The Mannâns

681. There are 5 or 6 coffee estates on the banks of the river itself.

Coffee Estates.

682. Near Vandanvatta the 8 colonies of Puliyars have cleared numerous detached valleys. Between this place and the Periyâr the Mannâns have destroyed all the good land, while to the West of the

Colonies of Hillmen.

same rive', which is the boundary between the Mannāns and Urālies, the last mentioned Hillm . nave left standing hardly a single stick of jungle. The Muthuvāns have been very destructive in all the forests beneath the slopes of Pārvathi hill (Nautachayee), on the Dhēvi ār near Mannān kandam, on the Mōthirappura river, again on the Kandanpāra river in the neighbourhood of Chennāyipāra mudi, and Chūtta vara (Shutta-wurray). They have also extended their clearings to the valleys of the Idiyara river, and Parishakkuttha thōda, where two camps of Vishavar are also settled.

683. On the High Range the planters have cleared for

Estates in the High Range. chinchona, tea, or coffee 3 or 4 square miles of land within this basin, but their fellings are insignificant compared to those of the Hillmen

684. In the low country, on the Vadakkanperin thōda and to the South of the Periyār, at one time or another, perhaps 20 square miles of land

Cultivation by low country people. have been felled by the villagers, and thus the 200 square miles are made up.

685. The third class, or grass land forest with large trees,

Area of park-forest small. is poorly represented on the Peermerd plateau, the elevation being too high, or the climate too moist, for this kind of forest.

686. In the neighbourhood of Kumili and Thēkkadi there are 10 square miles of park forest. Near Thodupura-Periyār there are some 7

Its distribution. square miles, and on the Periyār and the Vanjikkadava rivers and down the Kalār valley to the West of Rāmakkal, there are altogether 23 square miles of it. The other 60 are situated on the hills between the Kandanpāra and Parishakkuttha rivers, and North of the latter river, in the Idiyara valley, and on all the drier hills towards the low country.

687. The useless land extends over a very large portion of this river basin : it comprises all the hills and the greater part of the country on

Useless land. both sides of the Periyār, between Mlāppāra and the Cheyitthān mala ridge. North of the latter point it includes very nearly all the land to the West of the Periyār as far as Idukka vetta (Cullyka mala), and to the East of that it occupies all the ridges between Kalyāuappāra and the Chettiyan hill. To the North of Udumbanchōla, and to the West of Chāntha pāra, the hills are almost entirely covered with grass, and in the dry weather thousands of cattle come up to feed on it. North of this again, by far the major portion of the High Range consists of grass hills, the Eravi mala plateau especially being quite destitute of trees of any sort. Elsewhere the higher hills and many of the lower ones consist of bare rock, or are clothed with a scanty herbage unfit even for grazing, and the area of rivers, swamps, and flat rock must be not inconsiderable.

688. Within the basin of the Periyār the teak tree grows sparingly on the hills, but in fair abundance at the lower elevations. It is not found

Most valuable trees. Teak on the Periyār basin. anywhere above 3,000 ft., and land below this altitude is rare on the upper hill plateaux. At Thēkkadi this tree is met with, but the character of the timber is poor, owing chiefly to the fierce grass fires : it occurs here and there all down the valley of the Periyār, but is much rarer than I expected, and with the exception of a grove of very fine trees near the Pēkkānam estate the quality of the timber is poor. It is not till the Cheyitthān mala ridge is passed and we approach the neighbourhood of Thodupura-Periyār that this tree becomes abundant and good, but even here it is confined to the grass land on the banks of the river, and does not ascend the hills to any height. Some very fine trees may be seen near Madatthinkadava lower down the river, but unfortunately they are growing in such a position that it will never pay to work them down.

689. Eastward of this spot the teak tree is seen near Rāmakkal, and scattered all down the valleys of the Kalār, and the Vanjikkadava ār.

Teak found above Karunanal. At Pirinyānkūtta, where several streams meet, there are two groves of teak, one on the North and one on the South bank of the river, but this timber is also inaccessible and therefore useless. Some good trees are found on the Môthirappura river, but they cannot be worked out. It is not till we get to Karimanal that we can utilise any of the trees that grow, but not abundantly, in that neighbourhood.

690. Below the junction of the Dhêvi âr with the Periyâr teak used in former times to be numerous on the hills overlooking the river, and it *Teak in the Periyâr basin below its junction with the Dhêvi âr.* has been felled there for generations. The ridges about the Kandanpâra and Parishakkuttha rivers were famous for it in years past, but little now remains, for felling operations were carried on there and in the Idiyara valley every year until the present contract system was introduced, when the contractors, for reasons hereafter to be explained, gave up felling in the interior, and confined themselves to working out the teak in the forests near the villages.

691. The teak still standing is of small size, and represents what was rejected before, and this is scattered thinly about the country. Here *Very little large teak remaining* and there an enormous tree is met with, which was passed over at the time of previous fellings, from being too large or too badly situated for removal. A clump of 200 or 300 large trees are said still to be growing on the hills near the Kandanpâra river above Pendimâdakuttha, but a contractor who visited the spot to see if they could be brought down, told me that the difficulty of bringing them to the river is so great that he would not undertake the work under 10 Rs. a candy, double our present rate.

692. Blackwood is found in the same localities as teak, but though its distri- *Blackwood* bution is wider it is not nearly so common as that tree, for it is never gregarious. Below Karimanal it is found on the hills to the South-west of the river, and here and there on the grass covered hills in other parts, but it is no where abundant.

693. Sandalwood is entirely absent from this tract, and ebony is rare, but I *Sandalwood and ebony* have met with it occasionally in the moist forests at the lower elevations.

694. Kambagam is absent from the higher ground : like teak, it never grows *Kambagam* above 3,000 ft., and rarely in North Travancore above 2,000 ft. Below Karimanal it is not uncommon on all the river sides, but of late years, since ânjili has become so scarce, the merchants have taken to cutting kambagam for boats, and large number of trees have been felled lately.

695. Irûl or irumulla is abundant on the Vadakkanperin thôla, and on the *Irûl, thêmbavu and ânjili* banks of the Periyâr higher up, where its great weight has prevented the merchants from floating it down the river. The same remarks apply to thêmbâvu. Anjili is now very scarce in the lower forests on this river, as it surpasses every other timber for boats, and has been felled for this purpose wherever found. The valley of the Periyâr used to contain enormous trees of this species, but it has paid merchants to come up from Cottayam and cart the boats down the bandy road, so that there are very few trees left there now. Near Thêkkadi this tree grows well and has been much used by the Engineers for planking.

696. A demand for white cedar has lately sprung up, and the localities where *White cedar.* it grows, and where it can be got down are being rapidly denuded of this timber, but some trees still remain near Kuthira-kuttithâvalam, Erumukham, and the Vadakkanperin thôda.

697. Vênga is found near Thêkkadi and is cut for sale at the depôt there. *Vênga and ven-teak.* Ven-teak being a light wood is felled on the lower forests of the Periyâr, but the price obtainable being low, it has not paid the merchants to go far afield for it, and they were getting into the habit of cutting small trees of less than one candy, which could be obtained close to Malayâttûr, and which paid them better than large logs, when this was prohibited.

698. There are some other trees found in the forests of the Periyār basin which are occasionally felled, such as Shurali (*Hardrickia pinnata*), the Poon spar (*Calophyllum tomentosum*), Mala uram (*Pterospermum Heyneanum, and rubiginosum*), and others, but though the number of trees, is large there is but little demand for them.

699. The soil of the Periyār basin is good, and nowhere within the forest area does it contain laterite. In some parts it is exceptionally fine, and this is especially the case on the Peermerd plateau and on the banks of the main river. On the Cardamom hills a deep red loam of good character predominates, and on the High Range the Planters have no reason to complain of the want of fertility of their land. The Hillmen living in the basin of the river are little troubled by failure of their crops, which, in the sheltered valleys they occupy, present a luxuriant appearance.

700. The climate of the Periyār valley differs in its upper portions from that of the country to the West of it: the rainfall is less in the South-west monsoon, and heavier in the North-east, the total being much below that of Peermerd proper. As the river descends it flows more to the West, and the difference of climate then sensibly diminishes, and finally disappears altogether, as the river basin comes more under the influence of the South-west, and less under that of the North-east monsoon.

701. I have already explained that the Periyār valley is a very feverish locality, owing to its being shut off from the Western sea breezes which would temper the heat of the sun, and carry away the stagnant air in the hot weather, could they but reach its inner tracts. Except during the 4 fever months this country offers a very pleasant place of residence, especially in the monsoon, when the heavy rain of the Western Ghauts is diminished to a gentle drizzle.

702. I have already referred in para 47 to the general character of the Peermerd plateau and the High Range, showing that the former is an elevated tract of undulating grass land, broken by ridges and ravines, while the character of the High Range is still more rugged and uneven, the Eravi mala plateau alone excepted. As the river drains to the North-west the country ceases to have any resemblance to a plateau, and becomes broken up into deep ravines and valleys separated by high ridges.

703. Commencing from the South the elevated mountains which form the watershed of this river, and which must rise to a height of about 6,000 ft. first attract attention. The boundary between Travancore and Tinnevelly is in this part very marked, the very steep slope of forest towering up above Mlāppāra as boldly as the Western Ghauts rise above the low country. Kōtta mala the highest peak in these parts, rises to 6,400 ft., the Chēttur (Shaitoor) hill to the South of it is credited with 5,000 ft., and Chinga mala near Shivagiri hill with 5,100 ft. Kūttukal (Gootacal hill) must be nearly 5,000 ft. and being like the others in the middle of heavy forest, it is difficult of access.

704. Varayāttum mala (Vauragat hill) is a grass ridge rising to 5,170 ft., steep on all sides and frequented by ibex. Pālkulam mēda, Kumari kulam, Balakumari mēda (Walum Coorury mode), Pālkāchchiya mēda, and both Mangalam dhēvi hills are covered with grass, and rise to more than 4,000 ft. On the West of the river Kathira mudi is about 4,500 ft., and Kūttan mēda 4,200 ft. To the West of these two hills a tract of country extends consisting of undulating grass land which lies at an elevation of from 3,000 to 4,000 ft. In the centre of it a tongue of forest called Kōri kkānal, runs to the North. Māvadi mala a grass clad hill rises to 3,700 ft. or thereabouts, and the Padinyātta pāra plateau, on which is Mānammutti mala, and the area of which is 5 or 6 square miles of fine game country, is between 3,500 and 4,000 ft.

24

705. Proceeding Northward, Bâlarâmakatta (Balrunguddy) to the North-east of the Periyâr is reached. Its height is about 3,700 ft. which is also the approximate elevation of Aulcauty mode a hill overlooking the Plâkkâda estate, and now called Thondi-ât-mala. This is probably a corruption of Pon-eduttha mala, a name which Lieut. Ward ascribes to it, and in explanation relates a story to the effect that once upon a time the Pûnnyâtta chief discovered a treasure of gold on its summit. The Ottatthala mêda (Ootalûttamode), a grass hill overhanging Kumili, and at the top of a steep ascent, on the road from that place to Vandanvatta, is about 3,800 ft. To the North-west of this, Pâlkulam hill, grass-clad, with a very large boundary stone set up on its summit, is 3,700 ft., and the hill 1½ miles to the North-east of it which, (and not the hill 5 or 6 miles to the South-east) is called Cheyitthân mala is 4,250 ft. This is at a break in a ridge extending from Amratha mêda (Velnbaulay mudi) North-east to Râmakkal. Amratha mêda on the South-west is a rounded peak, and hill rising to 4,500 ft. The highest peak of the Thambi range (Culekote peak) now called Muppura mala is 4,900 ft., and continuing along the ridge beyond Cheyitthân mala to the North-east, Kunnûra mala (Antmode), and Puliyan mala (Peoli mulla) are reached each at an elevation of about 4,200 ft. The latter is a steep hill with grass on its Western and Northern, and forest on its Eastern and Southern faces. It is easily recognizable from having a flat summit covered with forest. The Vandanvatta Cardamom station is situated on a grassy knoll at 3,700 ft. Kambam mettu near the place marked Râmagerrycolum is 3,600 and Râmakkal mettu near the curious rock of that name is 3,200 ft.

High hills near the Periyâr crossing and Kumili.

706. North of the ridge above mentioned and to the East of the river, Kaly-ânappâra, already described, is about 4,000 ft. at its highest point, the long ridge on which it is situated being 3,000 ft. high at either end, Kâmâkshi mêda (Nachumbulla hill) is 3,300 ft., Kalkkûnthal near the place marked Chettyar rock 3,000 ft. Nayândi mala (Ninar malla), over 4,000 ft., is a cone of rock visible from a great way off. Udambanchôla is 3,300 ft., Mottayan mêda (Mootear mudi) 4,000 ft. Elayakâda thâvalam (Poolcuddy para) is 3,600 ft. Chânthapâra, the head quarters of the Pûppâra Cardamom Division is 3,450 ft. Thondi mala and Muraccha mala (Mool-chindy hill) must be about 4,500 ft. high, and the land between these higher hills lies at a general elevation of 3,000—5,000 ft. North of Thondi mala lies the Sûryyanala valley where the Panni âr takes its rise. It consists mostly of grass land, and lies below the steep slopes of the High Range. Elephants and bison abound here. Bodinâyakkanûr mettu at the edge of the ghats and at the upper end of this valley is 4,500 ft. high.

Peaks and high land on the Cardamom hills.

707. West of the Periyâr and North of the Aramana ridge the country is covered almost entirely with grass, except on the group of coffee estates at the foot of the Northern slope of this ridge. In this undulating grass land rises the bare peak of Lekshmi-kkunna (Lechmacoon peak) between the Chenkal and Perinthura rivers. The long Pâshippâra ridge lies just beyond the former river: more to the North again the cliffs that overhang the Vâgamanna (Bagamunnun Kaunum garden) Cardamom jungle rise to a height of about 4,000 ft., and are tenanted by at least one herd of ibex.

High hills to the West of the Periyâr.

708. Shortly to the North of this is the Vanjûram padi (Pearmoode hills), a grass hill reaching an elevation of about 4,200 ft., with a perpendicular cliff on the West and North, but a slope on the other two sides. The Urâlies who live in this neighbourhood have a tradition that long ago a Mahomedan established himself on this rock, and fortified himself there, making raids on all passengers travelling by the path that runs below the cliff, and throwing over it those who refused to embrace his religion. Whether there is any truth in this story or whether it is a legend which has grown out of the anecdotes related of Tippu Sultan, and his behaviour in other parts of India cannot now be ascertained, but the remains of a wall upon the hill give some grounds for believing that this place was, at one time, temporarily at least, inhabited.

Tradition connected with the "Bluff."

709. About a mile to the East of the "Bluff", as it is called by Europeans, the Bhîman kunna (Beeman kunu), a grassy cone, rises to about 3,800 ft.; from this point a strong ridge runs to the 'East as

The Bhîman kunna.

far as the Periyăr, at an elevation well above the surrounding valleys, and is called Sûryyan thanda. A little to the North of this a tolerably large tract of flattish grass land extends, and beyond this again the ground breaks up into ridges and deep valleys which make travelling very difficult.

710. North of the Chittăr, Idukkavetta (Cullyka mulla) is on the middle of

Hills North of the Chittăr.

a long grass ridge called the Gôpura thanda. It is an inaccessible pinnacle of rock with buttresses on its sides. To the East of it Pănyăva mêda a rocky grass hill at an elevation of about 2,500 ft. merits notice.

711. North of this again stands the great hill of Pălkulam mêda, which I have

The Pălkulam mêda.

already described. Its great height and the precipices on its Southern face distinguish it at once. Still farther North the Periyăr pours down the gorge below Mînirangiyăn kal, or as it is called by the Mannăns, Vălppuram mêda. On the North bank of the Periyăr great cliffs rise above the river, and stretch between Kutthirakkutti and Valiya kunna, or Thurava mêda, which separate the valley of the Periyăr from that of the Dhêvi ăr. Near the head of the last mentioned river Pettivara and Năbhikkunna (Nauvy coon peak), though not really high themselves, rise well above the surrounding country, and are visible from afar.

712. The tract of country marked in the map as " high rugged table land co-

The "high rugged table land "

vered with impenetrable forest" is not at all a rugged piece of country, the land is gently undulating, but it is covered either with forest or abandoned hill-cultivation, so that it is not easy to get a point of observation on it. Its elevation is about 2,500 ft. or a little more. The country to the North of this is much more broken.

713. Beyond this again the land begins to rise, and forms a steep slope or wall,

The Southern face of the High Range.

above which lies the High Range. This wall begins at the Părvathi ridge (6,170 ft.) on the West (Nautachayee), and extends to Karinkulam (Shulca mulla) on the East. I have already described the character of the land on the High Range, and the elevations of the more important points will be given in the Appendix.

714. West of Părvathi (Nautachayee) the ridge sinks, and is much less abrupt

Country West of the Părvathi ridge.

on its Southern face. Cheunăyippăra mudi (Chennapauray mude) is a hill of no great height, perhaps 1,500 ft., the whole of whose slopes are covered with abandoned clearings. Chûtta vara (Shutta wurray, a little to the East of it, is a rock precipitous on its Northern side, but covered with forest and sloping gradually to the South. Its shape is peculiar, and rising, as it does, abruptly out of the surrounding valley, it is a notable land mark visible from far out in the low country.

715. Westward from Anamudi a double ridge runs for several miles at an ele-

Country West of Anamudi.

vation of 5,000 or 6,000 ft., the Southern branch called the Gûdalûralatthanda (Kuddolauraythundu) sinks abruptly into the valley below, but the Northern divides, and one arm continuing to the West spreads into several ridges which form the watershed between the Kandanpăra and Parishakkutha rivers : the other branch bends to the North and passing through Churi mala (Shulee male) forms the Southern boundary of the Idiyara valley, and is called Thêratthanda. This is a lofty ridge of rock whose elevation must be about 3,000 ft. To the West of the Idiyara river, and evidently a continuation of Thêratthanda is Chokkan mudi, another rocky back bone which runs to the West, and finally merges in several high points which cluster about the Western end of the Idiyara valley.

716. On the Northern side of the valley just mentioned runs the boundary be-

Boundary between Travancore and Cochin.

tween Travancore and Cochin, which, starting from the trijunction station at Pannimădakuttha (3,800 ft.), follows the top of cliffs at an elevation seldom below 3,000 ft. and never sinks lower than 2,400 ft. for 15 miles. Indeed the boundary runs along the watershed except where it crosses 3 large streams the Kilitthôda, the Anamădan thôda, and the Manimăda thôda where they fall into the valley below. Further to the West the line of hills which divide the two States is not so elevated, but the boundary line is nevertheless clearly marked as far as the Ulanchēri hill (1,800 ft.). From here it descends straight to the river.

717. I have before me the names and elevations of all the important points on the boundary, and also of a great many minor hills scattered about this tract of country, but it will be more convenient if I collect the elevations together in the Appendix, instead of burdening this Report with more proper names, and as a large portion of this river basin to the West of Anamudi has been declared a Reserve doubtless the officer in charge will enter the names and descriptions of the hills and valleys in the more detailed map that will be needed when the Reserve is systematically worked, and I need not therefore particularise them further here.

Further details to be found in the Appendix.

718. Over so large an area as is drained by the Periyār within forest limits, and which amounts to fully one-fifth of the whole State, it is only natural to suppose that there must be scattered a large resident population, but, owing to the roughness of the country, and the rocky character of the hills, such a conjecture is incorrect, and would have been much more so before the commencement of the coffee enterprise. At the beginning of the century these hills were tenanted by Hillmen alone, who were dispersed over the country at a rate of no more than 2 to the square mile.

Sparse population of this river basin.

719. Even now, with the addition of the planters and their coolies, the density of population does not reach a higher figure than 3 to the mile, and it is unlikely that its numbers will be largely increased, at all events within our generation, for the greater part of the Peermerd plateau is better adapted for grazing than for agriculture or cultivation of any sort. Every year immense herds of cattle are driven up to the plateau from the Madura District during the hot weather, when the drought in the Kambam valley has killed down every green thing, and nothing remains for the animals to feed on. A charge of 3 annas a head is levied on the owners of these cattle, the total receipts amounting to between 2,000 and 3,000 Rs. a year.

Peermerd plateau much used for grazing.

720. The grazing of cattle dates back in all probability to a very distant period, and the curious lines of stones found on the hilltops in many places are probably boundary marks set up by the owners of the cattle to indicate the area apportioned out to different persons for grazing. These lines of stones, set up on end, are commonly seen in all the grass land of the plateau, but no where are they more numerous than in the valley of the Panni ār near Chānthapāra.

Curious boundary marks on the Cardamom hills.

721. One of the most curious of these is a large stone sunk in the ground near Pottankāda, about a mile to the East of the spot marked Talawary's cottages, and the same distance to the North of Urtyparae hill in the large map. I have described it in my diary of the 16th December 1899 : it is 15 ft. high, 10 ft. broad, and less than 1 ft. thick, and has a Tamil inscription on it, setting forth that it is the boundary mark of some land leased by a former Maha Rajah, but to whom or when I had not the time or patience to decipher.

A curious rock with an inscription.

722. Another very curious collection of stones is the ring of rocks situated on a ridge of grassland not far from the Māvadi chowkey, and overlooking the Periyār. It is frequently mistaken for a herd of elephants.

The Anakkal.

723. The population resident in the Periyār basin consists chiefly of Hillmen, but on the coffee, tea, and cinchona estates a considerable number of coolies, both from the low country of Travancore, and from British Territory are employed, while the number of the subordinates of the Forest, Cardamom and other Departments reaches a pretty high figure. Lastly, there are the Engineers and the laboures engaged on the construction of the Periyār dam.

The population.

724. The latter commence operations every year about the end of June, and
continue working until the fever season begins at the close of
March, when the work is stopped, and every one but a few care-
takers leaves the place for 3 months. Part of this population
when on the hills, is stationed near the site of the dam, and part
of it at Thĕkkadi, where the tunnel is being driven through the hill-side, while several
smaller camps are located along the stream which runs to the Periyār works What
is the number of persons employed in this work it is not easy to say, more espacially
as it fluctuates from day to day, there are 5 European Engineers, many European
and East Indian subordinates, and possibly 2,000 labourers engaged on various
branches of work. As however this population is migratory, and will finally disappear
in 5 or 6 years time, when the work is completed, I do not think it necessary to include it in the total resident population of this river basin.

*The population connect-
od with the Periyār pro-
ject.*

725. Coming next to the persons connected with the coffee and tea enterprises
we find in this valley that about 4,300 acres of land were sold
for coffee cultivation, and the greater part of this area was opened in coffee, forming 9 or 10 estates. Some of the land did well
for a time, but the peculiarity of the Periyār climate is that it is too forcing for coffee,
and fertile though the soil is, the trees overbear unless they are planted under shade,
consequently a great part of the coffee (which had been planted in the open) died
from overbearing. At the present time only 300 or 400 acres of coffee remain, but
tea has been planted to double that extent. These estates lie all pretty close together,
and with the exception of 2 or 3 a few miles back from the river, are clustered along
its banks, between the Peermerd cart road and the Aramana ridge. The number of
persons engaged on these estates is about 1,200.

*Population employed on
the estates.*

726. At different places on these hills there are stationed subordinates of the
Cardamom Department, who during the off-season generally
manage to get leave to the low country, but in the cardamom
season they are obliged to reside on the hills, and their numbers
are augmented by a detachment of the Nāyar Brigade which is sent up to prevent
smuggling.

*The Cardamom Depart-
ment.*

727. The stations of the Cardamom Department are Mlāppāra near Sangany
thāvalam: Vandanvatta North of Kumili and due West of Kambam: Kalkkūnthal 8 miles further North, and close to the Chettyar rock: Udumbanchōla North again, and near the garden
of that name: Elayakāda 4 miles North-east of the last mentioned place; and Chāntha
pāra which is close to the point marked Moulinpully garden in the map. Chāntha
pāra is the head quarters of the Pūppāra division. Besides the people stationed at
these places, Mr. Sealy, the First class Magistrate of the High Range, lives with his
office on the High Range during the slack season, but comes down to Udumbanchōla
when the cardamom crop begins to ripen.

*Stations of the Carda-
mom Department.*

728. I have included Mr. Maltby, the District Magistrate, with his staff, and
the Varukkappāra Vijārippukāran, and his staff in the resident
population of the Arutha valley, though the former goes out to
the Cardamom Hills when the season commences.

*Mr. Maltby's head quar-
ters at Peermerd.*

729. The cardamoms of the Western portion of the plateau, which ripen in
the month of Kanni, are collected by the Cardamom Samprethi-
ppilla, who comes up to the hills, and pitches his camp at Nag-
aram pāra, sending peons out in different directions to collect
the spice and form thāvalams. None of these officers, however, remain on the hills for
more than a couple of months, so they need not be included in the resident population.

*Collection of Kanni
cardamoms.*

730. Of the Forest Department there are about 10 subordinates stationed at
Kumili and Rāmakkal, the D. P. W. have perhaps 20 or 30, and
shopkeepers and others, not directly connected either with the
coffee estates or with the Periyār project, number as many
more, making a total of 60, and adding the Cardamom establishment (about 140), we get altogether 200 persons.

*Subordinates of the
Forest Department and
others.*

731. ` On the High Range the 13 or 14 estates within the basin of the Periyár
aggregate 1,500 acres still kept up, which support a population
of perhaps 600 persons, for cinchona does not require so much
labour for its cultivation as tea.

Population on the High Range.

732. Lastly, we have the Hillmen who are dispersed about the Peermerd
plateau in different tribes, which are again broken up into
numerous camps. As I shall describe these people in the
Appendix I will merely give their numbers now.

The Hillmen.

733. These different tribes recognise well known boundaries laid down by
tacit agreement, within which they are obliged to confine their
cultivation, and they never cross them into the land assigned
to any of the other tribes, without their permission.

Boundaries between the tribes strictly observed. \

734. The most Southerly of these tribes is the clan of Southern Mannáns,
who are confined to the country South of the cart road from
Peermerd to Kamham. They were at one time much more
numerous than they are now, their present number being not more than 60 or 70.

The Southern Mannáns.

735. Then come the Paliyar who number 250, and live between Puliyan mala
and Rámakkal on the North, the Periyár on the West, and the
Peermerd road on the South.

The Paliyar.

736. To the West of the Periyár, North of the Peermerd road, and South of
the Periyár after its junction with the Móthirappura river, are
the Urálies a tribe of Malayálam extraction broken up into
about 20 camps, and numbering 600 persons living within the basin of this river.

The Urálies.

737. To the East of the Periyár, North of the Paliyar and South of
the Muthuváns are the Northern Mannáns whose numbers are
about 600. Their Northern boundary is the Thondi mala
(Tonedu mulla) ridge, and the ridge running out to the West
through Talawary's cottages as far as the Móthirappura river, which then forms the
boundary.

The Northern Mannáns

738. North of the Mannáns are the Muthuváns who occupy the High Range,
the Anjináda valley, and the valleys to the West and South-
west of Anamudi. They number about 600 persons resident
within the Periyár basin.

The Muthuváns.

739. At Nériyamangalam there are 20 or 30 persons cultivating land, and at
various other places along the river bank where land has been
taken up, there are a few people living, the total number
amounting to about 100.

Low country people

740. Lastly there are three camps of Vishavar, two of them living on the
Idiyara river, and one employed by timber merchants, and
living in the jungles of . the Vadakkanperintbóda. They
number less than 100.

The Vishavar.

741. Thus the total resident population of the Periyár valley, excluding those
temporarily engaged on the Project, the cardamom ryots, the
sepoys of the Nayar Brigade, and those people who come up to
the hills for a short time only, and for some special purpose,
is 4,300, or not much more than 3 persons to the square mile.

Total population of the Periyár basin.

742. The 1,432 square miles in this river basin are, for purposes of Revenue
collection, under the Tahsildars of different Taluqs; for Magisterial purposes this area is subordinate to the 3 Magistrates
stationed at Peermerd and on the High Range; but for the collection of Forest produce the management is vested partly in the Conservator and Deputy Conservator
of Forests, and partly in the Superintendent of the Cardamom Department.

Divided supervision.

743. The existing arrangement is complicated, and requires simplification. The control of the forests in this area, and of the timber growing in them is in the hands of the Deputy Conservator of Forests, stationed at Malayáttûr, with the exception of the forest in the neighbourhood of Kumili depot, which, together with the depot itself, is under Mr. Maltby who again is under the Conservator of Forests, with reference to this one depot only. Then comes the anomaly; Rámakkal depot which is an offshot of the Kumili depot, and subordinate to it, is not under Mr. Malthy, the Cardamom Superintendent, but directly under the Conservator of Forests.

Complicated arrangement.

744. When Mr. Munro was in charge of the Cardamom Hills, he opened a timber depot at Rámakkal which was under him, and was quite independent of the Forest Department. Owing to a famine, and great scarcity in Madura, the demand for timber fell off for a time, and as the depot was off the direct road from Udambanchóla to Peermerd, and was troublesome, it was closed. Now that it has been re-opened, this depot has not been placed under the Cardamom Superintendent, although it is within 10 miles of one of his stations Kalkkûnthal, and is upwards of 140 miles from either Quilon or Malayáttûr.

Rámakkal depot formerly under the Superintendent of Cardamoms.

745. Undoubtedly, both Rámakkal and Kumili depots should be subordinate to the Superintendent of the Cardamom Hills, and it would be much better to lay down a boundary, and let each Department work out the forest produce growing inside the area assigned to it, whether it is timber or cardamoms or anything else. The line of the boundary I will propose later on.

These two timber depots should be under the Cardamom Superintendent.

746. Not only does the Forest Department have stations that must be occasionally visited within the jurisdiction of the Superintendent of Cardamoms, while permits to cut timber in the forests about Peermerd are given by the Conservator of Forests, but on the other hand the Cardamom Superintendent has control over gardens lying at a great distance from his head quarters, and therefore difficult of supervision, within the forests near Malayáttûr and Thodupura. Not long ago a timber merchant, holding a permit to cut white cedar logs, felled several trees growing in the cardamom gardens of Nériyamangalam, supervised by the Vijárippukáran of that place, who reported the matter to the Cardamom Superintendent, and thus a collision was imminent between the two Departments. Unless a definite area is assigned to each, such causes for unpleasantness may often arise.

Collisions liable to occur between the two Departments.

747. The Kumili depot was opened 3 years ago, and a Superintendent, 2 clerks and 4 peons were appointed. Timber is collected by a contractor, or contractors who convey it to the depot, and have it checked, after which they are at liberty to remove it, when they have paid the rate fixed, which is now about 5 Rs. a candy for teak and blackwood, and 2 or 3 Rs. for other kinds. The Rámakkal depot was opened within the last year, and there are 1 writer and 2 peons there. The conditions of working the forests are the same as at Kumili, but the rates are half a rupee lower.

The management of the two depots.

748. The timber felled in the valley of the Periyár is, for the depots at Kumili and Rámakkal, teak, and vênga, and a little blackwood. From the banks of this river, both at Peermerd and near Madatthin kadava, ánjili boats are conveyed to the low country. Above Malayáttûr a little teak and blackwood is felled, but the greatest demand is for white cedar and ven-teak logs, and ánjili and kambagam boats. Irûl, or irumulla as it is called there, is cut in the neighbourhood of the Vadakkanperin thóda, from which point sawn pieces can be brought in boats, but owing to the obstructions in the river already mentioned, they cannot be conveyed from higher up it. Chíni (*Tetramelrs uudiflora*) and payini (*Vateria indica*) boats are also cut in these forests. Here and there, wherever straight trees are found in suitable places, poon spars (*Calophyllum tomentosum*) are worked out by the merchants, who often realise large profits for them, if there happens to be a demand for wood for masts and spars. The nedunár tree (*Polyalthia fragrans*) is cut for the same purpose, though the trees are smaller, and therefore less valuable.

Timber operations in this valley.

749. The forests of the lower part of the Periyâr basin are not so rich in light timber as those of the Râni and Kulatthuppura, for two reasons : the Hillmen are much more numerous on this river. and have cleared miles and miles of good forest in the interior, and secondly the forests themselves have been worked for generations, and the timber that remains is what was not required formerly. On the Periyâr and Râni there is a good supply of heavy timber, which cannot be brought down till the water way is improved enough for rafts to pass, or till a road is cut, but the forests on the Kulatthuppura have been made accessible by road, and the heavy timber there is rapidly disappearing.

Forests of the lower Periyâr much exhausted.

750. On the upper part of the Periyâr basin, the forests contain very little saleable timber, the 3 species of wood in demand at Kumili and Hâmakkal being found only over a very limited area, while ânjili is almost extinct.

Forests on the upper Periyâr not rich.

751. There are many useful woods employed by the planters, but Government gets no revenue from them, as they are not royal timbers, and pay no seigniorage if they are felled on private land.

Some useful woods.

752. The greater part of the cardamoms gathered in Travancore come from the Periyâr valley. This spice has been grown for generations in gardens formed in the moist forest, under the natural shade of huge trees. These gardens are clustered round certain points, where stations have been formed, and which are situated at intervals along a line running parallel to the Travancore-Madura boundary, and some 5 miles to the West of it. They therefore lie to the East of the Periyâr. The cardamoms from these parts are collected by the Superintendent, and Assistant Superintendent, and are sent down to Alleppey via Pcermerd and Cottayam.

Cardamoms to the East of the Periyâr.

753. To the West of the same river there are few gardens, and the spice from these, and the wild cardamoms from the forest are collected by the Samprathippilla at Thôdupura, or by the Cardamom Vijârippukâran at Nêriyamangalam, and go from Thodupura to Alleppey.

Gardens West of the Periyâr.

754. Below Nêriyamangalam, and in the valleys of the Idiyara, Kandanpâra, Parishakkuttha, and Vadakkanperinthôda the cardamoms are collected by the Deputy Conservator of Forests, stationed at Malayâttûr, and here it is necessary to notice a curious mistake made by the Conservator of Forests in one of his Annual Reports to Government.

Cardamoms lower down the river.

755. In his Report for the year 1063 (1887—8) Mr. Vernède says with reference to the appointment of a Deputy Conservator. The old cardamom gardens "are situated near Nêriyamangalam, "and if due attention is paid to these gardens, I am confident " that they should be brought to pay better than the cardamom gardens in the " Koneyeur forests, where the soil is not so rich, though these gardens are bringing " in a large return, which is yearly increasing. In the year under review those " gardens yielded 701 tulams 2¼ lbs. while the gardens in the Northern Range " brought in only 4 tulams 8¼ lbs. "

Remark of the Conservator.

756. The gardens alluded to by the Conservator are Muttukkaunum garden, Naidiayra garden, and Punnikully garden which are marked in the map, and which, together with some others, lie about 4 miles to the East of Nêriyamangalam at an elevation of 1,400 ft.

The gardens alluded to.

757. Whether these can be worked up to yield better crops or not, I cannot say, as I have not seen them; very probably they can, as I am told they are never properly weeded, and the plants are treated simply as jungle cardamoms, i. e. the crop is gathered, and the weeds and bushes are beaten or cut down at the time of collection, and not touched again till the following year.

Their condition.

Tly vether this is so or not, it is only due to myself or whoever happens to be stationed at Malayāttūr, to say that these gardens *are not under the Forest Department at all but under the Nēriyamanga-lam Vijārippukāran who is subordinate to the Superintendent of the Cardamom Hills,* and who collects the crops and delivers it at Thodupura. The cardamoms gathered off the Nēriyamangalam gardens are therefore *not included in the 4 thulams mentioned above.*

759. As a matter of fact, the forests near Malayāttūr, or indeed any where in
Northern forests not likely to yield many cardamoms. the Northern Range, are little likely to yield a larger quantity of cardamoms than they do, so long as the present division of the land between the Forest and the Cardamom Departments continues, for the following reasons.

760. Cardamoms are rarely found growing at an elevation of less than 1,000 ft.,
Reason for this. and the extent of land over 1,000 ft., under the supervision of the Deputy Conservator, is limited, and what there is consists of rocky or grassy hills where cardamoms will not grow.

761. Not only so, but where there are forests at this elevation, as for instance
Another reason. on the slopes of the hills near Thodupura, and in the upper part of the Idiyara valley, the collection of the spice is entrusted to the Cardamom Samprathippilla of Thodupura, or to the Vijārippukāran already alluded to although the management of those very forests is vested in the Deputy Conservator.

762. The only places where cardamoms are collected by the Forest Depart-
The cardamom land in Northern and Central Travancore contrasted. ment in North Travancore are a few isolated hills, such as Eramukham, near the Vadakkanperinthōda, and certain places in the Idiyara valley, and outside it, where the subordinates of the Cardamom Department do not go. At Kōnniyūr there are numerous gardens, covering over several hundred acres, and lying at an elevation of about 1,400 ft., in the neighbourhood of Nānāttappāra, Pongampāra, and Vilanga-pāra (Shellycul), and there is a large extent of land above 1,000 ft., where new gardens can be opened out. Lastly, the officers of the Cardamom Department have nothing to do with these forests, so that no fair comparison can be drawn between those of Malayāttūr, and those of Kōnniyūr.

763. Such cardamoms as are collected by the Forest Department at Malayāttūr
System of collecting cardamoms and wax. are sent down to Alleppey, together with any ivory or wax that has been brought in. The system is to auction the collection of the produce in different localities, the bidder offering to deliver a certain quantity of cardamoms or wax at a fixed rate, and anything over and above that amount that he can collect becomes his property, but if he delivers less than the specified amount, he has to pay a fine. This system is very liable to abuse as I shall show later on.

764. The Cardamom Department collects a good quantity of wax and ivory on
Minor produce. the Peermerd plateau. A considerable revenue is also obtained from the collection and sale of gallnuts, or kadukka, the fruit of *Terminalia chebula,* chīyakka, the pods of a thorny bush (*Acacia concinna*), and other minor produce.

765. I have recommended that a large portion of the forests on the lower part
Land to be reserved. of the Periyār basin should be reserved, and the reserve has been proclaimed measuring 345 square miles. It will be described later on.

766. Various roads and paths cross the basin of the Periyār in different direc-
Roads and paths. The Peermerd road. tions. The most important is the cart road which runs from Cottayam to Peermerd, and thence across the plateau to Kambam, striking the river a mile South of the old crossing at Manimala-Periyār, and leaving Travancore at Kumili. This road is very much used for traffic, the whole of the Peermerd produce, and a good quantity of rice, and food stuffs passing along it. During the dry weather it is in fairly good order, but it gets very much cut up during the rains.

26

767. South of this, a bridle path runs from the 50th mile-po... ...up of
Other roads. estates near Padinyātta pāra, and a cart road... ...them
back to the main road at the crossing. A pa... ...cely been
cut from the Mount estate to the Māvadi chowkey, and is carried acr... the river and
on to the Project.

768. From Kumili a cart road has been made south for a mile to Thēkkadi,.
The path to Mlāppāra. and a path (which I hear is being widened into a cart road) is
carried on to the dam. Another foot path runs pretty level
from Thēkkadi to Mullayār, and on from there to Varukkappāra and Mlāppāra, but
this part of it is very uneven. A great portion of this path will be submerged when
the dam is completed, and the lake formed.

769. A very bad foot path runs through the forest from Mlāppāra to the
Paths at Mlāppāra. Shivagiri pass, and from there drops very steeply down to the
plains of Tinnevelly : there are also several paths through the
Cardamom gardens that have been opened in the neighbourhood.

770. North of the Periyār and west of it, a cart road runs from Peermerd
Paths north of Peer-merd. to the Perinthura group of estates, and from them a path was
cut out to the Cardamom Hills, via Thodupura—Periyār, and
Thōvāla, but this has been abandoned. From the low country
a path comes up from the Pūnnyātta valley, and strikes the cart road near Perinthura. Another path ascends the hills from Vellāmattam, and Thodupura, and
runs to Thodupura—Periyār. Farther north another foot path leaves Udambanūr,
and ascending the hills near Nagarampāra, crosses the river at Madatthinkadava, and
proceeds on to Pirinyānkūtta, and thence to Udambanchōla and Thēvāram mettu,.
and from there drops to the Madura low country.

771. From Kumili a path runs due north to Vandanvatta, over open grass
Paths on the Carda-mom Hills. country, and through a little forest. From here one branch
goes to Kambam mettu, and descends to Kambam, another
continues on to Kalkkūnthal, and from thence to Udamban-
chōla, Elayakāda, Chānthapāra, and Bodināyakanūr mettu, side paths running into
the Cardamom gardens at several points.

772. From Kōthamangalam on the West a path runs to Nēriyamangalam,
Nēriyamangalam path. crosses the river, and rising very steeply, near the Kuthirakkutti thāvalam proceeds level to Thōkkapāra, thence up and
down hill to the Mōthirappura river, and thence again to Pottan kāda and Chāntha
pāra, or Bodināyakanūr mettu. In former times all the cardamoms used to be
taken along this path to Nēriyamangalam, and thence to Thodupura, but since the
opening of the coffee estates on Peermerd they have been transported by the cart road.

773. From Bodināyakanūr mettu a good bridle path is being cut to Dhēvi
Paths on the High Range. colam, on the High Range, and from there other paths run to
different estates, the longest of them proceeding North-east,
and dropping down to the Thallayār valley.

774. Besides those mentioned, the Hillmen have numerous paths known only
Hillmen's paths. to themselves, which lead from place to place, and which are
sometimes covered with bushes if not much used, and at other
times are kept open, if there is cultivation in the neighbourhood.

775. In the forests of the low country, which are reached chiefly by river,
Low country forests destitute of paths. there are exceedingly few paths, the only ones being a path
from Kōthamangalam to Thatthakkāda and on to Urulapāra,
and another from Malayāttūr to the Vādamuri on the way to
the Athirappura waterfall, one branch of which runs to the Vadakkanperinthōda.
The rest of the country is destitute of good paths, and it is not often visited except
in the dry weather, when the people walk up the river beds, or along their banks.

776. When these forests are systematically worked, and an attempt is made
Paths much needed. to properly supervise felling operations, it will be one of the
first duties of the Forest Officers to make good roads. At
present, it is most difficult to get about, and the distance covered in a day is conse-

quently very small, but we shall not require cart roads, for it will be much more convenient to bring down the timber by water.

777. The planters on the High Range were at one time anxious to have a cart road cut right through the country from Malayáttúr to Thatthakkáda, and from thence up the Kandanpára valley to the gap South of Anamudi. From there it was to drop down into the Thallayár valley, and run from thence through the Anjinádá valley to Udamalapetta.

The proposed High Range road.

778. This idea has, I believe, been abandoned, as far as the Western portion of this road is concerned. It would cost an enormous sum of money to cut this road, as well as to keep it up, and the advantages to be derived from it are problematical. Application has lately been made to connect the Thallayár valley with the cart road from Udamalapetta, and as the distance to be traversed would n... ...ed 15 miles, nor the ascent be more than 3,000 ft., instead of 6,000 ft., t... ...est is much more reasonable, while the advantages that would accrue to the residents there are undoubted.

The Anjinádá road.

(17) The Kōttasshēri or Chālakkada river.

779. The Kōttasshēri rises in British territory, and is the most important river in the State of Cochin. It does not touch Travancore till it has approached within 3 or 4 miles of the Athirappura fall. At the point where it begins to mark the boundary between Travancore and Cochin it foams over a broad cataract, a grand river more than 200 yards in width.

The Kōttasshēri.

780. At this place the elevation is over 400 ft., and the river runs for four or five miles, sometimes spreading out in still reaches, and sometimes rushing over rocks, till the great fall is reached.

Its course.

781. Here the water dashes over a cliff, and falls upwards of 150 ft. into the pool below, its breadth at the time being nearly 250 yards. This is considered a sacred place, and there is a small shrine in a hollow of a rock above the fall.

The Athirappura fall.

782. Below this máda or barrier, the river runs almost due West for 10 miles, when it passes again into Cochin territory, and ceases to be the boundary between the two States.

Course below the fall.

783. A mile below the fall, above which a timber camp is always erected, another camp is annually stationed at Kalani. Here the river runs with a strong current, but is not impeded by rocks for about 4 miles. Two cataracts, called Chūran and Vannan, occur here, and are so bad as to prevent the floating of timber down this river. As the path along the river bank runs at some distance from the water at the places where these obstructions are situated, I did not inspect them, as I was on the march and did not learn of their existence till after I had passed them.

The Chūran and Vannan cataracts.

784. About 8 miles below the Kalani camp the Varamban cataract is reached. At this place a barrier of rocks crosses the river, which spreads out, and rushes down numerous channels separated from each other by rocks and islands. To float timber down here without directing its course would be to have it dashed to pieces, or wedged in among the rocks, so the usual plan is for the timber merchants to float their logs down to a little above the cataract, and then to get elephants to drag them down one of the side channels, where there is only a little water.

The Varamban fall.

785. Below the Varamban fall there is no obstruction, and the river flows with a strong current, but unimpeded, past Erāttamukham, where a watch station is situated, and where Travancore territory ends. Flowing west and then south, the river joins the Periyār at Elanthikkara, some 10 miles below Aluvā, and pours with it into the sea at Paravūr.

Below the Varamban.

786. The Kōttasshēri is not navigable above the Varamban cataract. Timber can be floated from the upper hills in Cochin territory to a short distance above the Athirappura fall, when it has to be landed, and dragged by elephants down the west bank of the river to Kalani, and then through the Vāda muri to the Pullāni thōda, the upper portion of the Vadakkanperin thōda, whence it can be floated down to the Periyār. The cost of dragging timber all this way is very great, and would be prohibitive, if the seigniorage levied by the Cochin Sircar were not so much lower than ours.

Only partially navigable or suitable for floating.

787. Whether the Chūran and Vannan cataracts can be in any way circumvented remains to be seen, and when we work these forests systematically it will be well to make a careful examination of these places, but the area of forest land on this river belonging to Travancore is insignificant compared with what belongs to Cochin, so it is to the interest of that State, more than of Travancore, to improve the waterway.

The Cochin State more interested in the improvement of the waterway.

788. The area of Travancore forests on the Kōttasshēri river is only 11 square miles, of which 9 are heavy forest, 1 secondary forest and cultivation, and 1 grass forest with big trees. There is no useless land here.

Different classes of forest.

789. Teak and blackwood are found, but sparingly, on the dividing ridge between this valley and the Periyār valley. Of ebony there is none. Anjili has disappeared. White cedar is fast being cut down, and will soon be exterminated like ānjili. It grows to an enormous size. Ven teak is found of good size, but a great deal of it has been cut.

Useful trees.

690. Timber trees of various kinds are found about half way between Kalani and Varamban for it.has not paid to work these forests from either end, on account of the distance the logs would have to be dragged, but, when prices go up, no doubt it will be worth while to begin felling here. The irūl is especially plentiful, and good in this forest.

Forest between Kalani and Varamban.

791. Kambagam trees are found along this river, and, wherever serviceable, they are cut down and hollowed out into boats, but the wood is not otherwise used, on account of its great weight. A few vēnga and thēmbāvu trees are seen on the grass hills within this area, but the weight of these trees has prohibited their removal also, as it has the felling of maniwarutha (*Lagerstræmia flos Reginæ*). Shurali (*Hardwickia pinnata*), and āval (*Holoptelea integrifolia*) for logs, and chīni (*Tetrameles nudiflora*), kolamāvu (*Machilus macrantha*), and payini (*Vateria indica*) for boats are sometimes felled.

Other useful trees.

792. The chief hills on this tract of land are the Ulanchēri hill (1,800 ft.), at its south-eastern corner, Nādakānimudi, and the Parayanpāra, about 2 miles to the east of Patticchērimudi. These are all on the ridge south of the river, and distant from it one or one and a half miles. It decreases in height as it runs to the west, and is not more than 300 ft. above the river, at its western extremity. These hills slope steeply at first, but from their bases to the river lies a level stretch of land with a very forcing climate and rich soil the consequence being that the forest trees are unusually tall and large with straight stems, showing an absence of wind.

Chief hills.

793. This river valley is for some unknown reason particularly unhealthy, the fever prevalent here assuming a very bad type, and from the end of November to the end of June the place is entirely deserted.

Very unhealthy.

794. There are no estates or villages, and no population beyond a few persons who live in the neighbourhood of Erāttamukham. This place receives its name from the occurrence of several islands in the river, a short distance above it, which split the stream up into numerous channels. A forest watch station is located here, with a Muthappēr and 3 peons under him, but they suffer much from fever.

No place of note.

795. Besides these forest officials, and the few people living at Erättamukham, there are no others resident within this river basin, that is to *Population small.* say, in Travancore. A few Hillmen (Vishavans) are employed by the timber merchants, to help them at their work, but they do not reside permanently on one side of the river and they disappear when the timber men leave the forests. The total population may therefore be set down at 20 persons.

796. At the Vädamuri, or the saddle on the ridge to the south of the river, at a point due south of Kalani, there are the remains of an old *Old remains.* ditch and mound, evidently constructed for fortification.

797. The forests of this river are under the Manyappra Aminadar, who lives at Kottamam, and visits these parts to measure timber, or to *Management.* check the logs brought through the country by holders of Cochin permits.

798. The quantity of Travancore timber worked out is not large, whether taken by the Vädamuri or down the river past Erättamukkham, *Timber operations.* but the quantity of Cochin timber passing the last mentioned place is considerable.

799. I have recommended that all the forests on the Köttasshëri river be included in the Malayättûr Reserve, with the exception of about a *Land to be reserved.* mile, or a mile and a half at the western end. The boundary line of the Reserve will strike the river at Varamban, instead of at Erättamukkham.

800. The elevation of this valley is low for cardamoms, but I noticed at the time of my visit in 1886, the occurrence of this plant at an *Cardamoms and wax.* altitude of not much more than 600 ft. These forests are given out on contract for the collection of cardamoms and wax, and the contractors secure a small supply of these articles from them.

801. The only path here is one which runs from Erättamukkham along the southern bank of the river, as far as the Cochin boundary at *The path.* Kardippära. Except in the neighbourhood of the Athirappura fall, it is almost quite level, but walking along it is tedious work, owing to the sticks or rollers laid across the path to facilitate the dragging of the logs. From Kalani a branch of this path runs south to the Vädamuri.

802. When we want to get at the timber east of the Varamban cataract, we shall probably have to cut a cart track for 5 or 6 miles follow-*Means of opening up the forest.* ing the present line. This would enable us to get out logs and pieces of heavy timber as far as Varamban, from which point they could be conveyed down the river with bamboos attached to them to make them float, or the pieces could be taken in boats.

(18) *The Pämbär or Anjinäda river.*

803. The sources of the Pämbär are at the head of the Thallayär valley, near Thallamala itself, and on the slopes of Vägavara, Chöttapära, *Sources of the Pämbär.* and Muthuvämala. Descending rapidly from 6,000 or 7,000 feet, the streams join together and flow N. N. E. down the Thallayär valley, the elevation being about 4,000 ft. From here the river runs for 7 or 8 miles, receiving on its way the waters of a large stream, that flows down from the Eravimala plateau, and of another which descends from Perumala.

804. The elevation at Näcchivayal is 3,000 ft., and the river is 30 yards wide and 2 ft. deep in the dry weather, with high banks showing *Course of the river.* that it rises considerably in the wet weather. From this point it continues to descend rapidly, and falls over a cliff, 4 miles below Näcchivayal. It then goes on falling over rocks, and 5 miles lower still it leaves the Travancore territory, where the Chinnär on the left and another stream on the right join it.

805. The Vattavadi river drains the valley of that name, rising in moist forest at an elevation of 5,000 or 6,000 ft. and after a course *The Vattavadi.* of about 15 miles, almost due north, unites with the Pámbär in British territory.

27

806. Owing to its torrential character the Pambâr and its branches are quite unsuited either for floating timber or for navigation.

Not navigable.

807. The area of this river is 130 square miles, of which 12 miles are covered with moist forest. This lies chiefly at the head of the valley, and on both sides of its upper part, and comprises about 9 square miles belonging to the Pûnnyâtta chief, and leased to the High Range Company. The rest of it is on the slopes above Perumala, south of Nâcchivayal, and in the valley between Kôttamala and Perumâlmala, and does not come within the Company's land.

Area of moist forest.

808. Of cultivation and secondary forest there are about 14 square miles, of which 10 are in the Anjinâda valley, 3 in the Company's land at the head of the valley, and 1 in the Vattavadi valley. Of these 14 square miles, 4 consist of estates, or the secondary forest growing up after land has been abandoned by the Hillmen, who have done a great deal of damage at the bottom of the Thallayâr valley. The rest comprises paddy fields, carefully cultivated and terraced, or the land occupied by the villages.

Secondary forest &c.

809. Below the village of Vattavadi, and below the Anjinâda plateau, there is a considerable extent of scrubby grass-land, with trees growing through it. Owing to the dry climate, the trees do not attain any size, and much of this low lying land from 3,000 ft. downwards, is too poor to grow good timber, but it has the appearance of forest, and here and there in the more sheltered parts trees of fair dimensions are found. Above the Anjinâda, and between it and the higher slopes, the trees are much larger, they grow close together, and their timber is more valuable. The area of the grass forests in both places is about 40 square miles.

Grass-forests.

810. The rest of this valley is covered with rocks, or with grass suitable only for grazing, and can never be utilised for cultivation of any kind. Its area is about 64 square miles, most of it lying at a high elevation.

Useless land.

811. The chief trees found within the basin of the Pâmbâr are firstly teak, which does not attain a large size, owing probably to the dry climate and rocky soil. Here and there clumps of this tree are found in the more sheltered parts. Above the Nâcchivayal plateau teak is absent, its range being confined to the lower part of the plateau and the dry slopes below it.

Teak.

812. Blackwood is found in considerable abundance above Anjinâda, and between it and the coffee estates in the Thallayâr valley. I was not able to inspect all the land, but from what I saw, and from the information I received, I judge that there is a good quantity of blackwood in this valley and that the trees attain a large size.

Blackwood.

813. Next to blackwood we come to sandalwood, which is found indigenous in Travancore in this valley only. Here the conditions are exactly similar to those on the Mysore plateau, where the tree thrives better than anywhere else. Sandalwood is found on the Anjinâda plateau over an area of about 12 square miles, and is particularly abundant near the village of Mamvûr, and on the Kilikkûda hill to the east of it. No use is made of this timber except by the Pûnnyâtta chief who is said to requisition 20 logs a year from this place. The people cut down the trees now and then, and frequently take the bark for chewing, as it is slightly pungent, while cattle and goats do their best to wound the stems of the old trees and to eat down the young. Thus the whole of this most valuable timber is allowed to run to waste, and no attempts are made to replace the trees that die or are cut down.

Sandalwood.

814. Vekkâli (*Anogeissus latifolius*) a very useful wood, thriving only on soils of free character is very abundant here. Vênga occurs in places and in the better parts of the valley, above the Anjinâda villages kadukka (*Terminalia chebula*), pâthiri (*Stereospermum xylocarpum*), and many

Other useful trees.

other useful but not valuable trees are found including kurangādi the Wynaad shingle tree (*Acrocarpus fraxinifolius.*)

No ebony &c.

815. Ebony, white cedar, punna and other trees belonging to the moist zone do not grow here.

816. One of the peculiarities of this valley is the occurrence of the male bamboo
(*Dendrocalamus strictus*), which is found in great abundance
The male bamboo. near Maravūr. It does not grow anywhere else in Travancore, the peculiar conditions of climate requisite to its existence being absent elsewhere. It attains a smaller size than the ordinary bamboo, and is very nearly, if not quite, solid. It is much used for spear handles, and for the shafts of carts, where strength and lightness are required. ⹂ \

817. The Anamudi hills, as I have already described them in the first chapter,
consist of a horse shoe of very lofty mountains, which form the
The Anamudi hills and High Range proper, and rise to an elevation of from 7,000 to
the Anjināda valley. 8,000 ft. This mountain range slopes very steeply down to the
Thallayār valley, which lies at an elevation of something over 4,000 ft. and in many places the hill sides are actually precipitous. Below the Thallayār valley the land slopes down easily to the north-east, the mountains on each side approaching each other until, between Pannivara and Pilāvavara, there is merely a narrow gorge. Below this the valley opens out again into the Anjināda plateau, a level stretch of land measuring about 4 miles long by 3 broad, and lying at an elevation of 3,000 ft. Below this again the dry grass hills slope more easily.

818. To the east of the Anjināda valley lies the Vatta
The Vattavadi valley. vadi valley, a narrow strip of land hemmed in on both sides by
high mountain ridges.

819. The soil of the Anjināda valley is excellent throughout the Thallayār
valley and the Anjināda plateau, and also on the hills around
Soil excellent. wherever the rocks and stones have any earth among them.
The rice of this valley is celebrated and is much relished, but unfortunately, on account of the want of roads, only just enough is grown to supply the wants of the people, as it cannot be exported.

820. The climate of this tract of country is peculiar, for while the rainfall is
much the same as in South Travancore, and measures in the
Climate and rainfall. lower part of the valley about 40—50 inches, the elevation is
much greater, while the valley is shut in from the influence of the sea breezes by high hills. In the upper part of the valley, the Thallayār valley so called, the rainfall is somewhat heavier, on account of the presence of forest there, and it probably ranges from 80—100 inches.

821. The lower slopes of land in the Pāmbār basin, the elevation of which is
below 3,000 ft., are very feverish, and with the exception of a
Low lying land. few wandering Puliyans no persons are so foolish as to live
there permanently.

822. There is no bad wind, but during the cold weather a steady breeze frequently blows from the north-east. On the higher lands frost
Wind and temperature. nightly nips the vegetation, and the thermometer has been
known to fall below 20° Fahrenheit. Even at an elevation of not much over 4,000 feet, in the Thallayār valley, frost often occurs to the detriment of the coffee or cinchona trees.

823. This valley is thickly populated for a mountain region, indeed I might
well have left the Anjināda plateau out of the forest area, but
Considerable population. that below it, and within Travancore territory, there is a large
extent of land covered with scrub forest, which may on closer inspection be found capable of yielding timber, and I have therefore included the plateau in the forest area.

824. In the Thallayār valley there are some 5 or 6 coffee and cinchona
estates, aggregating perhaps 700 acres, and employing on
The estates. an average throughout the year 400 persons.

825. Outside the Travancore Society's land there are the villages of Maravûr,
Nácchivayal, and Kárûr, and on the slopes above them, some
The villages. of them at a considerable elevation, are clustered the hamlets
of Kilandûr, Khandal, Puthûr, and Perumala.

826. These villages are picturesquely situated on the hill sides, the houses
being built close together on some flat piece of land, while the
Total population. paddy fields, carefully laid out in regular terraces, spread on
all sides. The number of persons in each of them varies from 100 to 600, and the
total population may be set down at about 2,000. In the Vattavadi valley there
are some 300 persons, and there are also 200 to 300 Muthuvâns in different parts of
the basin, who form colonies of their own, apart from the villages, so that 3,000
represents the whole number of people resident within this river basin.

827. The inhabitants of the Anjinâda are divided into 3 castes, the Vellálans,
who form the bulk of the population and live in the villages,
Different castes. the Puliyans or slave caste, whose huts are clustered on the out
skirts of the villages, and the Hillmen or Muthuvâns who dwell by themselves.
Besides these there are a few Maravars, Chetties, and others.

828. Each village is independent of every other, and forms a community of itself ruled over by a Mannâdi, or head man, whose office is here-
Government of the ditary. His right hand man is the village scribe or Nettâri, and
villages. to discuss questions of common interest the head man calls to
his aid a body of elders, who decide the matter in hand. Though the power of the
Mannâdi is very great, and though his office is regarded with respect, he is by no means
considered to be beyond the reach of caste law, and should he in any way depart from
the right path and outrage the feelings of his caste fellows he is summarily ejected
by the order of his council, and until his heir comes of age, his scribe manages the
affairs of the village.

829. The discipline exacted by the village council is severe, and instances are
not wanting where offenders have been expelled from their
Severe discipline. villages, and have been driven to commit suicide, in expiation
of some behaviour repugnant to the feelings of their neighbours.

830. Such a state of society seems peculiar in the present age, but it is justified by the isolation of the locality, and by the absence (until
Some such state of so- within the last few months) of any resident Magistrate, or other
ciety is required. officer appointed to enforce the law.

831. In olden times the Pûnnyâtta chiefs seems to have, occasionally at least,
resided within the limits of this valley, and a site of an old
Pûnnyâtta chiefs. palace is still shown, where the reigning family lived. At present, the authority of the chief is delegated to an agent, or Maniyakáran, who collects
the taxes, and remits them to his superior from time to time.

832. Mr. Munro in his pamphlet on the High Range considers that the population on those hills, and within the Anjinâda valley was at one
A larger population time much larger than it is now. There is evidence also to
formerly. show that when the track from Bodinâyakanûr to Nêriyamangalam was kept up, a certain amount of traffic passed along the path from the Anjinâda
valley across the High Range via Mûnâr to the Môthirappura river.

833. During the time that he was in charge of the Cardamom Hills, and while
he had authority over this region too, he frequently visited it,
The Kâders. and he speaks of a race of Hillmen called Kâders, as residing in
the low lying land below Anjinâda. I did not hear of them when I was making enquiries on the spot about the population, and possibly they may have moved to another locality within the last few years.

834. The forest area of the Anjinâda valley is supervised by the Assistant
Superintendent of the Cardamom Hills, but very little is done
The forests neglected there in the way of utilising the timber, in fact I may say no-
hitherto. thing. Hitherto, and before the appointment of an Assistant
residing on the High Range, the Superintendent of the Cardamom Hills had his head

quarters so far away from this tract, that he was not able to give it any attention, but, quite lately, the Assistant Superintendent has visited the locality, and has suggested that some steps should be taken to look after the forests, and especially the sandalwood there.

835. Sandalwood grows so freely in the Anjináda, that if any attempt is made
Free growth of sandal-wood. to cultivate this tree in Travancore it should be in this locality but whether it should be regularly planted out, or the existing trees merely be conserved remains to be considered. It is a slow growing tree.

836. Very little forest produce is taken out of these jungles, the want of a road
Forest produce. prohibiting any export of heavy articles. Teak of small size, blackwood, sandalwood, vênga, and vekkâli would be the forest products of this valley.

837. The climate is too dry for cardamoms and I believe that there is not
No cardamoms. much wax collected here. I understand that the Pûnnyâtta chief claims all the wax and honey obtained within his territory.

838. The only road, if it can be dignified by such a name, in the Anjináda val-
Roads and paths. ley runs from the British frontier at the Chinnâr up to Maravûr, and is continued up to the coffee estates in the Thalayâr valley. It is a mere bridle path, and is very rough in some places. * The Planters have applied for a cart road along the same line, and there is no doubt but that it would be of great advantage to them, and would also benefit the whole population of the Anjináda. Its length would be about 14 miles from the frontier to Maravûr, and 8 or 10 miles on to the coffee estates.

Other paths run from village to village, and connect them together, while yet others run up to the higher grass hills where no paths are needed, as it is possible to roam at will over the short and springy turf.

The total area of Forests.

839. In the preceding pages I have divided the area of forest land within each
Summary. river basin into different classes, with reference to the character of vegetation on them, and have estimated the resident population in each valley. I will now for the sake of convenience collect all the figures together in one table :—

Rivers.	Moist forest.	Secondary forest.	Grass forest.	Useless land.	Total.	Population.
Hanumannathi	¼	¾	6	8	15	None
Palli	2	6	20	10	38	350
Parali	10	16	30	15	71	800
Kôtha	30	20	30	19	99	500
Ney	4	20	24	12	60	500
Karamana	5	20	35	12	72	750
Vâmanapuram	23	40	70	10	143	2,700
Itthikkara	12	3	24	0	39	100
Kalleda	180	40	60	13	293	3,000
Thânuravarnni	4	4	5	7	20	50
Acchankôvil	70	40	59	20	189	300
Râni	367	60	20	20	487	2,000
Maniuala	16	110	30	12	168	2,500
Pâlâyi	5	52	30	15	102	4,000
Mûvâttupura	50	70	30	25	175	2,000
Periyâr	500	200	100	632	1,432	4,300
Kôttassbéri	9	1	1	0	11	20
Pâmbâr	12	14	40	64	130	3,000
Total......	1,299¼ Sq. miles.	736¾ Sq. miles.	614 Sq. miles.	894 Sq. miles.	3,544 Sq. miles.	26,870

* See para 778.

28

840. These 3,544 square miles, included within the forest line, are partly
Divided supervision. under the supervision of the Conservator of Forests, and partly
under the Superintendent of the Cardamom Hills, but the
boundary between the two jurisdictions is not very clearly marked, nor is the authority of each officer within his area kept distinct, as I have shown in paras 743 to 746.

841. With regard to the line, I think it very desirable that a well marked
boundary should be selected to separate the areas supervised
A line to be drawn be- by the two Officers. By well marked I mean a boundary
tween the two areas of following natural features, as far as possible, and I will give
supervision. my reasons for making some slight alterations in the nominal
line at present acknowledged.

842. The Superintendent of the Cardamom Hills should have under him the
whole of the Peermerd plateau, the Cardamom Hills, the High
Area supervised by the Range, the Anjinâda valley, and the basin of forest on the
Cardamom Superintend- upper portion of the Idiyara river between Pannimâda kuttha
ent. and Paruttha mala.

843. The boundary of this tract will therefore run as follows. Beginning in
the south at the high hills south of the Shivagiri peak, which
The proposed boundary are situated on the Travancore-Tinnevelly boundary, and which
between the two areas of form the watershed between the Periyâr and Râni rivers, it
supervision will follow that watershed to the west, and take in all the
Padinyâtta pâra plateau. From here it will run north along the western edge of
this plateau, and take in the Mount and the Granby coffee estates, situated just on
the edge of the cliffs. From the latter estate it will cross from the Periyâr basin to
that of the Arutha, and will continue along the edge of the Peermerd plateau,
westward, crossing the Arutha fall, and then turning northward, at the point where
the Peermerd cart road passes through a cutting at the 41st mile; it will take in
the Residency, and continue along the edge of the plateau till the top of the Amratha mêda is reached. Here the boundary will pass from the basin of the Arutha
to that of the Pâlâyi river, and will run along the edge of the plateau over Kudamurutti (Codamoortee hill), and Ponmudi (Punumudy hill). From this point it
will follow the watershed of the Periyâr, and the edge of the plateau as far as
Nagarampâra, situated near Nâdakâni peak, and will include a portion of the head
waters of the Mûvâttupura river. Then running down some miles to the north-
west along the Nâdakâni ridge, it will pass along the head of the Mullaringâda
valley, and drop down to the Periyâr at the Chîrppa mentioned in para 618. Rising
from the bed of the Periyâr at this place, the boundary will run to a point about a
mile to the east of Kallippâra, and crossing the Dhêvi âr will follow the watershed
between the Môthirappura and Kandanpâra rivers, first north and then east, as far
as Pârvathi mala (Nautanchayee peak). From here it will run to the point marked
Vengoya mudee in the map, and will cross the Pacchappulla thôda to the point
marked Aunaycolam, and will proceed on to Auveeray malc. From Auveeray malc
the boundary will run along the grassy ridge to the point marked Codinadee thundu,
opposite Pannimâda kuttha, which is the trijunction station between Travancore,
Coimbatore, and Cochin. The northern and eastern boundaries of the jurisdiction
of the Superintendent of the Cardamom Hills will coincide with the boundaries
between Travancore and Coimbatore, Madura, and Tinnevelly respectively.

844. The area supervised by this gentleman will be then confined to :—

(1) The basin of the Periyâr and its tributaries from its
River basins under the source down to the Chîrppa, and including such portions of the
Superintendent of Carda- Dhêvi âr, the Kandanpâra âr, and the Idiyara river (tributa-
moms. ries of the Periyâr) as lie above the crest of the hills.

(2) The basin of the Anjinâda river.

(3) So much of the basins of the Arutha, the Punnyâtta, and the Muvâttupura rivers as lie above the edge of the Peermerd plateau.

845. It will be seen, from what I have said, that this area is less than the
area over which the Superintendent of Cardamom Hills now
The present area of has supervision. His subordinates collect cardamoms on the
supervision would be lower slopes of the hills in the Arutha and Pambâr valleys as
somewhat reduced. far south as Lêkha mala (para 421), on the slopes of the hills

draining into the Manimala, the Thodupura, and Mūvāttupura rivers; and lastly in the forest above Nēriyamangalam where there are some old gardens (see para 753).

846. The jurisdiction of the Superintendent of the Cardamom Hills would
The parts excluded very remote from Peermerd. thus be slight'y reduced, but there is thus to be said in re-commendation of the change, that since the Cardamom Depart-ment was started as a separate Department some 25 years ago, no Superintendent has ever visited the localities named.

847. I believe also that I am correct in saying that Mr. Maltby would not be
This change suggested by Mr. Maltby. averse to the transfer of the control of the portions of forest to which reference is made from his Department to the Forest Department, and indeed he has already expressed a wish for it. *

848. But while relieving the Superintendent of Cardamoms of some part of
An exchange of work between the two Depart-ments. his work in places easily accessible to us, but distant from Peermerd, we should expect him in return to do some of our work in places conveniently near to him, but which the Con-servator could only visit at very long intervals. I allude to the charge of the Kumili and Rāmakkal timber depôts, which were under the supervision of Mr. Munro, when he was Superintendent of the Cardamom Hills, but are now managed by the Conservator of Forests, with some assistance from Mr. Maltby (see para 745).

849. Although the Conservator of Forests is primarily interested in the work-
The result would be a saving of labour. ing of timber, and only collects a few candies of cardamoms each year, and although the Superintendent of Cardamoms is only concerned in the collection and preparation of that spice, and cares nothing for timber, it seems to me that a better division of labour would be secured if each officer confined his operations to a certain area, but not so much to a single product, and if he endeavoured to make as much revenue as possible within that area, by the sale of timber, or by the collection of cardamoms, wax, or other produce. Much less travelling would then be required, and Mr. Vernède would not be expected to go to Rāmakkal, nor Mr. Maltby to Lēkha mala or Nēriyamangalam.

850. The line I have described has been so selected as to give to the Super-
Reasons for the selec-tion of the line described intendent of the Cardamom Hills those portions of forest, which from their elevation are less suitable for timber oper-ations (except in the neighbourhood of Kumili and Rāmakkal), while I have assigned to the Forest Department those parts of the hill slopes which are easily accessible to us, and most of which will fall within our Reserves. By thus laying down a definite line, all collisions such as were mentioned in para 746 would be avoided.

851. Two objections to the alteration proposed occur to me, one is that until
Objections to the alter-ation. we increase the staff of Forest Officers supervising the reserves, we shall not be in a position to manage the cardamom gardens which I propose to take from Mr. Maltby, so well as they are managed at present by the Thodupura Cardamom Sumprathippillai, but this diffi-culty will vanish as soon as we commence to work the Reserves as they are worked in British India, with a competent staff. The other objection is that it is convenient to have a part of the land that is supervised by the Deputy Conservator of Forests situated on the hills, in order that he may get a change there from time to time.

852. These objections are however overbalanced by the advantages to be
The objections over-balanced derived from a better division of area and of labour, and for the sake of argument I will suppose that this change has been carried out.

The measurement of the two areas of super-vision. 853. The total forest area of 3,544 square miles will then be divided out as follows between the two Officers re-ferred to.

* See para 74 of the Report of the Forest Commission in 1884.

, Area under Conservator of Forests ... ... 2,330 sq. miles.
'Area under Superintendent of Cardamom Hills... 1,214 Do.

3,544

This land does not all belong to Government, some of it falling within the territories of the Pûnnyâtta and Chenganûr chiefs, and some of it having been sold for coffee plantations, or leased to the British Government for the Periyâr Project, or otherwise alienated.

854. The Pûnnyâtta chief, who lives at the place of that name in the Mînacchal Taluq, claims two portions of territory within the forest area, one in the neighbourhood of his palace, and the other, comprising the High Range and Anjinâda valley.

Land claimed by the Pûnnyâtta chief.

\

855. The boundaries of the former, as I was informed when in the neighbourhood, are as follows: Starting from the Amratha mêda in the south-east corner, the line runs W. N. W. about a mile south of the place marked Kodunga yale, and strikes the ridge which forms the watershed between the Manimala and Pâlâyi rivers: this ridge it follows round to near Vengatânam, and then runs N. W. to the angle of the river, where the palace of the Pûnnyâtta chief is situated. From here it follows the river to Eerâttapêtta, and then ascends the Kavana river for about 6 miles. It then follows the easterly branch of this small river to its source on the high ridge dividing the Pûnnyâtta and Arakkulam valleys, and then runs along the edge of the plateau southwards to the top of Amratha mêda. The total area of this block is about 72 square miles. *

Description of the territory near Pûnnyâr.

856. On the High Range and in the Anjinâda valley the boundaries of the Pûnnyâtta territory are said to be as follows. Beginning at Karinkulam (Sholea mulla) in the south-east, it runs south-west through Kannivara and Kurukkacombu, to Chokkan mudi. Then turning north-west it follows the edge of the high land, and crosses the Môthirappura river (the Mûnnâr) where it disappears through the ridge. It then proceeds west to Pârvathi hill (Nautanchayee), and runs to Anakkulam in the valley below, passing through the hill marked Wonay pauray mudee. From Anakkulam it turns east, and comes up the Vailley mala or Nâval old ridge to Ana mudi. From here it runs all round the western edge of Hamilton's plateau missing Nilakkallôda, and, clinging to the high ground, it passes round the head of the deep valley below till it reaches Paruttha mala. From Paruttha mala, the boundary of the chief's territory is identical with that of Travancore, running north-east, and then south down to Karinkulam. This includes the whole of the High Range and the Anjinâda valley, and comprises about 250 square miles.

Description of the territory on the High Range, and in the Anjinâda valley.

857. The territory claimed by the Chenganûr chief is bounded as follows. Commencing at its south-east corner, a little north of Perumpâra mala, the line follows the stream running to the westward, which joins the Manimala river at Mundakkayam, and then follows the ridge, which encloses the Kûttukkal valley, as far as near Vengatânam, where the territory of the Pûnnyyâtta chief begins. Turning to the east, the boun-

Territory claimed by the Chenganûr chief

---

* I wish particularly to disclaim any accurate knowledge of the boundaries claimed by this chief. Those here given were supplied to me by a man whom I met at Eerâttapêtta, and represent the limits assigned by ordinary report to this chieftain's territory, but whether they are right or not I cannot say. The boundary of the High Range territory were read out to me by Mr. Turner from the deed having over certain portions of the chief's territory to the High Range Society, but the wording of the paper is vague, and the names given in it are not always easy of identification. The Conservator tells me that the chief now claims the basin of forest lying between Pannimêda kuttha and Paruttha mala, and on this account it has been excluded from the Malayâttur Reserve as proposed by me. Again, Lieut. Ward in his survey takes the western boundary of this chieftain's territory from the Mûnnâr to Anamudi in a line running due north, and thus shuts out a large extent of land now in the possession of the High Range Society. Then to, Mr. Munro in his pamphlet on the High Range gives a different, and much more natural boundary on the western side, more natural because it only includes land above 4,000 ft. instead of dropping down to the valley below as far as Anakkulam, but he is describing the High Range, rather than the land claimed by the High Range Society. I mention these facts merely to show that the extent of the land claimed by the chief is not accurately known, and that the boundaries here given are merely those that were told to me. I have made no attempt to verify them. It is highly desirable that the territories of this chief and that of the Chenganûr chief, as well as of any others who claim land within the forest area should be accurately defined in order to prevent disputes.

dary marches with that territory, until it reaches Amratha mêda. From here it follows the edge of the Peermerd plateau till the cutting on the road at the 41st mile is touched. It then drops down to the Panniâr, and follows it past the Arayan settlement at Kirakka (Kelekkay), until close to Perumpara mala. The area thus enclosed is about 47 square miles.

Land sold for coffee cultivation. 858. Then there are the 40,000 acres sold for coffee cultivation, and scattered about the forest area, so that they cannot be marked in the map, 62½ square miles.

The Periyâr Project. 859. The land leased to the British Government for the Periyâr Project works is 8,000 acres or 12½ square miles.

860. Some 30 or more years ago Mr. Huxham received a grant of 10 square *Mr. Huxham's grant.* miles, for the cultivation of coffee and other products in the neighbourhood of Pattbanâpuram. The boundaries of this grant are

East. A straight line running due North from Karuthaurutti waterfall.

South. The high ridge (Kôrakkunna) South of Mâmbaratthura.

West. A straight line running due North from Mâmbaratthura stream.

North. A line taken so as to include the ten square miles granted.

861. There are other small grants of land, such as the tract granted to the *Other small grants.* Kirakkumbhâgam Nambûthirppâd near Velliâmattam, to different pagodas, to the Arayans at Mêlakâva, and so on. These grants comprise all descriptions of land covered with heavy forest, with grass forest, with scrub, or occupied by paddy fields, and aggregate some thousands of acres. If we set down the total at 16 square miles we get a total area of 500 square miles within the forest area, which does not belong to Government.

862. Of this amount about 194 square miles fall within the area supervised *Area of Government land divided between the two officers.* by the Conservator, and 306 within the jurisdiction of the Cardamom Hill Superintendent, and deducting these from the figures already given, it will be seen that there are available in the former's charge 2,136 square miles and in the latter's 908 square miles.

These totals may be divided according to the different classes of forest as follows :—

The Conservator of Forests.

Moist forest.	980¼	sq. miles.
Secondary forest...	442¾	do.
Grass forest	504	do.
Useless land	209	do.
	2,136	

The Superintendent of Cardamom Hills.

Moist forest	221	sq. miles.
Secondary forest...	140	do.
Grass forest	42	do.
Useless land	505	do.
	908	

863. From these figures it will be seen that of the land belonging to the *Area under the Conservator of Forests.* Government within the forest area, rather more than two-thirds is under the Conservator of Forests, and of this nearly one-half is original forest, and nearly one quarter is grass forest containing useful trees, and finally only 209 square miles out of 2,136 are useless. The whole of the produce obtained here will find its way to the west with the exception of what goes to the Puliyara Depôt near Shencottah, and the trifling quantity raised on the Mahenthragiri slopes. It is from these 2,136 square miles of forest land that the Reserves will be selected.

29

864. Under the Cardamom Hill Superintendent the distribution of forest is quite different. Out of 908 square miles, 505, or more than one half, is useless land, being covered with grass, fit for grazing, but for the production of timber quite worthless. The grass forest amounts to but 42 square miles, and contains only small timber, small, that is, compared to what we get in the low country, and this kind of forest is confined to the banks of the Periyār. There are 140 square miles of secondary forest, which has been cleared by the Hillmen at one time or another, and finally 221 square miles of original forest are found About one half of this is situated at the sources of the Periyār, and besides being difficult of access, could not be much worked without endangering the water supply of that river : other patches are scattered about here and there, and, finally, the cardamom gardens occupy a considerable area.

Area under the Superintendent of Cardamoms.

865. Thus the forests under Mr. Maltby are never likely to yield a large revenue from timber, and this more especially because they lie at such an elevation that the cost of removing the timber therefrom would be prohibitive. Very little of the wood grown on the Peermerd plateau or Cardamom Hills will ever find its way down to the western side. At Kumili and in its neighbourhood there is a limited area of mixed forest yielding timber which can be sold in the Madura District at remunerative rates, and the supply will last for a year or two. At Rāmakkal some miles farther north, and also on the Travancore-Madura boundary there is a timber depôt, and the valley of the Kalār will yield a considerable quantity of wood for a good number of years to come, and far more than the forest near Kumili, provided bandy tracks are judiciously extended, and the timber yielding area is properly conserved. This wood will all be sold for removal to the Madura District.

The forests on the Peermerd Plateau and Cardamom Hills contain only a little good timber.

866. In the Anjināda valley, under the Cardamom Superintendent, some revenue may be secured by the sale of teak, blackwood and sandalwood, but how much that revenue will be it is impossible to say. The whole of the produce will be worked down to the Coimbatore District.

Some revenue may be obtained from the Anjināda valley.

867. I have spoken in previous pages of the system of hill cultivation followed in Travancore. With us it is called " Virippu " or " Malam krishi " : in Ceylon it bears the title of " China " : and in other * parts of India and Burmah it is variously named " Kumri ", " Podu ", " Jhum ", " Toungya ", and " Daya." The methods known by these different names are the same everywhere, and are familiar to every traveller in the East. As soon as the rains have ceased and the dry weather has regularly set in, the " Virippukāran ", as the Hill-cultivator is called, proceeds to some unoccupied part of the country, and selects a piece of waste land sufficient for his wants ; if he is a poor man he contents himself with a few acres, enough to supply him and his family with grain for a year, but if he is wealthy he selects a much larger piece, purely as a matter of speculation, for he knows he can always find a sale for the grain. If he can find a piece of virgin forest which has never been cleared before, he chooses that, knowing that the field will be much greater there than where the forest has been cleared, and the ground cropped before. Failing this, he contents himself with land covered with secondary forest, which has grown up after cultivation was abandoned, and the longer the land has lain fallow the better will be the return, anything under 10 years fallow being considered useless. Having decided on the scene of his operations the " Virippukāran " proceeds to cut down the undergrowth, trees, and grass growing on his piece of land. If the timber is very heavy the larger trees are sometimes left standing, but in this case they are usually girdled, so that they may be killed, and that there may not be any drip from the branches and leaves on to the ground below them. When the trees &c. have all been felled, they are allowed to lie for 2 or 3 months until they are thoroughly dried, and then they are burnt, and as soon as the ground has cooled after the fire, the ashes are dug into it, and a fence is usually erected round the clearing to keep out wild animals. The land is now ready for sowing, but the grain must not be sown until the first showers of the monsoon have fallen and moistened the ground. Three months after the sowing the paddy begins to ripen and is reaped, and the land is usually abandoned, and left fallow for 10 to 20 years, until the shrubs and trees on it have attained a good size. In some cases, after the crop has

Hill-cultivation described.

---

* Lands thus cultivated are called " Chārikkal. "

been reaped, the land is cleared again and sown with horse gram or some leguminous seed, and after this second crop has been gathered, the land is again cleared up, and another crop of paddy is taken of it. Curiously enough, this second crop of paddy is, I am told, often much finer than the first crop. In some places, and chiefly on the eastern side of the Hill-plateaux, rāgi (Eleusine coracana) is sown instead of rice, the reason being that in these localities there is a great deal of mist which causes the paddy to mildew, but rāgi does not suffer in the same way. In South Travancore tapioca, and various kinds of millet are also grown, but though the products raised are not the same in all cases, the system of Hill-cultivation is ·identical everywhere. Throughout Travancore the regular season for clearing is from December to June, when the seed is sown, and the crop ripens about September.

868. The return from this sort of cultivation is, as may be imagined, very good, for the people would never go to the trouble of clearing heavily-timbered land, often more than 10 miles from their homes, and of watching their crops and protecting them from wild animals, unless they were sure that they would, barring accidents, get a much better return than if they selected poorer land nearer home. In his Report on the Forests of Ceylon Mr. Vincent, quoting Dr. Roxburgh, sets down the yield at from 120—500 fold. * I have made numerous enquiries from "all sorts and conditions of men", but the largest authentic yield I have ever heard of was 200 fold, which Mr. Richardson tells me he obtained from a small plot of 4 or 5 acres on the banks of the Periyār some 20 years ago, and this he secured only by very thin sowing, at the rate of about 1 parra of grain to the acre. I am speaking of paddy, for rāgi yields more heavily than that plant, and 200 fold is not uncommon, but at the same time rāgi is a very exhausting crop and cannot be grown on the same ground again, without manure, except after a fallow of at least twenty years. Below 200 fold there are many different rates of return grading down from 150 and 100 to 80, 60 and 30 fold. Where the land is covered with forest which has never been felled before the yield should not be less than 100 fold, and on ordinary secondary forest land it should reach 40 to 50 fold. But some allowance must be made for loss by wild animals, birds, &c., and I therefore think that the estimate of 30 fold made by Mr. Painter, and given at para 80 of the Report of the Forest Commission in 1884 is a very fair one, and while making a very liberal allowance for losses of all kinds is sufficiently accurate to form a basis for calculations, as for instance in estimating what area of land should be given to Hillmen for their support. I shall revert to this subject later on. The returns from paddy land in the low country are generally about 10 fold, except when the fields are very heavily manured, or when two crops can be raised off them in a year.

Returns from Hill-cultivation.

869. This system of Hill-cultivation is most wasteful, and has destroyed immense areas of valuable forest land in the past, while its further continuance in the present prevents the growth of good timber. The land selected in the first instance for clearing and cultivation has always in Travancore been such as is covered by heavy moist forest, for the soil of the deciduous forests, in which the undergrowth is grass, is not suitable for growing grain, as it contains but little nitrogen. Thus the area of heavy moist forest has rapidly diminished, its place being taken by dense thickets of reeds, or by tangled masses of thorny creepers through which the plants of valuable trees have great difficulty in struggling. Where once there grew a forest of immense trees more than 100 ft. in height with a rich undergrowth of younger trees, there is now a wide expanse of lantana, reeds, thorns, and long grass.

Objections to the system of Hill cultivation.

870. Hill-cultivation is wasteful therefore because, in the first instance, large trees are felled and burnt to make way for the clearings. Taking 150 trees to the acre and assuming that only 10 per cent of these are saleable trees, these 15 trees would probably contain 100 candies of timber the value of which would be perhaps 1,000 Rs., and the loss to Government at least one-fifth of that sum. The tax at present levied by the Revenue Department on such lands is 2 fanams for every parra of 128 perukkams, which is equivalent to about one rupee per acre, so that Government, in permitting heavy forest to be felled for cultivation loses about 200 Rs. in their share of the value of

Direct money loss in the value of the timber.

* Para 78.

the timber destroyed, in exchange for 1 rupee realised by the tax collector, over a every acre of heavy forest land thus cleared. In the year 1888 I was called away from Malayāttūr to take charge of the office of the Superintendent of Cardamom Hills for 3 months, before I could name the boundaries of a certain forest, and so have it gazetted as a Reserve. On my return I found that the neighbouring villagers, many of them Cochin subjects who eventually escaped without paying any tax whatever, had felled and burnt upwards of 2,000 acres of original forest, causing thereby a loss to Government of about 2 lacs of rupees. It is true that the Hill-cultivators cannot always get heavy moist forest to fell, and they have therefore to content themselves with secondary forest, where the land has formerly been cleared and abandoned after cultivation, it is not fair therefore in estimating the loss to Government by " Virippu " to charge against it the whole value of the timber in the original forest, which took perhaps 100 years to come to maturity, but such part of the whole period for which the land has been allowed to lie fallow, say 12 years, or 12 per cent on the 200 Rs. mentioned above. Thus even when old abandoned land is cleared again after being left follow for some time the Government loses about 2 Rs. an acre for every year since it was last cropped.

871. Secondly, the clearing of the land, and its exposure to sun and rain entirely changes the character of the soil. The accumulation of leaf mould, the result of centuries of uninterrupted forest growth, is first of all burnt to ashes, and then the heavy showers sweep this most valuable part of the soil down into the rivers, silting them up, and interfering with the navigation of the backwaters. Even if the land is never sown but is allowed to revert into jungle at once, the whole character of the vegetation is changed, and in place of the original magnificent forest growth are found fast growing soft wooded trees of no use whatever. Generations, nay, centuries must pass away before the ground is again capable of supporting its former wealth of large trees.

Impoverishment of the soil.

872. In the next place this system of Hill-cultivation is wasteful because, under it, the same area of land will only support about one-fourth of the population which regular cultivation would feed. Much of the land cleared by the " Virippukāran " is steep and unsuited to permanent cultivation, but if the more level land were regularly occupied, it would for every parra of seed sown yield 10 parras a year or 120 parras in 12 years, while by Hill-cultivation the same area would yield a single crop of 30 parras.

Waste of land.

873. Fourthly, this system is objectionable because it encourage the people to be idle. They go out into the forest in the dry weather and make their clearings, and after a time they return and reap the crops, watching them for a short time before they ripen, and altogether they spend 3 months over the work, but for the other 9 months they have nothing to do, and these 3 months of work keep the cultivators and their families in food for a year.

Encouragement of idleness.

874. Further, Hill-cultivation is bad because those who follow it are exposed to the risks of fever or the attacks of wild animals, and the country is kept unhealthy to an extent that would not occur if the land were regularly cultivated. With easy work and an abundant supply of food it might be supposed that all the conditions necessary for a rapid increase of population were present, but as a matter of fact the population increases at a very slow ratio. To get to their clearings the people have to pass through rank scrub-jungle reeking with malaria; they have to sleep out in the open without proper shelter : sometimes the wild elephants which come to feed on the crops kill one or two : sometimes a man-eating tiger appears, and seizes a victim here and another there. At one time, when the crop is just gathered, they live in great abundance, at another they are starving for weeks together while the grain is ripening, and then they have to subsist on the seeds of the eentha (Cycas circinalis), the pith of the bastard sago (Caryota urens), or such roots as can be found in the

Great mortality among the Hill-cultivators.

---

* The size of the parra varies in different parts of the country, but in North Travancore the parra of paddy land is usually calculated at 64 perukkams and of waste land at double that extent or 128 perukkams, the perukkam being one square decout or 110 square feet, so a parra is either ⅓th or ⅔th an acre. In other parts of Travancore the parra is calculated at 40 perukkams, or about ₁'th of an acre, or at 80 perukkams ₁'th acre, and there are yet other rates.

jungle, while, should the crops fail, many die from insufficient food. Thus the average length of life of the Hill-cultivator is short.

875. Lastly, Hill-cultivation is to be condemned on account of the great loss of time occupied in going out to the clearings and returning from them day after day, and in conveying the produce, often for two days journey, to the virippukārans' homes. The crops too are often entirely destroyed by wild animals, and the labour of a whole season, and the means of subsistence for a year, are often lost in a single night. Thus, everything is against the system of Hill-cultivation, and it has nothing to recommend it.

Great loss of time and labour.

876. It is not easy to calculate the total area of land cleared each year for " virippu " on the hills, and in the low country. In the Taluq of Thodupura the revenue collected by the authorities is 19,000 fanams, which at 2 fs. a parra for Sirkar * land, and 1 fanam for jenmom land would give about 10,000 parras or say 3,000 acres. In this Taluq and those on the Rāni river there is more clearing done in the forests than in any others. Out of the 32 Taluqs there is much cultivation in 9 or 10, and none in the coast Taluqs. Perhaps therefore we may set down the number of acres that pay tax at 40,000 ; but a very large area of land escapes payment, as is well known, or at all events the money does not find its way into the Treasuries. Adding 10 per cent for this we get 44,000 acres cleared each year by the virippukārar. The Hill-men, roughly estimated at 8,000 (some of whom do not clear land, while others hold registered properties) possibly bring into cultivation 6,000 acres more. So that about 50,000 acres of land are thus annually cleared, and if we multiply this by 10 (the ordinary length of a fallow is 12 years, but often, on account of scarcity of land, the people have to clear it at shorter intervals) we get a total of 500,000 acres or 781 square miles which are devoted to Hill-cultivation, and all of which were originally covered with heavy forest. Compare this total with the 736¾ square miles mentioned in para 839 : the difference of 45 square miles is due to the fact that there is a certain amount of Hill-cultivation outside the forest area.

Total area of land devoted to Hill-cultivation.

877. The destruction of forest, chiefly in North Travancore, has been most extensive, and this more especially within the last 20 or 30 years. Mr. Munro tells me that in 1860 the forest extended from within 8 miles of Cottayam to the Peermerd plateau, and that during one period of a couple of years a block of about 60 square miles was cleared in one locality, and that in other places, as for instance to the south of Peruvanthānam, the original forest has all been cut down within a very short time. From every part of the country comes the same report. Not so very long ago Mīnacchal was a place surrounded on all sides by forests. Now they have all disappeared. At Vettikāt-mukku, there was large timber 20 years ago, as I was told, but now the whole country is covered with a low scrub not 3 ft. high, and scarcely a tree is to be seen. At Thodupura the Sirkar bungalow and cardamom treasury had, within the memory of man, to be protected by ditches from the incursions of wild elephants. Now that the forests have disappeared, the elephants never venture so far into the open country.

Great destruction of forest.

878. This practice of Hill-cultivation is carried on by two classes of people : (1) by the Hillmen who live in the forests, and have nothing to depend on but the animals they catch or shoot, and the grain they raise in this way, and (2) by low-country people who could find other means of subsistence if they liked. The former confine themselves chiefly to the hills and forests of the interior, and clear land which is often valueless to us, because it lies in such a position as to be inaccessible for the removal of its timber, or because it contains trees of but little worth. Further, there is a tendency among these people to settle down in the same places, and not to move about so much as formerly. This is a tendency which is by all means to be encouraged, and as the Hillmen are very good axemen, and are well acquainted with every track and by-path in the forests, their services can often be utilised by us. Besides this they, as a rule, only raise enough food for their own requirements, and their numbers do not exceed 8,000, so that the

Hill-cultivation practised by (1) Hillmen. (2) Low-country people. The meth·ds followed by the former.

---

* On jenmom land i. e land formerly granted to large land-holders on special terms, the Government demand is one half the usual rate, the land-holder taking the rest.

30

damage done by them is small in comparison with that effected by the low-country people. It is quite impossible to entirely prevent these Hillmen clearing land in the forests, and the only way to deal with them is to induce them to make their clearings in the less valuable land, where the timber is of no worth, and where its distance from the low-country would render the cost of working down the timber prohibitive, and to induce them to concentrate their operations in one place by allotting land to them. I do not consider it advisible to endeavour to tax or coerce them.

879. But the case is very different with the low-country people. They live in villages, far distant from the land they intend to clear, and when the season for felling commences, they travel from their homes to the land they have selected, and return every evening to them. Many of them are deeply indebted to Mahomedan traders, and borrow grain from them at exorbitant rates, to be repaid when the crop is reaped, and, when this happens, the quantity of grain left for those who have done all the work usually only just suffices to keep them alive till the following year. In fact the only people who really profit by this system of cultivation are the money lenders. The cultivators themselves follow in the steps their fathers trod : their ancestors cut down the forest and secured a living by raising grain in this way, and so do they. No necessity compels them, as it does the Hillmen. If they liked they could get other work, they could expend their labour on carefully cultivating land at their doors, they could find employment under the richer land owners, or they could move to other parts of the country, and our object should be to induce them to give up this system of forest clearing for some more fixed occupation.

Most damage done by the low-country people.

880. One or two rules have been published from time to time for the regulation of Hill-cultivation, but they are so vaguely worded that they are useless. The only well known order of Government is that no trees of 10 vannams or more quarter girth (say 1 foot diameter) shall be felled in the clearings, but this rule is easily observed but the spirit of it evaded by the axemen who leave all such trees untouched by the axe, but pile the branches round them and burn them, thus killing the tree . Further, the regulation alluded to provides no punishment for those who break it, and the culprits can only be proceeded against for mischief under the I. P. C. for destroying valuable timber, and many Magistrates refuse to convict. The Forest Act No. IV of 1063 permits no felling in the Reserves, a step in the right direction, but, with the small staff we have, it is difficult to enforce it.

Rules and regulations about Hill-cultivation.

881. The causes of this system of shifting cultivation carried on by the low-country people are (1) Poverty : many of those who make a living in this way possess no property whatever beyond a knife for cutting down small trees. They have no cattle nor ploughs, and they are therefore debarred from taking up suitable land, and cultivating it properly. (2) Want of communications. Those Taluqs such as Thodupura and Minacchal, where Hill-cultivation is much practised, are very badly off for roads, and, this, being so, it is not worth the while of the people to cultivate the land carefully, and raise large crops off it, because they could not get rid of their surplus. They therefore prefer to live from hand to mouth. (3) Want of education, whereby they do not understand the value of time, and this prevents them from seeing how much labour they lose in their wasteful methods. (4) The Puthuval system which permits any man to clear waste land, and settle on it without asking leave of any one. (5) The abundance of waste land, which makes it more profitable for a man to clear a plot, take one crop off it, and then abandon it for other land, than to work steadily at the same piece. (6) The difficulty of registration. It is the duty of the Proverthikārar and other low paid subordinates to collect the tax on these waste lands, which are often difficult of access, and which are not therefore likely to be often visited, and they make a considerable amount of money out of the work, by getting presents for underestimating the area of the land and so on, and they naturally throw every obstacle in the way of registration, which would mean the loss to them of a considerable income. (7) In the more *jungly parts wild

Causes which induce the people to adopt Hill-cultivation.

---

* The wild elephants seem to know exactly when the crops are ripening, and, regularly each year, shortly before the paddy is reaped, descend from the hills and raid on the fields, which then need a great deal of watching.

elephants and other animals destroy the crops, and the loss on hill-paddy land (i. e. un-irrigated land) where the labour expended is small, is much less than on the more highly cultivated land. Thus in the neighbourhood of jungle it is a great risk to open up new land for permanent cultivation.

882. It is impossible to put a stop to this system of shifting cultivation all at once without reducing a large number of people to starvation.

Hill-cultivation cannot be suddenly abolished. It is admitted on all hands that the system is a bad one, but the change must be made gradually; it cannot be entirely pro-hibited. Knowing as we do, what are the causes that induce shifting cultivation, and where its danger lies, we must take steps on the one hand to remove those causes, and on the other to reduce the damage to a minimum so long as the system prevails.

883. As regards the first cause, the poverty of the people. Although on the first blush it may appear a desirable thing to allow every one

Poverty of the people. The tax on Hill-culti-vation should be increas-ed to prevent the poorer classes clearing land. to be independent, and to have his own clearing, the independ-ence in the majority of cases is merely nominal, for by far the larger number of those who clear land are so poor that they have to borrow grain from others, to sow the fields, and often to feed themselves while they are waiting for the crops to ripen : thus these seemingly independent people are really much more the servants of others than if they obtained daily employment from them for a daily wage, and could work under this master or that as they liked. If a man wants to be independent he must have a little capital to help him over bad seasons, and the usual result of speculating without capital (and there is a considerable amount of speculation in clearing and sowing waste land) is that a man gets very much into debt. Instead of every one attempting to be his own master, it would be much better if the more intelligent alone cultivated the land, and the lazy and ignorant worked for them. The land would be more carefully tilled, and would yield better crops then, and the lazy and improvident, instead of often being on the verge of starvation, would get regular employment and food. To bring this result about, the best remedy is to raise the tax on Hill-cultivation, while re-ducing the area available for it. Considering the damage done to the land, and the large returns obtained from this sort of working an increased tax would be no hard-ship, and by closing certain lands and forming forest reserves, the available area would be reduced, and only the more wealthy would be in a position to clear land, and eventually they would cease to shift the locality of their clearings and would settle down.

884. As regards the second cause, the want of communications, this is being quickly remedied, and in Central and North Travancore, where

Great improvements in the extension of com-munications, and the spread of education. Hill-cultivation is chiefly practised, the people are fast settling down, as the net-work of road spreads over the country. The same may be said of cause (3), the people are becoming more and more civilised every day. In the most remote districts where a few years ago no one could read or write, schools are now springing up, and all but the very lowest classes desire to be educated. In one place I even found a colony of Hill-men paying a teacher to instruct their children in reading and accounts.

885. The custom of allowing people to go and clear what land they like with-out asking leave of any one is bad, but it is so widely spread

Hill-cultivation should only be allowed after pay-ing fees and taking out a permit. that it is difficult to know how to check it. A suggestion has been made that an order should be passed compelling every one who wishes to clear land to obtain permission, and to pay a fee, which would vary with the area of land to be cleared. After the fee had been paid it would still be necessary, as at present, to send an officer to check the area of cultivation, and to see that that the proper fees had been paid. If the person clearing the land were obliged to show his permit to any Government official who asked for it, whether Forester or Proverthikāran, a good control would be kept over this system. A copy of the permit should be sent to the nearest Forest officer to enable him to check the areas cleared.

886. An abundance of waste land encourages slovenly cultivation everywhere. A few years ago the yield of wheat per acre in America used to be 25 bushels to the acre. Now it has fallen to 8. This is because the area of waste land is so extensive that when a farmer found his farm becoming unproductive, he abandoned it and moved on to the more fertile land farther west, where a very little labour would give him a better return. New land is now becoming very scarce, and the farmers have not yet learned how to manure their fields, and select their seed. But in England, where land is scarce and valuable, the farmers could not make a profit unless they worked their farms highly, for there is no untouched land there, and the yield has thus increased to 35 and 40 bushels to the acre. Recent experiments in America have shown that the yield can be raised to 100 bushels an acre merely by carefully selecting the seed, and *Sir J. B. Lawes has proved the great difference in the yield of various kinds of wheat raised under exactly similar circumstances. Here then is a field for investigation as regards the possible yield of paddy. At the same time the area of waste land available for cultivation should be reduced.

Benefits resulting from selection of seed and high cultivation. The area of land available for cultivation should be reduced.

887. The difficulty of registration is much less than it was formerly, and, consequently, large areas of waste land are being registered, and the extent of the Hill-cultivation is being simultaneously reduced. Lastly, if the land occupied by the people was in compact blocks they would be able to combine and drive the elephants from their fields or catch them.

Hindrances to registration less than formerly.

888. To sum up, Hill-cultivation is universally condemned, but it cannot be altogether prohibited, and in the case of the Hillmen it can only be concentrated, not prevented. The cultivation of this sort by the low country people should gradually be reduced in area, and the people should be induced to register their lands by

Summary of remedies for checking Hill-cultivation.

(1) Increasing the tax on shifting cultivation.

(2) Forming forest reserves, and shutting off portions of the forest area from the operations of these men.

(3) Improving the communications.

(4) Introducing a system of giving permits for clearing land.

(5) Increasing the facilities for registration.

889. Grass fires do an immense deal of damage in Travancore. The extent of rock and grass land within the forest area is, as I have shown (in para 839) 894 square miles, and of grass forest 614 square miles, or more than 1,500 square miles in all. Every year, during the dry season, the whole of this grass is burned, except perhaps a few patches here and there. The fires occur in various ways, but usually they are intentionally lighted, sometimes by cattlemen who want fresh grazing for their cattle, sometimes by travellers who find they cannot move readily through grass 10 ft. high, and sometimes by native hunters who think thereby to secure game. Occasionally too the fire occurs accidentally by the rubbing together of bamboos, or by a spark falling from a passer-by, but the great point is that there are no rules for prohibiting fires, nor are any attempts made to put them out. Lighted in one valley they spread over hill and dale, and weeks afterwards are found travelling in some place 20 miles distant from where they first started. For the first 3 months of the year the open forests of the country may be said to be burning from one end to the other.

Immense damage done by grass fires.

890. That these fires do much damage to the forests is a statement that admits of no contradiction. In the first place, and even where there are no trees growing in the grass, as is the case on the higher hill-plateaux, the fires reduce to ashes all the valuable constituents of the soil, which have been brought up from below by the roots of the grasses, and the first showers wash these ashes into the ravines, so that the soil becomes yearly more and more impoverished. Some small proportion of this fertilising material finds its way into the swamps which are often taken up for paddy cultiva-

Fires impoverish the soil.

---

* Johnston's Agricultural Chemistry pp. 102.

tion, but the greater part of it is swept into the streams, and by them is deposited in the backwaters, or carried out to sea. The earth too becomes baked and hard.

891. In addition to this impoverishment of the soil, grass fires, in the second place, damage the forests by encroachment. Centuries ago the *Grass fires materially reduce the area of forests.* area of grass on the Peermered plateau was certainly much smaller than it is at present, but when the cattlemen from Tinnevelly and Madura discovered that they could obtain grazing for their cattle on our hills, they began to frequent them, and in order to make the grass spring up fresh and green they regularly burnt the hills as they do now. These fires sweeping through grass, often 10 ft. high, scorch the edges of the forest, and kill the bushes and trees, and it can easily be understood how such conflagrations, occurring year after year, should, in course of time, make very considerable encroachments on the extent of the forests. Any one who is at all sceptical on this point has only to notice the appearance of the trees on the edges of the sholas on Peermerd. Had there been no encroachment the trees would be branched to the ground, and would show no portion of their stems, but the trees are seen to be drawn up and unbranched, showing that they originally grew in the interior of a forest, with other trees outside them; and what has happened on Peermerd has happened in other places.

892. Thirdly, grass fires occurring in deciduous forest (and it is in this kind of forest that the most valuable trees, teak, blackwood and venga *Fires permanently damage the timber trees.* grow) burn the wood and scorch the trees. Where the burning does not penetrate very deep, the bark and sapwood are nevertheless permanently damaged, and it is a well known fact that when the bark of a tree is injured the heartwood suffers from sympathy. Then too the great heat caused by the fire makes the outer rings of the wood shrink, while the inner rings are unaffected, the result being that cracks or as they are called "shakes" of different kinds, "heart shake," "cup shake" and "radial shake" occur.

893. Lastly, the fires burn the seeds of trees and kill many promising young plants. Some trees indeed have such amazing vitality that, *Damage done to young trees and the destruction of seeds.* though they are burnt to the ground year after year, they send up shoots again and again, which are destined to be burnt every dry season, until in some fortunate year the fire does not happen to come there, and then perhaps the shoot manages to grow so high before the next fire occurs that it is out of danger. Even so, the timber is often injured by the damage done to the root, and this injury is inherited by the young plants which spring from it, and by their descendants too, so that the evil is far-reaching.

894. Some persons indeed, but their number is very small, contend that these annual grass fires do good to the forests, inasmuch as the grass *A class of people who say that grass fires benefit rather than injure the forests.* is burnt down, and manured with its own ashes, and they maintain that some seeds like those of the teak tree, which are enclosed in leathery capsules would not germinate unless they were subjected to great heat; and in support of their contention they point to the fact that hitherto fires have been allowed to rage through the forests, and yet the trees are living, and the forests which have not been worked are valuable, and they further say that without burning the grass it is impossible to travel about the country. It is hardly worth while arguing with people who make such statements, but it is sufficient to say that the fact that there is still timber in our grass forests in spite of the fires does not prove that the latter are harmless. In our unworked forests the trees grow fairly close together, and the shade thrown by their branches checks the growth of the grass under them, and fires do comparatively little harm, but, as soon as we begin to fell the timber and let the light in, the grass at once becomes vigorous, and the fires proportionately more fierce, the branches and dead wood left about still further intensifying them: and if we allow fires in the grass forests we are working, we permit injury to the standing timber, and the destruction of all future trees, or in other words we are content to see the conversion of our timber forests into waste land covered with grass alone.

§ See Fernandez' Manual of Indian Sylviculture pp. 431.

895. I have shown that about half the area of the country, or 3,544 square miles, out of a total of rather more than 7,000, is included in the forest area. Part of this land indeed is under permanent cultivation with coffee, tea or other crops, and there is a wide stretch of grass land on the hill-plateaux, amounting in all to 500 or 600 square miles, but this is quite made up for by the large extent of waste land covered with timber and scrub, which lies in detached pieces outside the forest line. Thus, quite 50 per cent of the area of the State may be set down as forest land. Comparing this with the figures given in Mulhall's Dictionary of Statistics of the area of forests in other countries we find that there are in

Large proportion of the land of the State classed as forest land.

				Ratio to total area	Acres of forest per 100 inhabitants
United Kingdom	...	...	3	6	
France	...	...	17	60	
Germany	...	...	25	75	
Russia	...	...	33	665	
Austria	...	...	27	127	
Italy	...	...	15	40	
Spain and Portugal	...	6	38	
Belgium and Holland	...	10	16	
Scandinavia	...	...	32	740	
Europe	...	...	27	220	
United States	...	...	8	343	
*Travancore	...	...	50	90	

Thus there is a larger percentage of the land of this State under forest than in any European country or in the United States, but, owing to the density of the population on the sea coast, the proportion of forest per head is not so high as in many other countries.

896. In para 81 of this Report I have drawn attention to the chief characteristic of the forests of this State, the marvellous number of species of trees. This peculiarity is undoubtedly due to the climatic conditions prevailing here, the abundant rainfall, the regular seasons, and the equable and high temperature. In northern latitudes or at high elevations the number of § different species is much less, for there the alterations of temperature are much greater, and but few trees are so robust as to stand these marked changes, but with us the conditions mentioned above favour growth, and not only do the trees themselves attain a larger size, but many delicate species can live and thrive, where they could not exist if the cold was greater, or the drought more prolonged.

Chief characteristic of the forests of Travancore the great variety of species.

897. Had our forests been thoroughly explored for timber, and was there the same demand for it for numberless purposes as in Europe, where every tree finds some use, this great variety of species would not be a disadvantage, but hitherto the demand for timber in Travancore has been so small that only the best trees, or as Dr. Hunter † calls them " the aristocracy of our timber wealth " are of any value. Consequently our forests. acre for acre, are less valuable at the present moment than those of most other countries, although, owing to their large extent, the total value is very considerable, and, as other trees become known, the value will increase year by year. A similar state of things exists in Ceylon, the Wynaad, and the Malabar Coast generally, and in his report on the Ceylon Forests Mr. Vincent (para 108) estimates that no more than 10 per cent of the trees have any present money value, even in the most favourable situations, while very often the useful trees do not exceed 2 per cent. In time, no doubt, the people will come to know the uses of the other trees, as the more valuable kinds become scarce and difficult to obtain, in fact I notice the change beginning already. A few years ago no one would look at a tree of the white maruthu (Terminalia paniculata), which yields a timber very much used in other parts of India, but now we frequently get applications for it, and no wonder, for it is a very common tree, and can be worked down at a trifling cost.

A very large proportion of the species is unknown, and therefore their timber has no money value.

---

* I have taken the present population of the State at 2,500,000 and the area of forests at 3,500 square miles.

§ Wallace's Darwinism Cap. ii pp. 33.
† The Imperial Gazetteer.

898. Comparing this state of things with what prevails in many European countries, it is easy to see why timber is so expensive in Travancore. Where whole forests consist of trees of the same or at most not more than 3 or 4 species, all of which are useful, and all of which are of the same age, the timber can be worked down at almost a nominal cost equivalent to about * 1 rupee per candy, even though the logs have to be delivered at a distance of 40 or 50 miles, for large quantities are brought down together. But in our case, though the distance may not exceed 20 miles, we have to travel perhaps some scores of miles to make up the required number of logs. We want for instance 100 logs of blackwood, and we may find 9 or 10 of these close together, then it is necessary to move on for several miles to get another 10, and on again over miles of country finding here 1 or 2 trees, and there 10 or 15, but nowhere many in the same spot, and to get these down to the river we must cut special tracks for each lot, all of which works entail great outlay. It is difficult to explain to any one who has not studied the question how thinly scattered through the forests are our more valuable trees. I have often been told by people who see our fine looking forests that the timber in this country ought to be very cheap. To them I say "come with me and I will show you that you are wrong." " In this forest the timber merchants have been at work and those stumps that you " " see are the stumps of sundry venteak and ạgil trees that have been cleared away ; " " those kambagam trees down by the river side are hollow, or they would have been " " felled long ago, and all these hundreds of trees you see around have no money " " value whatever." In a word, if the people would take all the timber we offered them, we could let them have it very cheap, but as they must have particular kinds, they must pay highly for them.

Thin distribution of the best kinds of timber makes them very expensive.

899. This thin distribution of the more valuable trees will increase the difficulty of treatment of our forests, for, unless very carefully supervised, the system of felling the better kinds and leaving the poorer will lead directly to the encouragement of the latter, and the disappearance of every valuable tree, as I pointed out in para 261.

This thin distribution makes forest work difficult.

900. In many parts of the forests which I have visited the reproduction is very bad, and Mr. Vincent, I see, makes the same remark with reference to Ceylon. The reason of this is that if left to nature the older trees do not fall as soon as they have arrived at maturity, and give place to their more vigorous juniors, but continue to live, and with their dense shade prevent the smaller trees growing. When these forests are taken in hand we should make it our business to fell and bring to market the trees as they become mature, and the light and air thus admitted will give the smaller trees encouragement to shoot up and fill their places. As matters stand now, there are very few medium sized and small trees in our moist forests, so that, when the big trees are felled, it will be a long time before we shall have anything large enough to take their places. I was very much struck with this state of things when visiting the forests on the Umiār (a tributary of the Chenthrōni river) in 1837. I found a large number of small kambagam trees (Hopea parviflora) of full size, and an immense quantity of seedlings under them, for the kambagam is a profuse seeder, but except in a few open glades, and on the river sides, where the young trees could get light and air, I found no trees between 5 and 100 years old.

Reproduction very bad.

901. In the last few paras I have been speaking of the trees in the heavy moist forests. The case is somewhat different as regards the grass land forests. where teak is often very gregarious, and where other valuable species frequently reach a considerable proportion of the trees, but in all accessible places these forests have been very heavily worked, and the timber that is left is not of much value.

The same remarksapply to the deciduous forests.

902. The timber grown in Travancore is brought down, as I have shown in describing the different river basins, partly by water, and partly by road. In South Travancore, owing to the smallness of the rivers, it is all conveyed by land, and lightness is not therefore such a *sine quà non* as in North Travancore. The woods used in the South are chiefly teak, blackwood, kongu, thēmbāvu and vēnga. But in

Chief timbers felled in North and South Travancore.

North Travancore, where the timber is transported by water, the heavier woods are·
in less demand, as they require bamboos to float them, and even when these are obtainable, they are often useless, because they cannot be attached to the logs on account of obstructions in the rivers. Thus the woods chiefly in demand in North
Travancore are teak, blackwood, white cedar, ãnjili, and ven-teak : punna and nedu-
nãr for spars ; and mango, kambagam and chīni for boats. Irũl is sometimes sawn
up, and conveyed by boats, as it is very heavy.

903. Most of the timber felled in Travancore is exported to foreign countries,
but the quantity annually required in the State is rapidly in-
Destination of the different timbers. creasing. Figures of the exports will be given in the appendix. Pondicherry and Tuticorin, and the country north, and
east of Madura take large supplies, chiefly of vēnga, kambagam and thēmbãvu.
Most of the teak and blackwood goes to Cochin, and is thence exported to Bombay
and other parts. White-cedar is largely used for oil casks, and is converted into
them for the export of oil from Cochin to Europe. Ven-teak, punna spars and some
of the common woods are sent across to Arabia, where in the drier climate of that
continent they last much longer than they would with us. In fact ãnjili, mango and
chīni are the only woods much felled which are almost exclusively retained for consumption in the country.

904. The most valuable timber trees of Travancore are the following :—
(1) teak, (2) blackwood, (3) ebony, (4) sandalwood, (5) ãnjili,
The list of trees taken from the Flora Indica, and brought up to date. (6) kambagam, (7) vēnga, (8) thēmbãvu, (9) white cedar, (10)
red cedar, (11) ven-teak, (12) jack, (13) irũl, (14) mayila,
(15) manjakkadambu, (16) pũvan, (17) manimaruthu, (18)
mango, (19) punna, (20) chīni, (21) pãthiri, (22) ilavu (23) karunthagara. Of
each of these I will give a short account before closing this chapter. Of these and
of the other trees indigenous in our forests a complete list, arranged according to
Natural Orders will be given in the Appendix. The names have been taken from the
Flora Indica of Hooker so far as it has been published, and from Col. Beddome's
Flora Sylvatica. In addition to those species which I have myself observed I have
included in the list the names of certain trees on the authority of Wight, Beddome,
and other skilled Botanists, who have examined the flora of our forests, and who
noted the occurrence of these species within our limits. Such lists are never thoroughly complete nor absolutely perfect, for new additions are constantly being
made, while the names of many species are from time to time changed, as our knowledge extends, or they are struck out altogether.

905. I have been at some pains to collect the native names of the different
species of trees, and this I have found a work of no small diffi-
The variety of native names for trees causes great trouble. culty. All except the very commonest trees have each at least
one Tamil, and one Malayalam name, and many of the Hill
tribes, Kãnies, Mannãns, Uralies, and others, know them by
names peculiar to themselves. Then again a certain word will be used in Malayalam for a particular tree, and the same word will be employed in Tamil to denote
quite another species, thus, pũ-maruthu is a name given to *Terminalia paniculata*
by the Malayalies ; and to *Lagerstræmia flos reginæ* by the Tamils, though they have
but little resemblance to each other. Of the less common trees a large number
have no special names, and it is not therefore easy to make a man understand what
tree is meant without going into an elaborate description of it.

906. The importance of having one name and one only, or at the outside two,
for each tree must be obvious to any one who considers the sub-
The elimination of the less common names very necessary for purposes of identification. ject, for the time lost in discussing its appearance for the proper identification of any particular species is very great. In
the list above mentioned I have given all the names I have been
able to collect, that after due consideration a selection of those most widely known
may be made, and that these may be adopted throughout the country. It is the duty
of those who have anything to do,'with the forests to endeavour to make the names
eventually selected more widely Vnown, to the exclusion of those which are local and
little used, that our many may be simplified. For it is very confusing to hear one
man talk of senjãl, another of ven-teak, another of vemvila another of vellei agu, and
another of vengalam, even though we happen to know that they all refer to the
same species, the widely distributed *Lagerstroemia lanceolata.* Without such knowledge

the number of species would be placed at even a much higher figure than is really the case.

907. The use of Botanical names is by many persons considered unnecessary and pedantic, but an accurate identification of our trees is a matter of the greatest importance, and this can only be obtained through the aid of scientific names. A Forest officer who is ignorant of the Botanical names of, at all events, the most important trees in his district will always be debarred from making use of the experience of others. There are thousands of able men who have made, and are making observations on trees and plants in all parts of India and the world, and in the absence of a common language, all the plants, trees &c that have been described are known, each by its own scientific name, and unless a man is conversant with this nomenclature (cumbersome though it often is) he is as much shut off from taking advantage of the results of these observations as if he were ignorant of the art of reading.

Necessity that Forest officers should know the scientific names of the chief trees.

The trees of Travancore.

908. *Tectona grandis.* (Linn, fil.) English: teak, Tamil: thēkku. Malayalam: thēkka. The teak tree belongs to the order Verbenaceæ. Its appearance is so familiar to every one in South India that it need not be described here, further than to say that the leaves are opposite, about 10 or 12 inches long, and nearly as broad, rounded, and with a pale under side. The foliage is sparse, and the whole of it falls at the same time, once a year in the hot season, leaving the branches bare. The teak tree flowers in June and July, and its fruit ripens in the dry weather, January to March. The flowers are white, about ½ inch in diameter, and are clustered in large bunches at the ends of the branches. The fruit consists of a nut enclosed in a spongy covering, which is further enveloped by the enlarged calyx, a loose case of one inch or more in diameter. The nut contains normally 4 seeds, but 2 or 3 are usually abortive, so that not more than one or two plants grow from each fruit. The only tree which is sometimes mistaken for teak is the *Dillenia pentagyna,* Tamil nāy-thēkku, Malayalam koda-punna, whose large leaves somewhat resemble the leaves of the teak, but on closer inspection the former can easily be distinguished by their pinnate venation, and their serrated or jagged outline, while the veins of teak leaves are distant, and their edges are regular. The flowers and fruit of these two trees are quite dissimilar.

909. The teak tree is indigenous in south-western and central India up to 25° north latitude and in Burma, Pegu, Java and Sumatra. In other parts of India north of 25°, it is cultivated, as in Bengal, Assam, Sikkim and the North West Provinces, and it has been introduced into Ceylon, but it is not indigenous in these countries.

Its distribution out of Travancore.

910. Teak grows best in a temperature ranging between 60° and 90°. It can bear cold greater than that indicated by these figures, but under such conditions it does not attain large dimensions. As regards moisture, the annual rain-fall should be not less than 50 nor more than 150 inches, and if there is a long period of dry weather, teak, owing to its habit of wintering and its ability to stand drought, will be found to form a larger proportion of the forest trees than if the rain-fall were more evenly distributed through the year.

The climate best suited to it.

911. The teak tree grows from sea level up to 3,000 ft. or in some cases to 4,000 ft, but in Travancore the former altitude is its highest limit, and, at this height, it does not thrive so well as a little lower down. The most suitable aspect is south and west.

Most suitable elevation for it.

912. This tree is found growing on all kinds of soils and in various localities, on sand-stone, granite, and lime-stones, on the steep hill sides, and on the alluvial land on the river banks. But, as Dr. Brandis * points out "under all circumstances there is one indispensable condition, perfect drainage, and a dry subsoil." When it occurs on a hard subsoil through which its roots cannot penetrate, the

Any soil suitable but there must be perfect drainage.

---

* Forest Flora of N. W. Provinces, pp. 336.

rapidity of its growth diminishes after a certain time, and the annual increment of its volume is very small. Under such conditions it becomes a question whether it is not better to fell the trees at an early age, as the yearly growth attained does not compensate for the loss of time.

913. It is only in open forests of deciduous trees that teak is met with.
Only found in the open and deciduous forests. During the hot weather it loses all its leaves, and growth ceases for a time. If it were found near evergreen trees which do not winter in the same way, and which retain their vitality and power of growth throughout the year, they would take advantage of its period of inactivity, and would occupy the land to its exclusion. Thus, teak is only seen in company with bamboos, or with trees that winter in the same way as it does, or are not aggressive, such as *Dalbergia latifolia*, *Pterocarpus marsupium*, *Terminalia tomentosa*, *T. paniculata*, *Anogeissus latifolius*, *Schleichera trijuga*, *Gmelina arborea*, *Stereospermum xylocarpum*, *Careya arborea*, *Phyllanthus emblica*, and others.

914. In Travancore the teak is usually found on the lower slopes of the hills
Teak most abundant at the foot of the hills. and at their feet, for here the extent of evergreen forest is small. Farther inland, such forest occupies all the valleys, and the teak is driven up the hill sides, where the soil is too dry to support evergreen trees.

915. In certain situations, this tree often forms the greater part of the forest,
Cause of its gregarious habit. purely, I believe, because it can stand a drought greater than would suit any other tree. But even teak does not thrive on slab-rock with a covering of about a foot of mould over it, and in such situations it does not attain a large size, but the trees often grow very close together. Such trees are called kōl-teak.

916. I have endeavoured without success to ascertain the derivation of this
"Kōl-teak." The derivation and meaning of the word. name. *The word kōl may mean a stick or pole of any sort, or it may mean the particular measure used for calculating the dimensions of timber. Thus, kōl-teak may be either sapling-teak, or teak of such small size that only a kōl length of timber can be cut from it. Neither derivation exactly agrees with the ordinary acceptance of the term. In the Proclamation of $\frac{1853 \text{ A. D.}}{1028 \text{ M. E.}}$ kōl-teak is defined as all teak under 10 vannams=(12½ inches) quarter girth, but this manifestly includes, not only all old stunted trees that would never attain larger dimensions, but also all young trees that certainly would grow larger. Again, the Report of the Forest Commission in 1884 (para 15) defines kōl-teak as "teak growing in poor soil and in the open country," and this properly defines it, but does not explain the derivation of the term. Kōl-teak is teak growing in unfavourable situations, such as on laterite, or on rock with very little surface soil, and, in consequence, so stunted that it would never attain large dimensions however long it lived, would never in fact be more than a pole.

917. The greater part of the teak found in the low country is kōl-teak because
Most of the teak in the low country is kōl-teak. there is so much laterite and rock on the dry hills, but the kōl-teak of North Travancore, growing on laterite, is much finer than the kōl-teak of South Travancore which grows upon a stratum of rock. Some good teak of large dimensions is seen in the low country, where, as on the banks of rivers, the soil is deep and fertile.

918. On account of its slow growth, kōl-teak has usually a different appear-
The timber of kōl-teak darker and heavier than the ordinary kind. ance to the fast grown teak of the better land. It is darker in colour, contains more oil, and is considerably heavier. This has led some people to speak of it as a species distinct from ordinary teak, and to call it "bastard teak." This is quite a mistake.

---

* It has been suggested to me that kōl-teak may be teak whose circumference is less than one kōl.

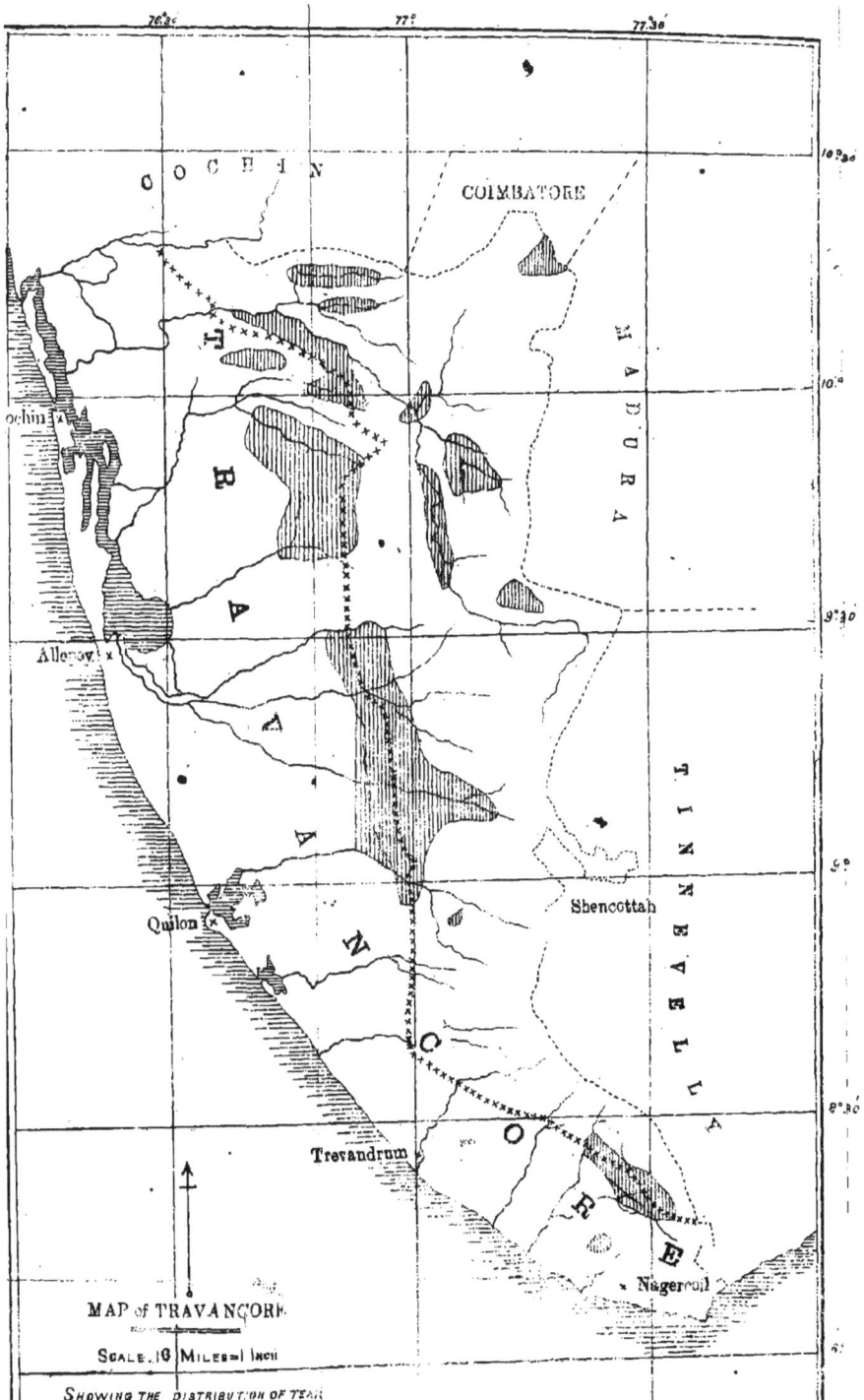

COIMBATORE

COCHIN

MADURA

TRAVANCORE

TINNEVELLY

Cochin

Alleppey

Quilon

Shencottah

Trevandrum

Nagercoil

MAP of TRAVANCORE

SCALE, 16 MILES=1 INCH

SHOWING THE DISTRIBUTION OF TEAK

919. The finest teak in Travancore is to be found on the hills at an elevation of from 1,000 ft. to 2,000 ft. The Idiyara valley, the sides of which run up to more than this elevation, used to be celebrated for the quantity and size of its teak. Dr. Balfour in his "Timber trees" mentions that a Mr. Edye about the beginning of the century felled a tree in this valley seven ft. in diameter at its base, and 26 inches in diameter at 70 ft., from its butt. This would give about 53 candies, or 900 cubic feet of timber. In the Trevandrum Museum there is a plank sawn from a tree felled in the same valley which is 4 ft. 3½ ins. across. The conditions to be found in the Idiyara valley are peculiarly favourable to longevity. The teak tree is there seen growing in thickets of the éetta reed (*Beesha Travancorica*), and it therefore does not suffer from fire as would be the case in grass land. The soil also is very fertile, and the climate forcing, while there are no violent winds. Thus those trees which, in other places and under less favourable conditions, would be burned or blown down, continue in this valley to grow, adding on a small increment to their size year by year.

Instances of specially fine teak trees.

920. It is particularly important to remember that this large size is not due to rapidity of growth, but to great age. *The plank I have mentioned as having been deposited in the Trivandrum Museum, shows 156 rings, and at least 50 more would appear if the plank had been sawn from the centre of the tree. These concentric rings which indicate the age with exactness, are in this plank never more than ¼ inch apart, and sometimes two consecutive rings are no more distant from each other than the thickness of a piece of note paper. This tree therefore must have been well over 200 years of age, and Mr. Edye's seven foot tree must have been a veteran of 4 centuries.

The large size of these trees due to great age.

921. The Idiyara valley and all the accessible forests of Travancore have been worked for teak for at least a century, and, knowing this, it is not surprising to find that there is little or no large teak left any where in the country except in very out of the way places, from most of which it will never pay to remove the timber. Of 3rd class and köl-teak there is a fair quantity. I will now show in what places there is still any teak left.

Very little large teak left in the country.

922. Beginning in South Travancore, there are first of all in the Hanamannathi basin a few trees, but they are so stunted and torn by the wind that they are of very little value, and may be neglected from the account.

A few small trees in the Hanamannathi valley.

923. Crossing over to the valley of the Palliár, we find a great deal of köl-teak on all the rocky hills bordering the cultivation. Though very numerous, these trees are all of small size. Few of them grow more than 20 ft. high, or have unbranched stems of more than 6 ft., and a girth exceeding 2 ft. at 5 ft. from the ground. This wood is all useful and saleable, but it is köl-teak of the poorest class, and can only be used for furniture or small work. In the neighbourhood of the Virappuli depôt, and on the hills near Black rock estate the teak begins to improve, and we here get for the first † time 3rd class timber between 8 and 10 vannams, in the centre of the log, or say 12 to 15 inches quarter girth at 5 ft. from the ground. The forests near the depôt which come within the basin of the Palliár have been heavily worked for some time past, and they contain but little teak timber of more than 10 or 20 years old.

Some small teak in the valley of the Palliár, but none of any size.

924. Passing next to the valley of the Paraliár, a good large quantity of teak of 3rd class is seen scattered about the low country at the foot of the hills, and especially in the long valley that runs down from near Mayilūnni, and joins the Parali on its northern bank. Teak also ascends the hills in the direction of the Māramala and

A good supply of 3rd class teak in the valley of the Paraliár.

* In a pamphlet by Mr. Bryce of the Bombay and Burmah Trading Co. ojﻌﻮ teak, he says " In " " a tree 8 ft. in circumference, I have counted 380 rings of annual growth, so that the great trees of 6 ft. dia. " " meter must have reached an age of several thousand years. "

† I find that the proportion of decrease from the butt to the centre of the log averages 20 per cent over a large number of logs measured. This corresponds with Mr. Gamble's measurements of standing trees, given in his Manual of Indian timbers.

Swâmikkûricchi estates. Four years ago, when I was at Vîrappuli, the Aminadar estimated that there were about 10,000 candies of fair sized teak in that neighbourhood, a supply sufficient for about 20 years at the rate of felling then carried on. This is I think a sanguine estimate.

925. In the valley of the Kōtha, and to the east of that river, a few hundred trees are found on the Ihacchammala hill, and below the Kalpadava estate, and some kol-teak is to be seen on the hills overhanging the village of Kulasōkharapuram. West of the Kōtha the teak tree entirely disappears, and is not found within the forest area (except as a planted tree in a few localities) all through the valleys of the Ney, the Karamana, the Vâmanapuram, and the Ittikkara rivers.

A little teak to the east of the Kōtha, but none in the valleys of the Ney, Karamana, Vâmanapuram, and Ittikkara rivers

926. Nor is there any teak, to speak of, south of the Kalleda river. It is not till that river has been crossed, and the basin of the Chālakkara reached, that teak is once more met with. A few teak trees of small size are scattered about the Kōrakkunua ridge just to the noth of the Kalleda river, and numerous trees are found in the vicinity of the road between Ottakkal and Punalūr. Crossing over to the Chālakkara valley, small teak is found in the neighbourhood of the place of that name, but all the way from there to the Ariankâvu pass the timber contractors have been cutting down teak for the Puliyara depôt, so that nothing remains but young trees of less than a foot in diameter. Up the Ambanâda river, and in the vicinity of Kōmarangudi, there are still some good trees, though the contractors have been at work here too. The Aminadar of Patthnâpuram estimates the total number of trees of 2nd and 3rd class growing within this valley at 1,000, a supply sufficient for about 5 years.

In the basin of the Kalleda no teak found to the south of the river except a little.

927. Kol-teak is found in the basin of the Thāmaravaroni thickly sprinkled over the stony grass-hills overlooking Shencottah, but it is of very poor class and small value.

Only kōl-teak in the Thāmaravarni valley.

928. The valley of the Acchankōvil was famous for its teak, and, as I mentioned in an earlier part of this Report, the Forest Department was hard at work felling there when Lieut. Ward visited the place in 1817. Ever since then there has been a steady drain on these forests, and there is scarcely a part of them where teak grows, that has not been worked at one time or another, by the Department or by contractors. On the Alappāda ridge some fine trees may yet be seen growing in places whence the descent to the valley below is declivitous, and whence contractors will never take them. If Government at any future time resumes the work of getting down the teak, this timber may be brought to market, but there are only a few hundred trees in this spot. On the stony slopes below Nāgamala teak trees of good size are found here and there, but a long way from the river. These forests were worked many years ago when the Department felled and transported its own timber, and now and then an immense log may be seen, which was left behind for some reason or other, and which the contractors did not think it worth while to bring to the depôt.[*]

The Acchankōvil valley famous for teak, but timber operations have exhausted nearly the whole supply.

929. There is said to be good teak up the Arimba thōda on the Kalār, a place that has been partly worked by the Forest Department in former times, and on the ridges to the west of Chōmpālakkara I saw a few good trees, but all these localities are most difficult of access, so the contractors will have nothing to do with them. In the valley below these ridges there are thousands of nice young teak trees of 3rd class, which would grow into good trees if they were left alone, but the contractors find that these trees pay them better than the larger ones, and they are not inclined to spare them to benefit posterity.

Good teak in other places that are difficult of access. Much immature teak cut by the contractors.

930. As I have already described, the valley of the Rāni river is clothed with heavy forest to within a distance of 3 miles of the forest line all along, so that with the exception of a few isolated patches, occurring in the middle of the moist forest, teak is not found except in this strip. These isolated patches are near Nāṇāttappāra, Nāladippāra, Thavalappāra, Nellikkal, Vampulialappāra,

Places in the Rāni valley where teak grows. Here too the contractors are cutting immature trees.

[*] See para 362.

on the Thêvara mala murippa, and close to Kirakka on the Panni âr. Most of them are difficult of access, indeed the very fact that teak still remains here is a proof that the Forest Officers of a former generation considered the timber to be too badly situated to allow of its being brought down at a profit. In the strip of mixed grass land and shifting cultivation, which lies between the forest line and the moist forests of the interior small teak occurs, and looks exceedingly healthy, where it has not been mutilated for manuring the fields. The trees themselves are small, because the contractors have cut, and are still cutting, everything that is worth having, and the fear is that they will go on removing the immature trees till they leave nothing but saplings four or five inches in diameter. Mr. Thomas (Assistant Conservator) estimates the total quantity of good teak over 18 inches diameter in the valleys of the Acchankôvil and Râni rivers at 10,000 trees, of which a great many are inaccessible.

931. The valley of the Manimala contains excellent soil admirably adapted to the growth of teak, but the same remark may be made about it that was made regarding the Râni valley. The con-tractors have felled all the good timber, and what remains is small, and requires to be left alone for 20 or 30 years. This small teak is very abundant to the south of the river, on all the stony hills lying within a short distance of the forest line. North of the river, the land is more level and the soil is retentive of moisture, and, except near Edakkôonam, teak is not found. From this place the contractors have felled all the trees of any size.

All the large trees re-moved from the Mani-mala valley.

932. From the Manimala river northward to the neighbourhood of Eerâtta-pêtta there is very little teak, but in the Pûnnyâtta Edavaga, which lies to the east of that village, teak is exceedingly abun-dant. The conditions here are most favourable to the growth of this tree, the land being rather steep, rich, and well drained, while there is an absence of bad wind. Consequently, both to the east of Eerâtta-pêtta and north of the Pâlâyi river, as far down as the village of Lâlam, teak is the commonest tree in the valley, in spite of the large numbers that are annually felled. Very little first or second class timber remains, but the third class teak is abundant, and, if properly conserved, would last for many years. At present, about 2,000 logs go to the Pârampura depôt every year, and a great deal is smuggled down to Cot-tayam, or is used surreptitiously for building, without our knowing anything about it, for the Forest Department is very short handed on this river.

Third class teak very good and abundant in the Pûnnyâtta Edavaga.

933. The Kurrinji ridge that divides the valleys of the Pâlâyi and Thodupura, grows excellent teak, and as these forests are at some distance from the rivers, some first and second class trees are still to be obtained there. The whole valley of the Thodupura river was famous for its teak some 30 or 40 years ago, and the Forest Department had a large establishment, and 40 or 50 elephants (so I was told) engaged in working out this timber. Even so lately as 20 years ago, Mr. J. S. Vernêde found fine first class teak at Mrâla, and on the Varippura, but all the good timber has disappeared now, with the exception of a thousand or so of 2nd and 3rd class trees on the Nyaralat-thanda, and on the ridge that divides the Velliâmattam from the Arakkulam valley. On the Perambukâda ridge, between the Thodupura and Vadakkan rivers, teak thrives, but the trees are young. Passing on to the valley of the Vadakkan we find teak absent, except in groves dotted about the country, and which I believe to be artificial. Some of these groves contain nice 2nd class trees, and they all deserve careful preservation. In the Mullamgâda valley teak is again found, and there are said to be lying there 400 fine large logs which were felled by a contractor, and which he has been unable to bring down. In the neighbourhood of Kôthamangalam extensive forests of teak occur, mostly of 3rd class, but containing a little 2nd class timber. The contractors have wasted a great quantity of this valuable wood, and the people in the neighbouring villages have burnt numbers of the logs for fuel, no one preventing them, and it is no uncommon thing to see logs that have been partly split up for firewood, with the remaining half chopped and hacked about. These Kôthamangalam teak forests are much mixed up with cultivation, both permanent and shifting, and the occupiers of chêrikkal lands do not scruple to lop the teak trees from base to summit to prevent the drip from them falling on their garden produce.

Good teak on the Kur-rinji ridge. Formerly there used to be much good teak in the valley of the Thodupura, but it has all gone now. Teak elsewhere in the Mavâtta-pura valley.

934. Next to the Múvättapuru river basin comes that of the Pefiyär. Coming down the river, the teak tree is found for the first time in the neighbourhood of Kumili at an elevation of 3000 ft. These forests have never been worked (as far as I can ascertain) until within the last year or two, and, though not extensive, they contain timber of full size, that is to say as large as it will grow at this high elevation. Along the river side teak occurs at intervals, but is not abundant till the neighbourhood of Thodupura–Periyär is reached. Before that, and near the Pèkkanam-estate there is a patch of about 100 very nice trees, and another patch, but not so good, near the Chenkara estate. At Thodupura-Periyär some capital teak is seen, and so it is all the way down to the Madatthin-kadava. On the right bank especially there is a clump of very large trees, which attracted the attention and admiration of Lieut. Ward when examining these parts in 1817. It is unfortunate that they are so situated as to be quite out of the reach of the contractors, for they would fetch a good price.

Teak found near Kumili, and down the Periyär at intervals.

935. Near the Rämakkal depôt, and all down the valleys of the Kalär and Vanjikkadava är teak is found, sometimes of good size. At Pirinyänkûtta, on both sides of the river there are groves of fine trees, and so it goes on; all down the river teak is met with at intervals, as also on the Môthirappuru, but with the exception of those trees near Kumili, or Rämakkal, the whole of this useful timber is of no value, because, every where above Karimanal, it is so situated as to defy removal at a cost which would be covered by its selling price.

Some teak at Rämakkal, and down the river at intervals.

936. Below Karimanal teak of enormous size was formerly abundant on the hills overhanging the Periyär, and even at the present time 2 or 3 trees are still pointed out from the river, which were too big to be dragged to the waterside by the contractors' elephants, and were therefore left to stand for generations as landmarks above the surrounding forest. On one of the highest and least accessible ridges adjoining Kuthirakkutti, and overhanging the Periyär, a contract has lately been given to a timber merchant to fell and deliver the last remaining teak trees still left in this neighbourhood. On the Idiyara river, on the Parisha kuttha, and the Kandanpära there now remain but few teak trees of any size. These forests were worked continuously, as already stated, until 20 years ago, and this valuable tree has been all but exterminated from them. Where hundreds of trees were annually felled, it is now difficult to find more than 3 or 4 trees of even moderate size in the course of a 15 mile march, and what is worse there are no small trees coming on to take the place of the older ones removed.

Entire disappearance of all the fine old teak from the lower Periyär.

937. It may seem curious that a tree should be found to disappear so entirely from a forest where it was once abundant, as the teak tree has done from the Idiyara valley, and I will therefore digress for a few minutes to show how such an event may very easily happen. Bearing in mind the great variety of timber trees in our forests, I will suppose, for the sake of argument, that the teak in this locality orginally amounted to 10 per cent of the trees there. After one or two cuttings of this tree, and no other, the percentage would probably not be more than 5, and supposing that all these trees shed their seeds equally, the chances of teak continuing to form a percentage of even 5 would be less than formerly, for it would be 19 to 1 against it. After further cuttings the percentage of teak would perhaps be reduced to only 1 in 100, and this percentage would fall yet farther in the course of nature even if no more trees were felled, for whenever a teak tree fell down the seeds of the other 99 percent of the trees would continue to spring up year after year, while there would be few teak seeds to compete against them.

The felling of one tree and no other in a forest leads in time to its complete extinction

938. Even in untouched forests there is a tendency among the species that compose it to change gradually in the course of time. Last year, when I was visiting the forests in the vicinity of Nériyamangalam, where there is a good quantity of white cedar of great age and large dimensions, my attention was drawn to the almost entire absence of any plants or seedlings of this tree. The explanation is probably that old trees, like human beings or animals, lose the power of reproducing themselves after a certain time, although they

A tendency among the trees which compose a forest to change, and for certain species, common at one time, to disappear.

continue to live, and when, after the lapse of perhaps a century from the time of reaching maturity, they fall down, their places are taken by other trees, and not by saplings of the same species. Thus, as in the case of the teak in the Idiyara valley, a species once plentiful in a certain place, may entirely disappear from it.

939. Above the Péndimáda kuttha on the Kandanpára river, and on its northern bank, there is a patch of several hundred large teak trees which were not felled before, because the cost of bringing the logs to the depôt would have been so great. Last year a contractor visited the spot with the object of seeing if the trees could be worked down at a remunerative expenditure, but after examining the ground carefully he decided that it would cost him 10 Rs. a candy to get the logs down. The contract was therefore not taken up, as the price offered was only 6 or 8 rupees.

A patch of large teak still remaining on the Kandanpára river in an inaccessible place.

940. On the hills between the Idiyara valley and the forest line teak of 2nd and 3rd class can be obtained, 20 or 30 trees growing in one place, and as many in another, on all the drier ridges of that tract. The total number however does not exceed 400 or 500. Near Malayáttûr itself the forest contains a good number of teak trees, for about 2 miles to the east of the village, after which the evergreen forest begins, and teak ceases. This is 3rd class timber, and 400 or 500 trees are felled every year from this locality.

Some small teak obtainable on the hills to the east of and near Malayáttûr.

941. On the Kôttashêri river the only teak trees left are some hundreds on the ridge overhanging the Varamban cataract. In the Anjináda valley teak is scattered all about the dry hills below the plateau, but it is of poor class.

Teak in the Kôttashêri and Anjináda valleys.

942. Outside the forest line, and beginning from the South, teak is found in abundance on the slopes of the Vélimala hills near Nagercoil, but it is chiefly kôl-teak, and some timber of the 3rd class. North of this, teak is not found, except as a planted tree in compounds or near pagodas until we get to the neighbourhood of Nílamayilum, for almost the whole of this country is occupied by gardens, is under cultivation, or is covered with short grass. Near the last mentioned place, kôl-teak is often seen growing on the dry hills, but it is of small size. Between Nílamayilum and Punalûr kôl-teak is abundant, and northward again as far as Kônniyûr it is a common tree within 4 or 5 miles and outside it, of the forest line, beyond which the country is quite open and treeless.

Occurrence of teak outside the forest line in the South.

943. Beyond Kônniyûr all the compounds contain teak, which grows well on this river about Kumala and Mala-ilappura. On the rocky hills near Rûni, Manimala, and to the south of Kânnyirapalli, at Thumbalakkáda, and all along the Pálâyi river as far west as Puliyannûr, teak may be seen. To the west of this country there is not much of this tree except where it is found in compounds, or near pagodas.

Northward of Kônniyûr.

944. At Kadanáda and on the long ridge running from from Mélakáva to Kúttháttakulam, teak is very abundant, and there is an extensive forest of this tree near Purappura, which I have recommended for reservation. At Mûváttapura, and along the river banks as far as Vettikátmukka, teak is seen on both sides, but it is only close to the river that trees occur. At one time all the open hills which are covered with low scrub, produced thousands of kôl-teak, but these trees have almost all been much felled, and burnt for charcoal, or taken to Cochin for sale, and the wood is almost exterminated from these parts.

North of Kadanáda and Kutthátthakulam.

945. Between Mûváttapura and Thodupura or Kôthamangalam there is some of this timber, thinly scattered along the road, but to the east of Kôthamangalam there are extensive forests of the composition of which this tree forms the greater part. West of this place and of Malayáttûr teak is only sparingly found, and it is much damaged by the cultivators lopping it in their compounds.

Near Mûváttapura and Thodupura.

946. It will be thus seen that the only places where there is teak of good size still remaining are the Châlakkara, Acchankôvil and Kallâr valleys, the Thêvaramala ridge on the Hâni river, the banks of the Pâlayâi river, the neighbourhood of Kadanâda and the Kurinji ridge, Purappura, the Thumbipâra ridge near Vêlliâmattam, the Mullaringâda valley, the forests ne r̲ Kôtbamangalam, and Malayâttûr, and the hills on the Periyâr, the Kandanpara, and Parishakkuttha rivers, and to the west of the Idiyara valley. The accompanying map will show at a glance the distribution of the tree through Travancore.

Only places where good teak is still found.

947. I estimate the total number of standing teak trees of good size, that is of the first, second, and third class above 8 virals (10 inches) quarter girth at 100,000 and of kôl-teak and smaller trees at about 400,000. I refer to timber that can be brought down at a profit, and I do not include such as grows on the Periyâr above Karimanal, and is inaccessible, saving what can be taken to Râmakkal or Kumili. These trees are scattered over about 400 square miles of country.

Estimate of the total number of trees still standing.

948. The number of teak logs sold by the Forest Department every year is not less than 12,000, (of which about 2,000 are below 8 virals quarter girth) if we include those sawn up and delivered at the depôts of Shencottah, Quilon &c. by contractors. There are also some 10,000 sleepers annually sold to the Bombay firm. Comparing these figures with those already given as showing the number of standing trees left in the country, it will be seen that the supply of good sized timber at the present rate of consumption will be exhausted in about ten years leaving nothing but immature and inferior wood. Besides this, there is a supply of felled timber lying in the forest sufficient for about two years.

At the present rate of consumption the supply of moderate sized teak timber is only sufficient for about 12 years.

949. That the quantity of teak still remaining in Travancore is very small is well known to everyone who has anything to do with the forests. The present Conservator Mr. Vernôde is well aware of this state of things, and has more than once noticed it to me, and in his Annual Reports for the last 5 years, as well as in former years, has urged the necessity of extending the plantations, in order to counterbalance the drain on the forests of the country by the annual felling of so many trees. To any one resident for a short time in one place, though knowing nothing of our forests, the rapid disappearance of teak from some parts is very apparent, and I have been frequently told that in such and such places the tree has been almost exterminated in a very short time.

The rapid disappearance of teak apparent to any-one.

950. Another proof of the growing scarcity of the timber is the increased cost of working it to the depôts, the contractors now refusing to undertake any contracts except at higher rates than formerly. If the quantity felled was not in excess of the annual reproduction, the contract rates should be less than formerly, for the contractors now have roads and means of exploiting their logs which were not available to them before.

The contract price for delivering teak at the depôts is rising.

951. Again, another sign that the teak tree is becoming scarce is the fact that the class of timber delivered at the depôts is annually deteriorating. Mr. J. S. Vernôde's diaries of 20 years ago are full of references to the large size of the logs he was working down, such as, "these logs were so large that they required 2 elephants apiece to drag them," "the logs felled this year each exceeded 5 candies (80 cubic feet)" "the logs brought down this year were so large that they could not be floated past Shêranellûr without the assistance of elephants to drag them over the shallows" and so on. In those days no teak was felled that was under 20 inches diameter in the middle of the log, that is, it was all first class timber. Now a days, nearly all the logs brought to the depôts are third class or under 13 inches diameter. Of the logs measured at the depôts of the Northern Range in 1064, 5,794 or 93 per cent were third class, 389 or 6 per cent were second class, and only 42 or less than one per cent were first class. Instead of the 5 candy logs

The inferior character of the timber brought to the depôts a sign that the good timber is nearly exhausted.

of the old days, the average seldom rises over 1 candy each, and logs of more *than 2 candies are quite unusual.

952. The timber in fact is now so small that purchasers object to it. In the Administration Report for 1868-9 the Chief Engineer is quoted as saying that "the quality of the teak supplied to my Department was all that could be desired," but the present Chief Engineer makes a very different statement, and complains that the logs taken for the work-shop are all so small that there is a very great wastage. Messrs Wallibhoy & Co., the Bombay contractors, have also remarked to me on the deterioration in the character of the logs. In a word, there can be no longer any doubt but that teak of good size will soon be very scarce, and that when it is wanted, (and for pagodas and religious purposes I believe that no other timber will do) we shall soon have to import it from abroad.

953. The question of the rate of growth of teak as bearing on the profits to be expected from plantations is one of the greatest importance. As the result of many experiments, it has been proved to a certainty that the concentric rings seen so markedly in teak wood exactly indicate the age of the tree, one ring being added on each year. In his Manual of Indian Timbers Mr. Gamble gives the diameter measurements of 29 trees with the number of rings in each, showing that the average width of a ring of wood is nearly half an inch, that is to say 2·62 rings go to the inch of radius : in other words teak increases at the rate of an inch in diameter in 1·3 years.

954. I have made measurements of a great number of trees growing in the forest, and I find that the average increase is certainly not more than ¼ inch of radius in a year, that is to say that the increase in diameter is 1 inch in 2 years, and it is usually much less. I find that, as a rule, unless the soil is particularly rich and deep, the rate of growth of the tree, which is rapid up to 40 years of age, sensibly diminishes after that time. Under ordinarily favourable circumstances, at 40 years old a teak tree should be 20 inches in diameter at 2 ft. from the ground, excluding the bark, and 25 inches at 60 years. These figures would show that the growth of teak is not so rapid as it is in other parts of India, but Mr. Gamble admits that his measurements were taken from particularly good specimens grown in plantations, and Dr. Brandis says that it generally takes a teak tree 100 years to attain a girth of 6 ft. at 6 ft. from the ground. The growth here may therefore be considered as much the same as it is elsewhere. As a matter of fact, the logs of 20 and 25 inches in diameter delivered at the depôts are more than 40 and 60 years old respectively, being probably double that age, but they have been exposed to forest fires, and have been distorted and retarded in growth accordingly.

955. As to the age to which trees may live, all depends on the character of the soil and the conditions that affect them. Under favourable circumstances, teak will live and go on growing slowly for 400 years or even more, but this is exceptional, and as the annual increment at an advanced age is small, most trees fall to the axe between 60 and 100 years old, by which time they should be about 2 ft. in diameter.

956. Although the teak tree, compared with some other trees, does not increase much in diameter from year to year, it shoots up vertically with very great rapidity. Under favourable circumstances it will rush up 10 ft. the first year, and 5 ft. a year for the next 10 or 11 years. So that at 12 years old it is 70 ft. high, and 6 inches in diameter. After that the tree makes but little upward growth, but continues to increase in diameter with great regularity.

---

* 1st class logs have a quarter girth at the middle of 15 inches or over: the 2nd class 12½ inches; and the 3rd class 10 inches. At the butt these measurements would correspond respectively to a diameter of about 2ft., 18 inches, and 15 inches.

34

957. The ash of teak timber forms only about 1 per cent of the weight of still
wood: this tree therefore exhausts the soil to a very slight de-
Teak exhausts the soil to a very slight degree. gree. Of this 1 per cent, ·22 of the ash consists of Magnesihe
·27 of Phosphoric acid, and ·33 of Silicic acid, the other corhji
tituents of the ash of plants including Potash, Soda, and Lime being poorly repré-
sented.

958. The weight of teak when freshly cut is about 55 lbs a cubic foot, and 40 lbs
when thoroughly * dried; so that the wood floats easily after
The peculiar properties of teak. being well seasoned, but large logs take 2 or 3 years before
they have parted with their moisture. The value of † P is set
down at 600 so that for cross strains teak is not so strong as many other timbers, as
blackwood for instance. The great value of teak lies in the fact that it does not warp
or twist, that owing to the essential oil found in it, it is not eaten by white ants, and
that it possesses a straight grain, and is easily worked, and finally that it can be obtained of large size.

959. These numerous good qualities make it available for a great number of
purposes, and teak is probably the most generally useful wood
Uses of teak. Its great utility. in the world. For shipbuilding and the backing of ironclads
it stands unrivalled, on account of its great powers of resistance.
It is very extensively used for railway carriages, for house building, and for furniture of all kinds, and lastly it is employed in the construction of bridges, and other
works in the open air. As, in our moist climate on the west coast, teak wood does
not maintain its strength more than 7 or 8 years, it seems to me a waste to expose
this valuable timber, which is already becoming scarce, to the destroying influences
of sun and rain, when, if kept under cover, it would probably last for hundreds of
years.

960. Teak now sells in Travancore at from 10 to 17 Rs a candy, or say 10
annas to 1 Re a cubic foot. Elsewhere, the price is at least
Price of teak in Travancore. double this, but Travancore teak is now a days mostly so small
and so much damaged by fires, which cause heart and radial
shakes, or by borers, that it is less valuable than the same wood grown in other
countries.

961. Teak found in the low country of Travancore suffers much from a borer
which tunnels irregular holes of the diameter of a lead pencil
Teak much riddled by a species of borer which only attacks trees that have been lopped. in different directions through the wood. This boring is entirely due to the abominable system of lopping the leaves of this
tree for manuring paddy fields. In May and June, when the
teak is in full growth, the owners of the neighbouring paddy lands strip the trees of
all their leaves, leaving nothing but snags all the way up the stem, and a tuft of a few
leaves at the top. If this lopping were performed in the dry weather, when the tree
was resting, the damage might not be great, because the branches which had been
cut back would then possibly sprout again when the season of rest was over, but,
as the lopping is practised when the trees are in full growth, the sap, trying to find
an exit, instead of forming buds at the ends of the branches, expends itself in sprouts
thrown out from the sides of the old branches which then die back : and, as this lopping
goes on year after year, a large extent of dead wood is left in the places where the
old shoots once grew, all up the tree. This dead wood is seized on by a certain
species of moth, which Mr. Cotes of the India Museum, Calcutta thinks to be identical with the coffee moth (Zenzera coffœophaga), as a place of deposit for its eggs,
and from these eggs the borers hatch out and tunnel into the wood, which in bad
cases thus becomes thoroughly riddled, and is then quite useless for any purpose except for rough out of door work, posts and so on. These borers may be found in the
lopped teak trees any time between June and the following February, after which
they pupate, and finally, the perfect insect emerges about one year after the egg

---

* The weight of water is 1,000 oz. or 62½ lbs. per cubic foot, and the weight of teak being less than this
it floats readily.

† The letter P is used to denote the comparative strength of timber, and is found by the formula
$P = \dfrac{l \times w}{b \times d}$ = where, l = length in feet of the scantling used, w = weight producing fracture, b = breadth, d = depth of
scantling, both in inches.

which originated it was laid; but this requires further confirmation, as I have not been able to thoroughly trace the successive stages of its growth. That the lopping of the trees is entirely answerable for this great destruction of valuable timber can easily be proved, for no signs of any such boring are ever to be seen in trees growing in the forest at a distance from cultivation, and which are therefore not lopped. The annual loss to Government must amount to many thousands of rupees, all of which could be saved by prohibiting the baneful practice of lopping.

962. Owing to the rapid growth of the teak tree, the utility and value of its timber, the facility with which seed can be collected and plants raised, and above all its natural habit of gregariousness, which enables us to raise a larger number of these trees in a given area than would be possible with many other trees, no tree in India is better suited for cultivation in plantations than teak. In the year 1844 a commencement was made with the planting of teak at Nelambūr in Malabar by Mr. Connolly the Collector of that District, and up to 1836-7 about 3,500 acres had been planted. The plantations have already paid off all the original cost with interest, and it is calculated that in the year 1986 when all the trees will have been felled the profit to Government will amount to 423 lacs of rupees. With such a splendid example before us, it cannot be said that the suitability of teak for plantations is any longer doubtful. Teak plantations have been opened in Travancore to the extent of 1,150 acres, but they have not proved an entire success for reasons which I will explain hereafter. As the earliest plantations were only opened in $\frac{1867 \text{ A.D}}{1042 \text{ M.E.}}$ none of the teak growing on them will be large enough to fell for another 40 years, though a certain quantity of small timber in the way of thinnings may be brought to market.

Suitability of teak for plantations. The Nelambūr plantations and those of Travancore.

963. As regards the yield of teak in plantations Dr. Brandis writes in his Forest Flora. "It is estimated that the teak plantations of "Burma, when mature, will contain at the age of 80 years "about 60 trees per acre, measuring on an average 6 ft. in "girth and yielding 3,000 cubic feet of marketable timber, "which, with the thinnings, is expected to amount to a mean annual yield of 47 "cubic feet per acre. (Report on the revised plan of working the Burma Forests of "February 1868). The natural teak forests not being pure or compact, do not "distantly approach to this yield. As an instance of a particularly rich forest, I "may quote Col. Pearson's survey of a sample acre in Abiri, stocked with 18 large "trees, containing an aggregate of 22 tons or 1,100 cubic feet of timber. Most of "these trees however were probably more than two centuries old." In Travancore it is very unusual to find more than 10 or 12 teak trees of good size to the acre in natural forest, and the areas over which such a large number occurs are very limited. I do not include the kol-teak forests which contain perhaps 100 trees or even more to the acre none of which are more than 8 inches in diameter.

Yield of teak in plantations and in the natural forests.

964. (2)—*Dalbergia latifolia*. (Roxb.) English, blackwood or rosewood. Tamil, thotbagatti: eetti. Mal. vitti: eetti. Blackwood is a handsome tree with light foliage (which it loses about February–March) and a cylindrical stem and rusty bark. The flowers are white, small and pea like, and about ⅓ inch in diameter, and they appear in great profusion in April, at the time that the fresh foliage bursts forth. The seed begins to ripen in October. An allied species, *D. paniculata*, is sometimes mistaken for it, but a close inspection shows that the difference between the foliage of the two trees is distinct, while the wood is utterly dissimilar, that of *D. paniculata* being quite white.

Blackwood. Its appearance.

965. Blackwood likes a temperature slightly cooler than that in which teak thrives, as pointed out by Dr. Cleghorn in his § Gardens and Forests of Southern India, although it is not very particular, and will grow well on the plains, but it reaches its greatest size and height at an elevation of over 1,000 feet. It does not ascend the hills above 3,500 feet. It requires a wet climate with a rainfall averaging from 50 to 150 inches. As regards soil, blackwood grows best on a free rich soil where teak

Most suitable temperature, elevation and soil.

would also thrive, but unlike that tree it attains large dimensions, though its growth is slower, on the stiff laterites where teak would languish. On the other hand it cannot grow on a slab-rock like teak.

966. Blackwood, like teak, is found in the deciduous forests with an under-growth of grass. In a similar way too it suffers every year *Blackwood found in the deciduous forests* from the severe fires which sweep through the grass forests in the dry weather, but, thanks to the great vitality of the small plants which shoot up again after they are burnt down, and to the large numbers of seeds which are annually produced by this tree, as well as to the fact that the spreading roots send up shoots all along their length, young blackwood trees are to be met with in great abundance in every grass forest in this State lying below 3,500 feet. Indeed, so abundant are they, that, if the grass fires were prevented, this tree would be one of the commonest in the country. Outside Travancore, blackwood is found all along the west coast, in Central India, and parts of Bengal and Sikhim and the Andaman Islands, but not in Burma or Ceylon.

967. In Travancore the blackwood tree is met with in all the grass forests *Distribution of this tree through Travancore and outside it.* under 3,500 feet, but nowhere is it very abundant, nor does it ever reach a percentage of more than 2 or 3 of the trees in a forest. The places where I have seen most of it are on the Coffee estates near Blackrock in South Travancore: below Mârn-mala estate, and, occurring with teak, as far as the Kôtha river and a little to the north of it. Beyond this, blackwood becomes less common through the valleys of the Ney, the Karamana, the Vâmanapuram and Itthikkara rivers, but it does not entirely disappear as teak does here, and a certain number of logs are felled in these localities every year. Blackwood becomes more common in the Kallada valley, and in the Acchankôvil valley, and near Kônniyûr fine trees may be seen. North of this, blackwood is found at intervals all through the country, but nowhere is it common except on the rich clays of the Manimala river, the Pûnnyâtta Edavaga, the Thumbi-pâra ridge on the Thodupura river, and near Kumili on the Periyâr.

968. It is not easy to estimate the total number of trees in the State. From *Estimate of the number of trees in Travancore.* the above it will be seen that blackwood is far more widely dis-tributed than teak, but it is not nearly so gregarious, and the number of trees in Travancore is not so large. If I may hazard a guess, I should put the total of trees above 18 inches in diameter at 10,000, scattered over 800 square miles of country. Until lately the demand for blackwood was not large, but in the last year or two there has been much more application for it, and we now sell about 750 logs a year, at which rate the supply would last for some 12 years.

969. Blackwood attains a height of 80 ft. with a girth of 12 or 15 ft., but it is *Rate of growth.* of slow growth, especially at first. Dr. Brandis gives 5 to 9 rings per inch of radius, which would be equivalent to a growth of about one inch in diameter in 4 years, thus a tree 2 ft. in diameter would be about 100 years old. But there is a considerable thickness of white and useless sap-wood, so that to get a plank of 2 ft. in breadth the tree would have to be from 120 to 130 years old at least.

970. The power of blackwood to resist a transverse strain is very great, and *Weight and strength.* the value of P is given by Dr. Brandis as 950: it is therefore half as strong again as teak, but on account of its higher price and comparative scarcity, blackwood is seldom used in Travancore for building. The average weight of seasoned blackwood is given as 52 lbs, but this must have been obtained from weighings with very well dried wood, for with us blackwood will only float when there is a considerable amount of sapwood still adhering to the log.

971 The colour of blackwood is blackish purple, and the timber possesses a *Uses.* fine grain and takes a good polish. It is largely used for furni-ture, for doors and windows, and for boxes &c. In other parts of India it is employed in boatbuilding, and for gun carriages, agricultural imple-ments, ploughs, cart wheels &c.

972. At the present time blackwood is valued at about 16 Rs a candy in the log, equivalent to about one rupee per cubic foot. The greater part of the timber sold in Travancore is exported to Bombay, where, under the name of Bombay rosewood, it finds a ready sale.

Present value.

973. The blackwood tree grows readily from seed, but it is of slow growth, as already stated. It is therefore less adapted for regular planta-tions than many other trees, more especially as it is not a gre-garious tree, and might refuse to grow freely if planted closely together. It would do well for avenues, or for belts, or as scattered trees, mixed with teak, provided the faster growing teak was not permitted to dwarf it. If only grass fires were prevented, blackwood would at once become a very common tree, for it seeds profusely, and the grass land is full of young plants waiting only for a little encouragement to push their heads above the long grass.

Not suited for planta-tions.

974. (3.) *Diospyros ebenum*. (Kœnig.) Eng. ebony. Tam. karinthâli. accha. Mal. kaïu. mushtimbi. karinthâli. karingâli. This valuable tree grows to a height of 80 or 90 ft. with a girth of 6 to 8 ft. Its bark is exceedingly black, smooth, and often covered with grey lichen. Its flowers are small, and yellowish green : male and female are found on different trees. They appear at the commencement of the dry weather, and the fruit, which is a pale green round berry, about ½ inch in diameter with the enlarged calyx almost enclosing it, ripens in June and July. The leaves are tough and leathery, elliptic and pointed. There are many trees of this genus with similar foliage, and most of them with black stems, but the wood of the others is entirely white, and the flowers and fruit are dissimilar.

Ebony. Its appearance.

975. Ebony is only found in the dense moist forests at elevations varying between sea level and 2,500 ft. It requires a considerable rainfall, and is never found on the outer hill slopes where the climate is at all dry.

Best elevation and rain-fall.

976. This tree is found outside our limits in the South Indian Peninsula and Ceylon, and it is planted in North India. In Travancore, ebony is nowhere common, but is sparingly scattered through the forests from the south to the north, wherever the elevation and climate suit. It is said to be common at the foot of the Ashambu hills. It is met with all through the forests near Ponmudi both to the south and the north. I have found it in the Chenthrôni valley, but where I have seen it in the greatest abundance is in the dense forests between Nellikal and the Shabarimala pagoda on the Pamba river. From this locality it has never been felled.

Distribution out of Tra-vancore and through the State.

977. It is not easy to estimate the total number of trees in the country : the number cannot be large, but scarcely any logs of this timber are ever felled, so that the supply is not decreasing. When any ebony happens to be required it is usually procured from the forests near Shencottah, and taken to the depôt there.

The supply not large.

978. Ebony is a tree of slow growth. The structure of the wood is exceed-ingly indistinct, and the annual rings have not been counted, so that no exact information is obtainable on this point, although it is known that the tree does not shoot up fast. Probably more than 100 years would elapse before an ebony tree attained a diameter of one foot.

Rate of growth.

979. When young, the ebony tree consists purely of white wood, with a faint streak of black running up through the centre, and even after it has attained maturity, there is usually a considerable thick-ness of white sapwood to be cut through before the black heart is reached. It is not easy to find trees with a heart whose diameter exceeds one foot. The character of the wood of ebony is well known to every one. Good specimens are intensely black, but inferior wood is sometimes streaked with brown. It has a very compact texture and fine grain, and the pores are so minute as to be invisible. The value of P is 720, and the weight of a cubic foot is 70 to 80 lbs. when dry, unseasoned 90 to 100 lbs.

Appearance of the wood, and its weight and strength.

980. Ebony is used chiefly for ornamental work, furniture, inlaying, mathe-
matical instruments, rulers &c. It is very much in request in
Its uses. Not suited for plantations. Its value. China for chopsticks and large quantities * are exported to that
country from Ceylon. Its advantages are strength, durability,
fine appearance, and close grain. The price of ebony in Travancore is nominal,
but when it is procured by request it sells at about double the price of teak, or
from 10 Rs. up to 25 a candy. On account of its slow growth it is not adapted
for plantations, but it may be grown as an ornamental tree.

981. (4) *Santalum album.* (*Linn.*) Eng. sandalwood. Tain. and Mal. chantha-
nam. The sandal is a tree of small size, and never attains a
The sandalwood tree. Its appearance. height of more than 20 ft., or a girth exceeding 3 feet. Its
foliage is light, drooping, and graceful. The leaves are pale
green,\ small and opposite. The stem is dark rusty black, often deeply cracked.
The flowers are small brownish purple turning to straw colour, and appearing from
March to July. The fruit is a black one-seeded berry, the size of a pea, ripening
in the hot weather January—March.

982. Sandal requires a very dry climate with a rainfall of from 20—50 inches,
and an elevation of from 1,000 to 3,000 ft. if it is to have any
Most suitable climate, elevation &c. value. In Travancore, the tree grows freely in the plains, and
shoots up rapidly. In Quilon especially, it has run wild, but
the wood is quite scentless there, and is therefore of no use. The temperature which
suits it best is one that ranges between 70° and 90°; as regard soil, the tree does
not seem to be particular, as it grows well in apparently poor and dry situations.
Though an evergreen, the sandal is found in the deciduous forests, where the prolonged drought causes other trees to lose their leaves. It is often seen growing in
company with the ordinary bamboo.

983. In Travancore, sandal is only found wild in the Anjinádo valley, in the
extreme north east corner of the State. Here the tree
Distribution through Travancore and outside it. thrives and reproduces itself naturally without any assistance,
and the villagers as a rule respect the trees, but do not hesitate
to remove the bark, which is slightly pungent, and may be eaten with lime. A
small patch of sandal, self sown from a planted tree, was discovered about 3
years ago on the hills overhanging the township of Shencottah, but the tree is
certainly not indigenous there. Outside Travancore, the sandal tree is found in
Mysore, and the drier parts of the Peninsula, in the island of Timor, and in East
Java. It is cultivated elsewhere in India, but there the wood has very little
scent.

984. The area of distribution in Travancore being so small, the number of
trees in the country is also very limited. No trees are felled
Total number of sandal trees in Travancore small. by the Government, but the Pûnnyátta chief, who owns the
Anjináda valley, cuts down, and makes use of, some 20 or 30
logs yearly, although, properly speaking, he has no right to do this without permission, any more than he has a right to utilize the other 3 royalties, teak, blackwood,
and ebony.

985. The sandal in its natural state is a tree of slow growth. Gamble gives
5—10 rings per inch of radius, so that a tree of 10 inches in
Rate of growth. diameter of which only about 7 inches of heartwood would have
any scent, must take some 40 years to attain this size, and probably the time required would be much longer. Maturity is reached in from 60—100 years, by which
time the trees should have a diameter of one foot of heartwood, but except in very
favourable circumstances, sandal stops growing after it has attained a diameter of 8
inches.

986. Sandalwood is dark yellow and heavy. The weight varies between 55
and 62 lbs. a cubic foot, and the value of P is about 870.
Weight and strength. The outer sapwood is much lighter in colour than the heartwood, and has little or no scent.

---

* The ebony of commerce is obtained from several different trees: D. ebenum, D. exsculpta, D. melanoxylon, and D. ebenaster from India and Ceylon, and D. reticulata from Mauritius.

987. Sandalwood is valued entirely * on account of its scent. It has been exported from India to China from time immemorial, and the § almug wood brought from the land of Ophir to Palestine in Hiram's ships for king Solomon is usually supposed to have been sandalwood. Large quantities of oil are distilled from the wood (the roots being the most useful for this purpose), and the greater part of it is exported from Bombay. Sandalwood is also used for inlaying, for carved work, boxes, &c., and it is used in the religious ceremonies of the Brahmins, at their funerals, and is burnt as incense.

Uses of sandalwood.

988. The value of sandalwood is, on account of the great demand for it, very high being about ½ rupee per lb. The whole of the wood used in Travancore is imported, for the small quantity felled in the Anjinàda valley is, as I have shown, taken by the Pùnnyàtta chief for his own consumption.

Its money value.

989. Though the sandal is a slow grower, it shoots up comparatively fast at first in plantations, and where everything is done to assist it. It is frequently planted together with the casuarina, and the combination is said to be a good one. Extensive plantations of it exist in Mysore. In Travancore an attempt at a plantation was made some years ago at Quilon, where 50 acres were opened, but the site selected contained very poor soil and was much exposed, while the elevation was no higher than sea level, so that the experiment was doomed to failure from the first. If sandal plantations are to be a success in Travancore they must be opened in some spot where there are three conditions present (1) good soil, (2) elevation not less than 1,000 ft., (3) light rainfall. The only places where all these conditions exist are some parts of the Pecrmerd plateau, the Anjinàda valley, or the hills overhanging Shencottah. As regards the yield of plantations, Beddome expected to secure at least 150 maunds of heartwood per acre when the trees arrived at maturity, which he hoped would be in 25 years, but the stems would have obtained but a very small diameter in that period.

The sandal well adapted for plantations if suitable land is selected.

990. (5) *Artocarpus hirsuta.* (Lamk.) Eng. ânjili. Tam. and Mal. ayani : ânjili. Next to the 4 royal timbers ânjili is probably the most useful wood in the country. This tree presents a very handsome appearance, with its grey smooth bark, and glossy dark green foliage. It attains a height of 100 to 150 ft., and a girth of 15 ft. The flowers are monœcious i. e. they consist of distinct male and female borne on the same tree, and they open from December to February. The fruit is a round prickly ball, containing numerous seeds immersed in an edible pulp, and about 4 inches in diameter : it ripens from May to July. The ânjili is never bare of leaves at any time.

Anjili. Its appearance.

991. Anjili is found wild only in the moist forest from sea level up to 3,000 ft., about which limit it is seen to attain its greatest dimensions, and to thrive best. It is not found in the deciduous forests except on the banks of rivers. It is also very much cultivated through the low country, where its fruit is eaten by the poor. The soil best suited to it is a yellow loam. The rainfall must exceed 60 inches.

Most suitable elevation, soil and class of forest

992. This tree is indigenous on the western ghauts from Canara southward, but is not found out of India. In former days it was very abundant throughout the State, but there has always been so great a demand for it for boats, that it is not now easy to procure even in the most remote forests. From one end to the other of Travancore I cannot call to mind a single place where I have found many ânjili trees together, and indeed, so high is the value placed upon this wood, that any timber merchant who knew of the existence of 100 trees of this species within reasonable distance of the low country would make quite a fortune. Merchants sometimes spend 1 and 2 months dragging down a single ânjili boat from the interior, so valuable is it.

Not found out of India. Its increasing scarcity. Distribution here and elsewhere.

---

* Besides our tree, other species of the same genus *S. freycinitianum* and *S. paniculatum*, which are found in the Sandwich islands, yield sandalwood.

§ l. Kings. x. 2.—Others say it was red sandars wood. *Pterocarpus santalinus.*

Anjili is to be found in the low country scattered through private gardens, where it has been planted for its fruit, but these trees never attain the dimensions reached by the Anjili of the hills, and they are therefore of little value for boats, though well adapted for house building. In the low country Anjili thrives best in the red loams of the neighbourhood of Mundakkayam on the Manimala river.

993. It is impossible to estimate the number of trees of this species, but large
Quantity felled annually.
trees are certainly very scarce. Our books show that permits are taken * out for some 2,000 candies a year. This number would certainly be quadrupled if the trees were available.

994. The Anjili is a tree of rapid growth. There are no statistics available to
Rate of growth.
show how fast it does grow, nor can any information be obtained by counting the rings, for, owing to the structure of the wood, the rings are very indistinct. Trees of immense size are sometimes met with on the hills, with clean stems for a height of 90 ft., and a girth of 15 ft. and more. Such trees are invaluable, but unfortunately they are often found in places from whence they cannot be removed. The Anjili probably reaches maturity in 25 to 40 years attaining a diameter of 2 ft.

995. Anjili wood is bright yellow turning to brown with age ; it is very straight
Its valuable properties.
grained, and free from knots, it takes an excellent polish, but owing to the straightness of the grain, it is not so ornamental as jack and other knotty woods. It does not warp or alter its shape with changes of weather : white ants eat the sapwood but do not touch the heart : it is light, tough, strong and quickly seasoned, and if kept under cover lasts for centuries. The value of P is 750, and the weight of a cubic foot of unseasoned wood is 48 lbs., and of seasoned from 34-42 lbs.

996. As already stated, the chief use to which Anjili is put is boat building.
Its uses.
Dug outs of 30 to 40 inches in diameter sell for 400-600 rupees each. If well rubbed with fish oil, Anjili boats will last for 30 years or more. Large boats are built of 2 inch planks of Anjili which are sewn together with cocoanut fibre rope. It is also much used in ship and house building, where its smooth surface gives it a very clean appearance.

997. The price of Anjili is about 12 rupees a candy in log. It is usually sold in
Its present value.
planks 3 inches thick, § which sell at the rate of 300 rupees per codge of 20, equivalent to 20 Rs. a candy. Almost all the Anjili felled in Travancore is used in the State, little of it being exported, indeed we appreciate it far too highly to lose much of it, and in other Districts like Tinnevelly where there are few navigable rivers there is no demand for it.

998. On account of its rapid growth Anjili is well suited for plantations.
Well suited for plantations.
Planted with kambagam (*Hopea parviflora*) it would out-grow and dominate it, and if mixed with teak it would outpace it while teak was wintering in the dry weather. It might profitably be planted in belts running through the teak plantations, which the Anjili would shelter, or mixed with the red cedar (Cedrela toona). We have no data for calculating the quantity of timber which could be grown on a given area in a given time, but it would undoubtedly be very large, and looking at its high value the profits of such plantations would be considerable.

999. (6) *Hopea parviflora.* (Bedd.) Tam. kongu. pongu. nam pongu. Mal.
Kambagam or kongu.
Its appearance.
thambagam. kambagam. A very lofty forest tree more than 100 ft. high, and up to 15 ft. in girth, with a rusty brown bark, marked with longitudinal cracks about 1 inch apart. The foliage is dark green and very glossy, the leaves being small and drooping. A young kambagam tree growing away from other trees is exceedingly ornamental,

---

* In 1859 Mr. Crawford, Commercial Agent of Alleppey, estimated the exports at 4,000 candies which was "only a moiety of what is annually felled." Cleghorn's Gardens and Forests.

§ Writing in 1859 Mr. Crawford says the value of Anjili in log was 8-10 rupees per candy, and a codge fetched 120-170 rupees. Comparing these figures with the above it is seen that the price of planks has doubled, but of logs has only increased 50 per cent in 32 years. The inference is that large trees, from which planks are cut, are very scarce, but small trees are not so. Cleghorn's Gardens and Forests.

owing to its regular pyramidal shape and dark foliage. The flowers are very small and cream coloured : they are borne in great masses at the ends of the branches in February and March. The fruit is the size of a small pea, black and shining with two straw coloured wings about 1½ inches long attached to it, and it ripens in May and June.

1000. The kambagam is found only along the sides of rivers in the open country, and in the dense moist forests between sea level and 3,000 ft. It never ascends the hills above that limit. Nor is it a tree of the deciduous forests. It above all things likes plenty of moisture, and may often be seen growing in swamps, and places where trees like teak, which require thorough drainage, would die instantly. This tree is found most abundantly at an elevation of from 1,000 to 2,000 ft., where it frequently forms a considerable percentage of the trees of the forest. Many people consider that the presence of kambagam is a sign of poor soil, and it certainly thrives where many trees would not grow, but I have also met with it on good soil. The rainfall must be not less than 100 inches.

Elevation, soil, and class of forest best suited to it

1001. Kambagam is found at suitable elevations all along the hill slopes in South Travancore, but it has been very much cut on all the coffee estates for building purposes, and the quantity remaining is not large. Farther north its place is taken by an allied species (*H. racophlea*) which possesses timber of equal value, but does not reach the same size, anything more than 15 inches in diameter being uncommon. In the Kulatthupura and Chenthrōni valleys and outside them, and in some places near Ariyankāvu, kambagam is still to be had in fair abundance, but large fellings for the Puliyara depôt during the last 5 years have depleted these forests very much. On the Ambanāda river trees are numerous. Farther north, kambagam has been much felled for boats in the absence of ānjili, and anywhere near a river there is scarcely a sound tree left, but doubtless there is still a good quantity of this timber scattered through the moist forests. Not long ago the forests near Mundakkayam contained much of this timber, but the D. P. W. have purchased large supplies from there. Outside Travancore, kambagam is found in Tinnevelly, Malabar and South Canara, but not out of India.

Its distribution through Travancore, and elsewhere.

1002. Permits are given for about 500 boats, each year, and at the various depôts &c. something like 8,000 candies are annually disposed of.

Quantity annually felled.

1003. The kambagam is a tree of slow growth. Unfortunately, no data are available to show how many rings go to the inch of radius. It attains a diameter of 3 ft., a tree of this age being probably some 200 years old.

Rate of growth slow.

1004. Kambagam is a heavy, close-grained wood of a yellow colour turning darker with age. It has a straight grain, but does not split easily. White ants do not eat it. Its weight is 62—63 lbs. a cubic foot so that it does not float. The value of P has not been estimated, but it is probably not far off 1,000.

Appearance and properties of the wood.

1005. This wood is used for building of all kinds, and it would answer excellently for sleepers. For beams it is especially useful, but when sawn into planks it is liable to crack. It is used, as already stated, for dug-out boats, which last 30 or 40 years, but, owing to the weight of the wood, boats made of it sink if upset. To use it for such purposes is a great waste of fine wood.

Its uses.

1006. The price of this timber is now about 12 rupees a candy in log, or 12 annas per cubic foot. A great deal of kambagam is exported to Tinnevelly and the Eastern Coast, and an increasingly large quantity is used in Travancore.

Present value.

1007. This tree is less suitable for plantations than many others on account of its slow growth, but as it is a free seeder, some attention to conservancy and thinning will cause large numbers of small plants to come on and take the place of the parent trees. It

Not suited for plantations.

36

might be planted on a small scale on the river banks in the teak or other plantations, and be put out along the roads as an avenue tree, but regular plantations of it would hardly pay.

1008. (7) *Pterocarpus marsupium.* (Roxb.) Tam. vĕngei; Mal. vĕnga.

Vĕnga. Its appearance. A large tree with a rough black bark, not specially noticeable, and sparse glossy foliage: not by any means an ornamental tree. It grows to a height of 80 or 90 ft., and attains a girth of 10 ft. Its stem is seldom straight or clean, being generally branched at 12 or 20 ft. from the ground, and it is therefore not easy to get long scantlings of the timber. The flowers are bright orange, pealike, ½ inch across, appearing profusely in July, and very sweet scented. The fruit, a flat circular pod, ripens about the end of the year.

1009. Vĕnga likes an elevation rather higher than that of teak, but it is found in the plains as well as on the hills, ascending the latter to the *Most suitable elevation, soil and class of forest.* height of 3,500 ft. It is found in the deciduous forest in company with blackwood, thĕmbāvu, and other trees. It prefers a stiff soil, and, curiously enough, is found in abundance where teak is scarce or absent, and is scarce where teak is common.

1010. This tree is distributed through all our grass forests below the elevation given above. It is very common and attains its largest size *Distribution here and elsewhere.* in the valleys of the Ney, the Karamana, the Vāmanapuram, and the Itthikkara rivers. Elsewhere it is not uncommon, but it does not reach so large a percentage of the trees. Outside our limits, vĕnga is found in South India, Mysore, Bengal, Bombay and Ceylon.

1011. In spite of the large number of trees felled every year vĕnga is still a very common tree, and if proper attention were paid to conser-*The supply of vĕnga abundant.* vancy, and the suppression of grass fires, this tree would become exceedingly abundant, as its light seeds are carried to great distances by the wind, and the small plants are very hardy, and, though annually burnt to the ground, shoot up again with the first showers. In the Nedumangāda Taluq thickets of young trees are seen in many places along the road sides where the destroying fire has been unable to cross the road, and the small trees have thus been able to grow.

1012. No one seems to have examined the wood with a view of ascertaining the rate of growth of the vĕnga tree. As far as I can ascer-*Rate of growth &c.* tain, it is a moderately fast grower. It does not attain any very great size, 2 ft. in diameter being about the largest I have seen. It probably attains maturity in about 60 to 80 years and grows to double that age.

1013. Vĕnga wood is in colour a dirty yellow, darkening with exposure. It is durable, seasons well, and takes a fine polish, but should not be *Properties and uses of the wood.* exposed to sun and rain, which cause it to crack, and lower its value very much. It is apt to warp if sawn green, does not work easily, and stains chunam yellow. It is much used for building of all descriptions, including cart and boat building, also for sleepers. Brandis says it makes beautiful furniture. The weight of a cubic foot is 65 to 70 lbs. unseasoned, and about 56 when thoroughly dry. The value of P averages 800. From wounds in the bark an astringent red gum-resin exudes which is exported under the name of kino or dragon's blood for medicinal purposes.

1014. The price of vĕnga in log is about 8—10 rupees a candy. Some 4,000 candies are brought to the depôts or sold on permit every year. *Price and destination.* The greater part of the vĕnga felled in Travancore is exported to Tinnevelly and Madura, where it is considered, next to teak and blackwood, the most valuable tree. In Travancore it is but little appreciated.

1015. Vĕnga is not suited for plantations. It is only a moderately fast grower: it prefers a stiff soil: it has a long tap root: and it is not *Not suited for plant-ations.* gregarious. All that is required to encourage it is to keep down the grass fires.

1016. *(8)—Terminalia tomentosa* (Bedd) Tam. karumarutha. Mal. thembávu.

Thembáva. Its appearance. A very lofty deciduous tree often 70 or 80 feet to the first branch, 120 feet high, and up to 12 feet girth, often buttressed. The trunk is very straight, and the black bark is deeply cracked in longitudinal lines about 2 inches apart. Foliage thin, and not ornamental Flowers small, greenish, ¼ inch in diameter, numerous, on drooping stalks in January to April. Fruit one seeded, about 1 inch long, with 5 wings or ribs, ripens the following dry weather.

1017. The thembávu is founded only in the deciduous forests at the foot of

Elevation and kind of forest best suited. the hills. It does not ascend them above 1,000 feet. It does not like the stiff clays where vênga thrives, but it is found in the greatest abundance in the drier subalpine tracts of South Travancore. It is not found in the dense moist forest.

1018. The area over which the thembávu is found in Travancore is limited,

Distribution through Travancore and elsewhere. because it is confined much more to the foot of the hills than many other trees. Common all through South Travancore as far as the Kôtha, it then becomes scarce for a time, but is found again in some places and not in others, in quite an unaccountable way. In the poor forests about Kôttûr it is a stunted tree not 50 ft. high. North of the Acchankôvil river it is less abundant, but it is seen again in parts of North Travancore. Outside our limits it is found throughout the Madras Presidency, Bengal, Bombay, and Ceylon as well as in Burmah.

1019. The quantity of thembávu sold annually must be about 6,000 candies.

The supply not very great. This is fully as much as can be spared in a year without unduly reducing the supply. The rate of growth of this tree is proba-bly slow, but the annual rings are so indistinct as to prevent the forming of any conclusions. Maturity is probably reached in 80–100 years, but the larger trees, 4 ft. in diameter, must be over 200 years old.

1020. The wood is hard, dark brown, mottled with darker streaks, often near-

Properties of the wood ly black, and somewhat resembles walnut. Its weight is 60 lbs. a cubic foot when really dry, and 70–80 lbs. when unseasoned, and, as wood seldom is really dry, thembávu does not float. The value of P is 860, but the peculiarity of thembávu is that it is of very unequal strength, being some-times very weak, and at other times sustaining an unusually heavy strain. The wood does not season readily, and is apt to warp and crack. Its grain is course and curly, and it is not easily worked. It is used for building, sleepers, and furniture, and out of Travancore for ship and boat buildings. It makes good fuel, and potash is found in the wood in large quantities. Its leaves are much lopped for manuring the fields.

1021. The value of thembávu is about 6 to 8 rupees a candy in log The

Price and destination. greater part of the timber felled is used in Travancore, for, while vênga is not so much esteemed with us, the people of Tinnevelly much prefer it to this wood : some thembávu however is exported.

Not suited for planta-tions. 1022. Thembávu is not suitable for plantations, but it de-serves attention and protection from grass fires.

1023. (9) *Dysoxylum malabaricum* (Bedd.) Eng. white cedar. Tam. agil.

White cedar Its ap-pearance. Mal. vella agil. The white cedar of the West Coast is a very lofty tree growing to a height of more than 120 ft., and attain-ing a girth of 18 ft. The bark is light coloured and covered with curious warty excrescences which at once distinguish the tree. Its foliage is sparse and not ornamental. The leaves are pinnate, and the leaflets about 5 inches long. The flowers are borne in long drooping racemes and appear in abundance in February—March. The fruit is shaped like a Turkey fig, and consists of a rough yellow capsule containing 4 large seeds which ripen about the commencement of the rains.

1024. This tree is only found in the moist forest and never ascends the hills

Most suitable elevation and class of forests. above 2,000 ft. The white cedar of Peermerd is quite another tree (*Ileynea trijuga*.) It seems to attain its greatest size at an elevation of about 1,400 ft., and it is a common saying among the people that cardamoms thrive best where the ånjili and white cedar are found. I have not noticed that it has a predilection for any particular kind of soil, but it likes plenty of moisture.

1025. 1he white cedar is only found on the West Coast of India from Canara southwards to Cape Comorin. With us, it is scattered through all our moist forests at suitable elevations, from the south to the north of Travancore, and, before the demand increased, it must have been at one time very common in the northern forests. So many trees are felled every year at the present time that it is becoming scarce, except in the forests that are very difficult of access. I have found it in abundance on the Kuthirakutthi and Eramugam forests near Malayåttûr, but in a few years it will have been exterminated from those parts. It is probably most common iu the interior forests on the Râni river, where it is difficult to get at. In South Travancore it is not unfrequently met with, but is seldom felled because of the remoteness of the place of its growth from rivers by which it could be floated down.

Distribution here and elsewhere

1026. The number of trees still to be had in the country is not large: if the timber was easily obtainable three or four times the amount now sold would find a ready sale in Cochin, but the cost of working it down is so high that the greater part of the white cedar used in Cochin is imported from farther up the coast. Mr. John Grieve of that place has kindly supplied me with the following statistics of the annual imports of white cedar into Cochin. From Calicut 13,000 candies.

Number of trees in the country not large Great demand for white cedar at Cochin.

„	Trichur 2,000	do.
„	Malayåttûr 2,000	do.
„	Alleppey 4,000	do.

21,000

1027. The cedar is a tree of moderately fast growth. I have counted the annual rings, and find they run from 6 to 20 to the inch of radius or say 12, on an average, so that these large trees of 5 feet diameter must be upwards of 360 years of age. Five hundred years is probably the limit of its life. In its younger stages the tree grows more rapidly, but its reproduction in the forests where the large trees are now found is very bad. No attempt has ever been made in those forests to let in light, and the young plants which have from time to time sprung up have long since been killed out by the heavy shade of the parent trees, so that after the present trees are felled it will take a long time for another generation to attain its full size.

Rate of growth and reproduction.

1028. The wood is pale yellow with a smooth, somewhat silky vein, but not much marking. It is sweet-scented, and easily worked, but, unfortunately, it has a great tendency to contract and expand with changes of weather, so that, although it is sometimes used for furniture and house fittings, it is not well suited for these purposes. Its weight is about 35 lbs per cubic foot; the value of P has not been determined, but the wood is not strong.

Appearance and properties of the wood.

1029. The chief use to which white cedar is put is the construction of casks for the conveyance of cocoanut oil to England. From 1 to 1½ candies of this wood are required for a ton of pipes, equivalent to about a ton of oil.

Chief use.

1030. The price of white cedar is now about 8 to 9 rupees a candy for logs, but at the time of the Cochin fire in January 1889, when so many casks and so much oil were destroyed, the price rose to 10 rupees a candy and over. Travancore has a serious competitor in Calicut, as I have shown: from thence to Cochin the freight is 1 to 1¼ Rs. a candy, and the supply of wood in the forests of Malabar is said to be still considerable.

Price.

1031. Cochin takes practically the whole of the white cedar felled in Travancore.

Destination.

1032. This tree is of sufficiently rapid growth to warrant its being tried in mixed plantations, but the timber is required for a special purpose, and as the supply in the Cochin market is likely to run short before many years have elapsed, it is probable that some

Not desirable to plant it though it would grow well.

-other timber may be employed for the construction of oil casks, so that long .before the trees we plant can grow sufficiently large to fell, the timber may have ceased to be in demand, or the cocoanut oil trade of Cochin may have passed away. Steps should therefore be taken to conserve this tree rather than to propagate it.

1033. (10) *Cedrela toona* (Roxb.) **Eng.** red cedar. Tam. santhana vêmbu : mathagiri vêmbu : Mal. thêvathâram : vedi vêmbu.

Red cedar.
Its appearance.

The red cedar is a handsome tree, attaining an immense height and size, 120 ft. × 15 ft. girth. It has a rough brown bark, and a pretty foliage of red, or pale green, pinnate leaves. The old leaves fall about the end of the year, and the young leaves and flowers quickly appear, so that the tree is not long destitute of foliage. The flowers are small, sweet scented, and borne in abundance. The fruit is a hard capsule about 1 inch long, containing numerous winged seeds, ripening during the monsoon.

1034. This tree thrives at all elevations from sea level up to 3,000 ft. It requires a rich soil, and a rainfall of not less than 100 inches. It is found in the moist forests only, but seems to prefer those which are more or less open, where there is plenty of light. It does not grow in the deciduous forests.

Most suitable elevation and class of forest

1035. The tûn has a wide distribution : it is indigenous in Java, Australia, Burmah, Bengal, Oudh, and in South India. In Travancore it is met with in suitable localities throughout the country, .being less common in .the south. At present, it is most abundant in the valley of .the Periyâr on Peermerd, and on the Peermerd estates. At one time it was probably abundant through the country, as young trees are frequently met with, but there has always been a demand for the timber, and at present it is not easily procurable. The number of trees sold each year is very limited.

Distribution here and elsewhere.

1036. The red cedar is a tree of fast growth. Experiments show that 3 to 9 rings go to the inch of radius, so that, under favourable circumstances, it increases in diameter at .the rate of ½ inch in a year, or a tree 60 years old would be 40 inches in diameter.

Rate of growth.

1037. The wood is coarse, red, and sweetscented, and closely resembles the wood used in England for cigar boxes. It has an open grain, with very distinct annual rings and large pores. The weight of a cubic ft. of seasoned wood is 33 lbs. and the value of P varies between 420 and 560. It is durable and is not eaten by .white ants.

Properties of the wood.

1038. Red cedar is highly valued for furniture, house building, and carving. It is used for tea-boxes, and shingles, and is hollowed out for boats. It is employed in the manufacture of cigar boxes, and is exported to England from other countries where it is found. Its flowers yield a red or yellow dye.

Its uses.

1039. Its present value is nominal, but is about 10 Rs. a candy though the tariff rate published in the almanac is 12 Rs.

Value.

1040. The tun is well suited for plantations, on account of its rapid growth, and the facility with which it can be raised from seed. It would grow well with ânjili, and there is no fear that the demand for it would cease.

Well suited for plantations.

1041. (11)—*Lagerstrœmia lanceolata*. (Wall.) **Eng.** ven-teak. **Tam.** venthekku : vemvila. **Mal.** senjal. Hillmen. vengalam : venda : vellei râvu : vellil âgu : vehlla.

Ven-teak.
Its appearance.

A large forest tree with a smooth, very pale bark scaling off in thin flakes not much thicker than paper. It has a straight cylindrical stem with very few branches, and attains a height of 100 feet with a girth of 10 feet. The leaves are bluish-white beneath, and fall off in the dry weather. The flowers are small and appear in April to May. The fruit is a capsule rather larger than a pea, ripening in July to August, and containing numerous small seeds.

1042. The ven-teak is found only in the moist forests, like red cu.

Most suitable elevation and class of forest. ring those which are more open. It attains its large the forests of North Travancore on the Kōttasheri an rivers, at an elevation of 200 or 300 feet above sea level, but it ascends tl over 3,000 feet.

1043. This tree is found only on the West Coast from Bombay to Cape (

Distribution here and elsewhere. and nowhere out of India. In Travancore it is met suitable elevations from the extreme south to the limits of the State, near which latter locality it is most abundant.

1044. There is a considerable demand for ven-teak, mostly in planks

Destination and quantity sold. are exported to Arabia for ship and house building. are also exported for the same purpose. Permits are g: about 6,000 candies a year.

1045. Its rate of growth is not fast. Gamble gives 10 rings to the :

Rate of growth. radius, so that in 100 years a tree would only be 20 inc. diameter. It attains a size of more than 3 ft. so that if measurements are correct, ven-teak must often attain an age of more tha centuries before it begins to decay.

1046. Ven-teak wood at a distance resembles teak, being of much the

Properties of the wood. colour, but the annual rings which are so well marked in latter, are much less distinct in the former, nor has it the culiar odour of teak. It is eaten by white ants. Its weight is 57 lbs. and the v of P probably about 600, but it has not been exactly ascertained. If exposed to weather it soon rots. It is used for furniture, house building, and packing ca: and for shingles, and stakes on the estates, for its straight and even grain make easy to split.

1047. Its present value is about 6 rupees a candy. As already stated, t

Value and destination. greater part of the timber is exported, only a small proporti of it being consumed in the country.

1048. On account of its slow growth it is not suited for plantations, but it

Not suited for plantations. still a common tree through the country, so that if steps ai taken to conserve it, there should be no difficulty in accurin its reproduction.

1049. (12) *Artocarpus integrifolia.* (Linn.) Eng. jack. Tam. and Mal. pilāvu!

Jack. Its appearance. A medium sized tree, in a wild state never attaining a height of more than 80—90 ft., or a diameter of more than 2 ft, but, as a cultivated tree, it often reaches a diameter of more than 4 ft. As a wild tree it is very far from ornamental, for, growing as it does in heavy forest, where the trees surrounding it force it to shoot up, it throws out few side branches, and their place is taken by knots and excrescences from which the fruit grows. Its foliage is dense and throws a heavy shade. The male and female flowers appear in January to March. The compound fruit sometimes nearly 2 ft. long and 10 inches in diameter, which is much eaten by the people, takes 3 months to ripen. The English name is derived from chakka the native name for the fruit.

1050. The jack is only found wild in the moist forests at elevations of from

Most suitable climate, soil, and elevation. 1,000 to 4,000 ft., and it requires a rainfall of not less than 50 inches. It seems to like a rich red soil. When the seeds of trees cultivated in the low country are taken to the hills and planted there, the resulting plants often do very badly, if the elevation exceeds 2,000 ft., although the wild tree may be seen growing at a yet higher altitude, showing that the tree has become acclimatised to a greater heat, and will not thrive without it.

1051. This tree is indigenous in the forests of the Western Ghauts but does

Distribution here and elsewhere. not occur wild elsewhere. It is not abundant in Travancore, being met with chiefly in the forests to the south of the Agasthiar peak. It is very much planted all over Burmah, Ceylon, India and the Archipelago.

1052. Practically, no wild trees are felled for timber in this State, but the cultivated jack is cut down in large numbers, and is much used locally. It is not easy to estimate the total number of candies used per year, as much of the timber is grown on private properties, and is used on them, so that it does not pass through our hands.

Quantity of timber felled.

1053. Jack is a tree of rapid growth, and will run up to a height of 6 ft. in a year, but after this its progress is slower. The structure of the wood is homogeneous, and the annual rings too indistinct to be counted, as is the case with the anjili. The growth may probably be at the rate of an inch in diameter in two years, but this is only on good soil, and the tree probably grows very slowly after it has attained a diameter of 2 ft., so that the very large trees mentioned above must be a couple of centuries old.

Rate of growth and maximum age.

1054. Jackwood is yellow when young but darkens with age, and often resembles mahogany. It has a fine grain which takes an excellent polish, and the wood of old trees is usually very well marked. Its weight is about 42 lbs. a cubic foot, and the value of P is 780. A yellow dye is obtained from the wood, and a milky juice, which may be worked into bird-lime, flows from the stem.

Appearance and properties of the wood.

1055. The jack tree is planted as an avenue tree for shade, and in gardens for its fruit, or as a support to the pepper vine, although its shade is rather too dense for that purpose, unless other trees of lighter foliage are interspersed with it. The wood is used largely for furniture and house building, though it is not employed for beams, or where a great strain would be imposed on it. From other countries it is exported to Europe for cabinet work, turning, and brush backs. The fruit is eaten by the people, and at certain seasons forms a large portion of their diet. Both the pulp that encloses the seeds and the seeds themselves are thus employed.

Its uses.

1056. The price of jackwood is about 12 rupees a candy in log. All the wood felled is used in the country, and it is not exported except to a very limited extent.

Price and destination.

1057. Although from its rapid growth it might be thought that the jack tree would be suited for plantations, it is so largely planted in gardens, and along roads, that it would not be worth while to form regular plantations of it, because there will always be a good supply of the timber available.

Not suitable for plantations.

1058. (13) *Xylia dola'riformis.* (Benth.) Tam. irūl: Mal. irumulla. A large tree growing to a height of 80 ft, and attaining a girth of 6 or 7 ft. with mottled brown bark, and glossy foliage which it loses in the dry weather. It is not an ornamental tree, especially as it remains for several months destitute of leaves. The flowers are minute, and are borne in globose heads about ¾ inch in diameter. They appear when the tree is bare, and the fruit, which consists of a hard flat pod containing 6 to 10 seeds, ripens about August.

Irūl. Its appearance.

1059. The irūl is found in our forests at elevations from sea level to 1,500 ft. It is not met with in the dense moist forest, but prefers either the deciduous grass forest, where it is often seen in company with teak, or the open moist forest where from some reason or other the vegetation is not rich, and where it has room to spread. It is often very gregarious, forming almost pure forests to the exclusion of other trees. The rainfall must be not less than 100 inches.

Most suitable elevation and class of forest &c.

1060. This tree is not found in South Travancore south of a line drawn from Quilon to Stencottah. It is very abundant in the grass forests near Kōnniyūr, from whence it is not much felled, and it is also a very common tree in the open moist forests round Malayāttūr, and elsewhere in North Travancore. Outside our limits, it is found in Arakan, Burmah, and the Philippine islands.

Its distribution.

1061. The number of candies of this wood felled each year averages about
1,000, and the total supply of the country is maintained,
Quantity felled. though in some parts the demand for this timber is rendering
it scarce.

1062. The rate of growth has not been ascertained. It is probably fairly fast,
especially at first. The wood is dark, red, hard, heavy, close
Properties and uses. grained, but not easily worked. It contains a resin which
makes it very durable. Its weight is 70 lbs. a cubic foot unseasoned, and 5? when
thoroughly dry. The value of P is 950. It is used for boat building, sleepers,
posts, carts, house building, telegraph posts, and agricultural implements. It lasts
a long time under water, but is liable to crack and warp if exposed. But for its
great weight, which prevents its being floated without the aid of bamboos, it would
be much more largely used than it is at present.

1063. The price of irûl is about 7 to 8 Rs. a candy. Most
Price and destination. of the timber felled is used in the country.

1064. On account of the large size and abundance of its seeds, this tree could
be easily propagated, and it would probably do well in mixed
Suitable for plantations plantations with teak. These two trees naturally grow to-
and should be carefully gether, and both winter at the same time. Protection) this
preserved. valuable tree would ensure its reproduction. All out
Mundakkayam it is very common, but when their annual clearings are made the
Hill-cultivators millions of young seedlings are cut down and burnt.

1065. (14) *Vitex altis·ima.* (Linn. f.) Tam. shambagapâla: mayila. Mc·,
Mayila. Its appearance. mayilella. Hill-men. meilâdan. This tree grows to a height
of 80 ft., and attains a girth of 12 ft. It is an ornamental tree
with pale trifoliate leaves, small lavender-coloured flowers, which appear in April
and May, and a black drupe the size of a pea, which contains 1—4 minute seeds,
ripening in June—July.

1066. The mayila is found, like the irûl, either in deciduous grass forest, or,
more commonly, in the more open moist forest. Both of them
Most suitable elevation require a certain amount of light and air, and cannot exist in
and class of forest the thick, dark forests of evergreen trees. It is to be seen at
all elevations between sea level and 3,000 feet. It does not seem to be particular as
regards climate, for it is found both in the drier tracts where the rainfall is small, and
in the moister regions of North Travancore. It is more common perhaps in the
former.

1067. This tree is widely distributed throughout Travancore at suitable elevations from the extreme south to the north: it is abundant in
Distribution here and the forests near Shencottah. Outside Travancore, it is found
elsewhere. on the West Coast, and, doubtfully, in Bengal. A variety of the
same species is indigenous in Ceylon.

1068. The aggregate of trees felled each year is not large, amounting to about
50, of 2 candies each. This small number is not in excess of
Number of trees annu- the annual increase in growth of the existing trees, so that we
ally felled. are not trenching on the capital of our forests expressed in
terms of this tree.

1069. Its rate of growth is only moderately fast. Experiments show that 8–9
rings go to the inch of radious, so that the tree increases in
Rate of growth. diameter 1 inch in 5 years, and the very large trees 4 ft. and
5 ft. in diameter, sometimes seen, must be 250 to 300 years old at least.

1070. The wood is hard, durable and flexible, with a coarse grain, is light
brown in colour, and does not split, or warp. It is highly
Properties and uses of esteemed in parts of India for buildings, carts &c. and it is
the wood. Value &c. liked in Tinnevelly, but in Travancore the people prefer other
trees. Its weight is 63 lbs. unseasoned, and 53 lbs. seasoned, and the value of P
is 700. Mayila is worth about 8 to 10 rupees a candy in log.

1071. This tree is not suited for plantations on account of its slow growth,

Not suited for planta-tions. but it deserves careful preservation and protection from fires and indiscriminate felling.

Manjakkadambu. Its appearance. 1072. (15)—*Adina cordifolia.* (Hook. f.) Tam. and Mal. manjakkadambu.

A large, deciduous tree, growing to a height of 100 feet, and reaching a girth of 18 feet. Leaves pale green and heart-shaped, falling in April to May. Flowers minute, yellow, sweet scented, borne in globose heads about an inch in diameter, in June and July. Fruit a small capsule. Seeds minute, ripening from December to March. Decidedly an ornamental tree.

1073. This tree is found in the low country at the foot of the hills, but does not ascend them above 1,000 feet. It is found only in the *Most suitable elevation and class of forest.* moist forests, preferring those which are rather open. It is particularly abundant, and reaches a very large size, in the forests on the Acchankövil river, near Könniyür, and in North Travancore. Outside our limits, it is found in North India, Burmah and Ceylon.

1074. About 1,000 candies are felled annually, which does not reduce the total quantity of this timber in this country, as the annual increase *Annual fellings.* amounts to more than this, for the tree is a common one.

1075. The rate of growth has not been ascertained as there are no distinct annual rings, but it is probably fairly fast. The wood is *Rate of growth, pro-perties and uses.* yellow, seasoning to nut-brown. It is close grained, smooth, light, and polishes well. It is said not to be eaten by white ants. It is in great request in other parts of India for furniture, boxes, turning &c. but it will not stand wet. In Travancore it is not much used, but bedsteads made of it are said to have a very soporific effect. Its weight is 50 lbs. unseasoned, and 42 lbs. seasoned, and the value of P is 625.

Value and destination. 1076. The value of manjakkadambu is about 8 or 9 Rs. a candy. About one third of what is felled is used in the country, the rest being exported.

1077. This tree is not suitable for plantations. The minuteness of its seeds would make it very difficult to propagate, and it is not of suffi-*Not suited for plantations.* ciently rapid growth or high value to make it worth planting.

1078. (16)—*Schleichera trijuga.* (Willd.) Eng. Ceylon oak. Tam. pûvan. Mal. *Puvan. Its appearance.* pûyam. A handsome, spreading tree with a sturdy much-branched trunk, attaining a height of 80 feet and a girth of 12 feet. The leaves fall in February, and the new foliage that comes out in March is tinged with pink. The flowers are small, yellow, bisexual, and appear with the young leaves. The fruit is dry, the size of a nut, often prickly, and contains 1 or 2 seeds, ripening in May—June.

1079. The pûvan is found scattered through the deciduous forests at elevations between sea level and 2,000 feet. It is also much plant-*Elevation and class of forest.* ed in the low country. It is not a gregarious tree, but its distribution through Travancore is very general. It is more common perhaps in the south. Elsewhere, it is found throughout India, Burmah and Ceylon.

1080. Its growth is probably slow, but this has not been ascertained. Old trees are probably not less than 3 centuries old. The wood is *Rate of growth, pro-perties and uses.* very strong and durable, reddish, hard, and heavy, and seasons, and polishes well. It is used for sugar and oil mills, for carts, and for building: a lamp oil is made from the nuts, and the lac insect frequents its branches. The weight of a cubic foot is 67 lbs, and the value of P is about 900.

1081. The price of pûvan is about 6 to 8 rupees a candy in log, but it is sel-*Price and destination.* dom used except in certain localities for any thing but oil mills, for which purpose there is a considerable demand. All the timber felled is used in the country.

1082. The pūvan is not suitable for plantations. Its growth is too slow, and
Not suited for plantations. its value too small to warrant expenditure in this direction. Still, it is a valuable tree, and should be preserved.

1083. (17)—*Lagerstræmia Flos-Reginæ.* (Retz.) Tam. pūmaruthu: Mal. mani
Manimaruthu. Its appearance. maruthu: nīr maruthu. A tree of very ornamental appearance on account of its handsome pink flowers. It attains a maximum height of 80 feet and a girth of 12 feet, but, more often, its size is very much less. The bark is pale and smooth. The flowers appear in immense masses in March and April, and cover the trees with a clothing of pink. The fruit is a rounded capsule ¾ inch long, containing numerous seeds which ripen in July to August.

1084. The manimaruthu is found in the dense moist forests, and does not
Elevation, class of forest, and distribution. occur in the grass land. It likes plenty of moisture, and may often been seen during the monsoon standing in water 4 & 5 feet deep when the floods are out, and it is none the worse for the excess of water. It ascends the hills to a height of 1,500 feet, but is most abundant at an elevation of 300 feet. It occurs in South Travancore, but does not attain the large size it reaches in the north. It cannot be called a very common tree, though it may be abundant locally, but it is scattered all through the State. Elsewhere, this tree is found from Assam to Malacca, in the Deccan Peninsula, and in Ceylon. It is much planted along avenues and in gardens, on account of the beauty of its flowers.

1085. The number of trees felled annually does not exceed 200 or 300 of 2
Annual felling and rate of growth. candies each, which is not in excess of the annual growth. The rate of growth is rapid showing 4–7 rings to the inch of radius, and the large trees mentioned above are probably from 1½ to 2 centuries old.

1086. The wood is pale red, tough, and very durable under water, but
Properties and uses. it decays under ground. It does not receive in Travancore the attention it deserves. It is used for building, ship building, boats, canoes and casks &c. Its weight is 48 lbs unseasoned, and 38 lbs seasoned, and the value of P is 750.

Value and destination. 1087. The value per candy is 4 to 5 rupees, and all the wood that is felled is used in the country.

Not suited for plantation. 1088. The manimaruthu is not suitable for plantations.

1089. (18)—*Mangifera indica.* (Linn.) Eng. mango. Tam. and Mal. māvu. A
Mango. Its appearance. lofty forest tree with a very straight trunk, the bark of which is curiously creased. It attains a height of 100 feet, and a girth of about 10 feet. The foliage is dark green and ornamental. The minute flowers appear in masses in February to March, and the fruit ripens from May to July.

1090. The mango is found wild in our moist forests at all elevations up to
Elevation, class of forest, and distribution. 2,000 ft. It is not found in the grass land except on the banks of rivers. In the low country and all over India it is very much planted for its fruit, which is much improved by grafting, the wild fruit being hardly edible. Outside our limits, the wild mango is found on the Western Coast, Sikhim, the Khasia hills, and Central India.

1091. A considerable quantity of mangowood is felled every year in Travan
Annual fellings and rate of growth. core, but it almost all comes off private lands, and it is not easy to estimate the amount. Permits are taken out for some 3,000 candies a year, and for 500 boats, and about as many more boats are stamped, after seigniorage has been realised. The rate of growth is not very rapid, the annual rings running from 6 to 10 to the inch of radius. Large trees are therefore upwards of a century and a half old.

1092. The wood is pale, with curious shining plates which, if they are large, improve its appearance much. It bears the action of salt water well, and trees are hollowed out for canoes, which with care will last 10 or 12 years. The timber is brought down chiefly in the form of planks, 12 virals broad and 3 thick, which are sawn up and used for packing cases, tea chests &c. Its weight is 55 lbs. unseasoned, and 42 lbs. seasoned, and the value of P is 600.

Appearance and properties.

1093. The present value is about 4 Rs. a candy. Most of the wood felled is exported to various places, only a small quantity being consumed in the country.

Value and destination.

1094. Mango trees are very much planted along avenues and in gardens, and there will always be a good supply of wood from these sources, and, as its value is so low, it is not worth while forming regular plantations of it, when there are so many trees of higher value available.

Not suited for plantations.

1095. (19) *Calophyllum tomentosum.* (Wight.) Tam. and Mal. punna. malam punna. A tree of very large size, its straight, unbranched stem reaching a height of 120 ft. or more, and a girth of 10 ft. Its bark is yellow, and deeply cracked longitudinally, giving the tree the appearance of having parallel lines running from base to summit. The leaves are dark green and very glossy, and the tree is very ornamental when young. The flowers are white and about ¼ inch in diameter, and appear in January and February. The fruit is a one-seeded nut, one inch long.

Punna. Its appearance.

1096. The punna tree is only found in the dense evergreen forests from 300 feet elevation up to 400 ft. It will not grow where the rainfall is less than 100 inches. It thrives on very poor soil, and attains a large size where no other tree would succeed. It is often found growing in thickets of the eetta reed. It is found all through Travancore, where the above conditions of elevation and climate exist. Elsewhere, it is only found on the West Coast, and in Ceylon.

Elevation, class of forest, and distribution.

1097. A very small number of trees is felled every year the total being not more than 50 to 100. There are tens of thousands of trees in the country, but the straight tall trees are eagerly sought after, and cannot be very easily procured. The rate of growth is rapid, and runs from 2 to 4 rings to the inch, so that the large trees are something under a century old.

Annual fellings and rate of growth.

1098. The wood is reddish, loose-grained, and long fibred. It splits easily, and may be used for shingles, but does not last. It is very elastic, but is not capable of sustaining a great strain. On the coffee and tea plantations it is used for reepers and packing cases, and for rough planking, furniture &c. Its chief use is for spars of vessels, its great length, straightness, lightness and elasticity making it most suitable for this purpose. A single spar sometimes realises 200 or 300 Rs. Its weight is about 38 lbs., and the value of P has not been ascertained, but it cannot be high.

Properties and uses.

1099. No fixed price per candy can be set on this timber, as its value entirely depends upon the local demand there may be for it. It is not suitable for plantations though it would, no doubt, grow rapidly.

Value. Not suited for plantations.

1100. (20) *Tetrameles nudiflora* (R. Br.) Tam. and Mal. chīni. A very lofty forest tree, only second to the cotton tree in size, attaining a height of 120 ft. and a girth of 15 ft. It has a smooth bark and cylindrical stem, often furnished with buttresses. The branches are very spreading, and the foliage is pale green and ornamental. It loses its leaves in the dry weather, and at that time the branches are loaded with the small bisexual flowers, not more than ¼ inch in diameter, hanging in strings from the ends of the branches, and from every joint in them. The seeds are very minute, and are enclosed in small capsules ripening in the monsoon.

Chīni. Its appearance.

1101. The chíni is found only in the moist forests, from sea level up to 2,000 feet. It seems to require a heavy rainfall, as it is not found in the drier parts of the hill forests. It is met with in Travancore from the extreme south up to the Cochin frontier in the north, but it is not at all common in the southernmost parts of the country, whereas in the forests of the Cottayam Division it abounds. Outside our limits, it is indigenous in Sikhim, the Western Ghauts, Burmah, Tennaserim, the Andamans, and Java.

Elevation, class of forest, and distribution.

Rate of growth.

1102. The rate of growth is very fast, but how fast has not been exactly determined. Large trees are probably nearly 200 years old.

1103. The wood of this tree is dirty white, and somewhat resembles mango, but the shining plates are absent. The wood too is much lighter, of finer grain, and takes a good polish. Its weight has not been ascertained but it probably does not exceed 25 lbs. The wood is not strong, nor durable, and white ants eat it. The chíni is used exclusively for dug-outs for which its lightness makes it very well suited. If rubbed with fish oil and used in salt water a boat will last from 8 to 10 years, but, ordinarily, it is very short lived.

Properties and uses of the wood.

1104. No fixed value can be placed on this wood; but boats sell for about ⅓ the price of ánjíli, or from 30 to 100 rupees, according to size. This low value makes the tree unsuitable for plantations, but the seeds are freely carried about by the wind, and trees spring up in numbers in all kinds of places, so that if they were carefully preserved, the chíni would become much more common than it is. The large honey bee (A. dorsata) is very fond of making its nest on this tree, and therefore it may not be commonly felled, and this has had the effect of saving the tree in years past. Permits to fell this tree for boats are only sparingly issued.

Value not suited for plantations.

1105. (21)—*Stereospermum xylocarpum* (Wight.) Tam. male uthi: sirakora: malealantha. Mal. páthiri: vedang konnan: edang konna. A tree of only moderate size, up to 80 feet high and 8 feet in girth. Foliage pale green, drooping, and deciduous. Flowers very large, white and fragrant, appearing in March to May, when the tree is leafless. Fruit a pod often 2 feet in length, covered with tubercles and containing numerous winged seeds which ripen towards the end of the monsoon.

Páthiri. Its appearance.

1106. This tree is found in the deciduous forests of South Travancore and in the more open moist forests of the north. It ascends the hills to a height of nearly 4,000 feet, but is there stunted and dwarfed by the cold. It is a small tree in the south of the State, but grows to a larger size in the forests of the Periyár. Outside our limits it is indigenous in the Deccan Peninsula and Khandeish, but has not been discovered elsewhere.

Elevation, class of forest, and distribution.

1107. The number of trees sold every year is not large, on account of their comparative rarity and small size. The rate of growth is moderately fast, 7 to 8 rings occurring to the inch of radius, so that a tree 2 feet in diameter would be about a century old.

Rate of growth.

1108. The wood is light brown, close grained, tough and elastic, and takes a good polish. When fresh cut it has a greenish tinge. Its weight is 47 lbs. It is used for cabinet work, and in South Travancore small trees are felled for bandy poles &c.

Properties and uses.

1109. Its price is about 10 rupees a candy. What little is felled is used in the country. It is not suitable for plantations, but it should be carefully preserved on every opportunity.

Value and destination.

1110. (22)—*Bombax Malabaricum*. (DC) Eng. cotton. Tam. and Mal. ilavu. The largest tree in the country, often seen from a distance of 2 or 3 miles towering high over the other trees in the forest. It sometimes § reaches a height of nearly 200 feet, and a girth of 20 feet. The stem is thorny when the tree is young, and it is always rough. Large

Cotton or ilavu. Its appearance.

buttresses spring from the stem. It is unbranched for the greater part of its height, the crown being formed of immense horizontal branches springing out at the summit of the tree. The foliage is deciduous. The flowers are large and showy, 3 inches in diameter, and generally red, but sometimes white. The fruit is a capsule 3 or 4 inches long, containing numerous black seeds surrounded by silky cotton.

1111. The cotton tree is found both in the deciduous forests, and in the moist forests from sea level up to 3,000 feet, and is evenly scattered all through the country, but is not gregarious. Elsewhere, it is found all through India, Burmah, Ceylon, Java and Sumatra.

Class of forest and distribution.

1112. Its rate of growth is very fast, the number of rings to the inch varying from 3 to 7. Even at this pace it must take the largest trees two centuries to attain their full size. The wood is very coarse, white in colour, with a rough grain, and decays very rapidly. It is used for boats and occasionally for packing cases, its straightness and large size rendering it suitable for the former purpose, but the boats do not last long even in salt water, and insects attack it readily. Its weight averages 24 lbs. when seasoned, and the value of P is 650. The large honey bee suspends its nest from its branches, and the cotton tree may not therefore be felled for fear of interfering with the supply of wax. The gum is collected in North India and used medicinally.

Rate of growth, properties, and uses.

1113. The value of this timber is nominal. The tree is not suitable for plantations, but it is frequently planted for the support of pepper or vanilla. It grows easily from cuttings.

Value &c.

1114. (23)—*Albizzia procera.* (Benth.) Tam. vāga : Mal. vāga. karinthagara. A moderate sized tree with a rough white bark blotched with black, and a spreading head. It attains a height of 60 or 70 feet and a girth of 8 feet. It loses its foliage in the dry weather. Its small white flowers, clustered in globose heads, appear in May and June, and its flat brown pods ripen in November to January.

Vāga. Its appearance.

1115. The vāga prefers light jungle, and springs up rapidly in abandoned clearings. It does not ascend the hills above 1,000 feet. It is less common in South Travancore, and its wood is not much used there, the people preferring an allied species (*A. o loratissima.* Benth.) In Central and North Travancore the tree is far more common and is frequently used. Outside our limits, it is found over the greater part of India where the climate is moist, in Burmah, and the Philippine Islands.

Elevation, class of forest and distribution.

1116. The growth of this tree is very rapid, 2 to 6 rings to the inch of radius having been counted. The wood is dark brown, and resembles bad specimens of blackwood. It is straight, even-grained, seasons well, and the heart wood is durable. It is used for furniture, boxes, rice-pounders, agricultural implements, bridges, and houseposts. It splits easily and is burnt for charcoal. Its weight averages 46 lbs., and the value of P is about 800.

Properties and uses.

1117. Its price is about 6 to 8 rupees a candy. Being of such rapid growth, it might be put out in plantations if its value were higher, but as it very readily spreads through the agency of birds and animals, it is preferable to protect and conserve it, rather than to go to the expense of forming plantations of this tree.

Value &c.

1118. In addition to the 23 trees mentioned above the following are used and have some market value. In some parts of the country one tree will be commoner and will be preferred to others for certain purposes, while in other parts the choice is given to yet other trees. Of the less known trees I may mention *Anogeissus latifolius* (Wall) Tam. vekkāli : Mal. marakānjiram. This tree is common in the drier districts of South Travancore, and on the Peermerd hills on the side nearest to Cumbum, and again in the deciduous forests near Kōnniyūr. The wood is dark-coloured and strong, and is used for bandy poles and agricultural implements. Its gum is very valuable. *Mesua ferrea*

Other useful trees employed in the low country.

(Linn) Eng. ironwood. Tam. and Mal. nāngu, is exceedingly heavy, hard and durable. But for its great weight, it would be more commonly used for building. Unfortunately, it is very much riddled by a species of borer, and it is very unusual to find a tree that is sound all through. The wood gives out a very great heat when burnt, and makes first rate charcoal. *Poly-lthia fragrans* (Benth) Mal. neduvār, is a straight tree abundant in the forests of North Travancore. The wood is light and very elastic, and is very well adapted for masts and yards. About one thousand trees are sold for this purpose every year. *Hardwickia pinnata* (Roxb.) Mal.shurali, a very large tree of all the moist forests is exported to Kurrachee. *Odina wodier* (Roxb.) Tam. uthi. Mal. kalasan, a tree of only small size, has a light reddish wood, that is very useful for furniture and house building. A tree called shokala in Tamil, which I have been unable to identify, is used at Shencottah for the manufacture of bandy wheels. A species of *Eugenia*. Mal. thavaram, is used for boats. It is a large tree growing in the moist forests of North Travancore. The ordinary payini tree *Vateria indica* (Linn) is sometimes cut for boats. It is very abundant in the moist forests. Two species of *Pterospermum*, *P. rubiginosum* (Heyne), and *P. Heyneanum* (Wall). Mal. malavūram, are felled for building and boats. The former especially is said to be a very good wood, and it is an exceedingly handsome tree. *Machilus macrantha* (Nees) Tam. and Mal. kolla māvu, is a moderate-sized tree of light wood growing in the moist forests. It is much used for common boats. The famous "upas" tree of Java under whose branches it was certain death to stand, is now identified with the ara-ānjili or aran-thella, Mal. *Antiaris toxicaria* (Leschen) a tree of the dense moist forests reaching an immense size. Its wood is light, but not strong or durable. It is used for boats and has been tried for tea boxes. The bark is particularly tenacious, the inner portions of it being connected by transverse strands of fibre, and the hill people take advantage of this peculiarity to reverse it, and to make portions of it into coarse bags by sewing up one end. Another tree of immense size *Holoptelea integrifolia* (Planch) Mal. āval, common in the moist forests of the north, is sawn into planks or fashioned into boats. The wood is said to be light and fairly durable, if smoked. The same remarks apply to *Lophopetalum Wightianum* (Arn) Mal. vengkotta, another very large tree of the moist forests. *Alstonia scholaris* (Brown). Tam. and Mal. mukkampāla, is another large tree with a milky juice. Its wood is white, very light, but not durable. It is used for rough planking. *Oroxylum indicum* (Vent) Tam. ālāntha Mal. palaga payāni, a tree of moderate size, is occasionally cut into boats. *Terminalia paniculata*. (Roth) Tam. marathu Mal. pūmarathu, a large tree, and one of the commonest trees of the deciduous forests, possessing a strong durable wood which is much valued in other countries, is used to a small extent for building. The country people have a tradition that tigers will never attack cattle kept in sheds made of this timber. *Chickrassia tabularis* (Adr Juss) is not a common tree in our moist forests, but it occurs here and there and reaches a large size. It supplies the Chittagong wood of commerce, a pretty pink wood of great utility for furniture, and of considerable value. Its properties seem to be unknown in this country. Two very elastic woods, which may be used for carriage shafts, spear handles and such purposes are *Miliusa velutina* (H. f. et T.) and *Bocagea Dalzelii* (H, f. et T.) both known as kāna kayitha in Malayalam. The former is found in the deciduous forests, the latter occurs only in the moist forests. Another tree the wood of which is used for axe handles, and is preferred for this purpose beyond all others *Grewia tiliæfolia* (Vahl). It is known in Tamil as unu, and in Malayalam as chadiccha. A species of *Dipterocarpus* which has not yet been identified, and is called kār-ānjili, is a tree of immense size, growing in the moist forests of North Travancore. It has a distant resemblance to the ānjili, out its leaves are much larger, being more than 18 inches in length and 12 broad. It is felled for boats. *Bridelia retusa* (Spreng) mullu vēnga, a tree of the deciduous forests, yields a hard and heavy wood, much valued in other countries and employed to a limited extent in Travancore. Lastly, *Trewia nudiflora* (Linn) a moderate sized tree possessing light wood, is used for carving, the images put up in Roman Catholic churches being commonly made of it. This completes the list of trees used for their timber : out of the large number of others which are indigenous in the country many are used for rough house building, for posts, or for the construction of jungle-wood roofs, but they possess no commercial value, and are used only by the poor, or for temporary buildings.

1119. The planters living at elevations between 1,500 and 4,000 ft., use woods other than those which I have mentioned above, in addition to *Timber trees used in the hills by the planters.* those whose range extends from the plains to the hills. Foremost among these are the two species of illupa or pāla, *Dichopsis elliptica* (Benth) and *Chrysophyllum Roxburghianum* (G. Don) which yield a reddish brown timber with a very straight grain, easily worked when young, but hardening with age. These trees make the best shingles of any, lasting about 10 years. A sticky, milky juice exudes from them which has some commercial value. They are abundant in the moist forests. *Pœciloneuron Indicum* (Bedd) pūthangkōlli. Tam, a large tree yielding a hard and heavy reddish wood occurs in moist forest up to 2,000 ft. The wood is used for building and is very durable. *Cinnamomum Zeylanicum* (Breyn), a large and common tree on the Peermerd plateau, yields a dull white wood bearing some resemblance to mango, and useful for rough planking and building. The ennay or oil tree *Dipterocarpus turbinatus* (Gœrtn f.), grows in the moist forests up to 3,000 ft., and yields a soft resinous wood which is useful for reepers, but decays rapidly with exposure. Two species of *Hemicyclia*, *H. Venusta* (Thwaites) and *H. elata* (Bedd) possess a hard white wood which might be used for turning. They are sometimes employed for building. A very beautiful red wood, suitable for furniture but not strong, is obtained from *Gluta Travancorica* (Bedd), a very large forest tree, confined to the extreme south of the Peninsula, and ascending the hills to an elevation of 4,000 ft. *Nephelium longana* (Comb) Tam., kātta pūvan, has a hard yellowish red wood, which is very suitable for building if cut in large scantlings, but it is liable to crack if sawn thin. The Wynaad shingle tree *Acrocarpus fraxinifolius* (Wight) is found only on the eastern side of the Peermerd plateau, and at the upper end of the Anjināda valley, where the climate is dry. Its wood is pink and splits easily. It is very useful for shingles, as well as for building and furniture.

1120. From these and other trees various products are obtained. Dragon's blood, or gum kino is collected from the vēnga tree, *Pterocarpus* *Trees yielding gums and resins.* *marsupium* (Roxb) from exudations caused by incisions in the bark. Dammer or "kunthirikkam" is the gum of *Canarium strictum* (Roxb) a lofty forest tree which is abundant from sea level up to 4,000 ft. The beautiful "payini" tree *Vateria Indica* (Linn) which is so much planted in gardens and along avenues for the fragrance of its flowers, yields a useful gum suitable for a varnish. The lofty *Ailantus malabaricus* (D. C.) possesses a sweet smelling gum called "mattipāla" which is burnt as incense. The tree is common in the forests of North Travancore and has been planted in the low country. Another very large tree *Hardwickia pinnata* (Roxb) possesses a gum which is said to be as useful as copaiba. No attempt has ever been made to collect the gamboge which flows from *Garcinia morella* (Desr) and *G. Wightii* (T. Anderson) which is an excellent pigment in both cases. The former is found in Central Travancore, and the latter is abundant on *river banks in the North. Neither of them are large trees.

1121. Of trees yielding useful fruits I must mention first of all the tamarind, *Tamarindus Indicus* (Linn) which is doubtfully indigenous, but *Trees yielding useful fruits.* has sprung up spontaneously on the dry hills of the south. The nux vomica tree, *Strychnos nux vomica* (Linn) is common in grass land at the foot of the hills, and its produce is largely collected. Two species of myrabolams are gathered from *Terminalia chebula* (Retz) and *T. belerica* (Roxb) the former especially being in good demand. It is a local tree, and is most abundant on the Cardamom Hills. The latter is much more common but its fruit is less valuable. The fruit of *Hydnocarpus alpina* (Wight) is used for making lamp-oil, and the tree itself has been largely planted in gardens for this purpose. A medicinal oil is extracted from the curious fruit of *Samadera Indica* (Gœrtn) which is good for rheumatism. The tree is indigenous in North Travancore. A medicine is procured from the long pods of *Holarrhena antidysenterica* (Wall) which is efficacious in cases of dysentery. The fruits of *Spondias mangifera* (Willd) and *Phyllanthus emblica* (Linn) are pickled, and that of *Baccaurea courtallensis* (Muell) is eaten raw. The only wild fruits that are at all worth eating are produced by two species of *Eugenia*, which grow in our moist forests at an elevation of 2,000 ft. and over.

---

* A few pounds of lac are collected each year by the Forest Department and are sold at Alleppey. I do not know what tree produces this article in Travancore, but we have many trees frequented by the insect, *Schleichera trijuga* (Willd). *Butea frondosa* (Wall) and others.

The Hillmen, however, eat the produce of many trees and shrubs, of the lofty *Flacourtia cataphracta* (Roxb), of *Clausena Indica* (Oliv), of some species of *Zizyphus*, of *Aporosa Lindleyana* (Baillon), of *Eleocarpus serratus* (Linn) and others, but curiously enough they do not seem to appreciate the nut of *Buchanania latifolia* (Roxb) which is excellent.

1122. Under the name of "Ponnanpū", the mace of a species of nutmeg [*Myristica Malabarica* (Lamk)] is collected by the poor in the hot weather and sold in the bazaars, but to get this product the people usually lop off all the branches and thus permanently ruin the trees. The small fruit of *Calophyllum Wightianum* (Wall) is gathered for oil in the same way. Other products collected from the jungle are the buds of two species of *Eugenia* under the name of "vellelenji" and "vennyāra", the unripe fruit of a Lauraceous tree (*Phœbe*) or (*Actinodaphue*) which is called "varrachil", the pods of *Acacia concinna* (D. C.) "pulinjikā" or "chīyakā" for washing, and the fruit of *Sapindus trifoliatus* (Linn) for the same purpose. The leaves of "thāli" *Actinodophue Hookeri* (Meiss) are also used instead of soap. A small quantity of the bark of some species of wild cinnamon is collected, as well as the wood of the sappan *Cæsalpinia sappan* (Linn) a small tree which has been introduced into, and has become naturalised, in the country.

Some special products collected in the country.

1123. Of poisonous and medicinal trees and plants the nux vomica tree already mentioned stands first. A small tree growing on the banks of the backwaters *Cerbera odollam* (Gœrtn) yields a very poisonous fruit often used in cases of suicide. The famous *Datura fastuosa* (Linn) is a common weed. A very poisonous and acrid juice exudes from *Sapium insigne* (Benth) a small tree growing on the upper hills, and from *Holigarna ferruginea* (Marchand) a lofty tree found both on the slopes of the hills up to 3,000 ft., and in the low country. The sap on exposure to the air turns as black as tar, and when it falls on the body raises large blisters. The Hillmen use the fruit and bark of many trees and shrubs for poisoning fish; foremost among these are the seeds * of the common croton *C. tiglium* (Linn), and the berries of the creeper known as "Cocculus indicus" *Anamirta cocculus* (W. and A.) and the fruit of one or two species of *Randia* (Linn), and the bark of *Lasiosiphon eriocephalus* (Decaisne) a small tree with pretty yellow flowers growing on the higher hills. Many other barks and fruits are thus employed, but, as Mr. Hooper has pointed out to me, the fish are affected rather by their astringent than by their poisonous properties. A small tree with yellow wood and prickly leaves *Berberis Nepalensis* (Sprengel) is found on the High Range and yields the principle called berberine which is largely used in medicine. The Muthuvāns have discovered for themselves the valuable properties of its fruit and bark, and they esteem the latter a sovereign remedy for snake bite. I have been in correspondence with a Bombay firm of Chemists regarding several plants yielding drugs which are obtainable in Travancore, but a list of these would occupy too much space in this Report. I may however mention that the roots of the common sensitive plant *Mimosa pudica* (Linn) which has spread all over the waste lands of North Travancore, and is such a dreadful pest to bare-footed travellers, on account of its prickles, have their value. Doubtless, as our knowledge extends, we shall find that there are many useful products in our forests for which we may obtain a sale.

Poisonous and medicinal trees and plants.

1124. Of indigenous cycads the most important is *Cycas circinalis* (Linn) which produces a great abundance of spherical fruit nearly an inch in diameter each containing a kernel. The Hillmen, and the low country people too in some parts, carefully collect these fruits, and either dry the kernels and keep them in that state, or convert them into an insipid flour which is baked into cakes. The fruit is known as "chanangkā" or "eenthakā", and it forms the staple food of some Hill races for several months together. Of palms the most useful is *Caryota urens*, the bastard sago, from which a toddy is extracted. In times of scarcity these trees, which are planted about the low country, are felled, and the pith is mixed with water, and the resulting fluid is strained, and a flour is prepared from it. This tree is scattered all through our moist

Indigenous cycads and palms.

---

* The croton of commerce is not wild, though it is very largely cultivated in gardens.

forests at the lower elevations. The mighty talipot *Corypha umbraculifera* is doubt-fully indigenous, for, though it is found all through the forests near Nāgamala, I believe that the seeds were scattered about by Mr. Huxham, as thatching grass was unobtainable on the large grant of land made to him. I have met with this tree nowhere else in our forests. The Hillmen make toddy from one or two other species of forest palms, and eat the tender tops of the wild arecanut (*A. Dicksoni*)?

1125. Of bamboos and grasses the most useful is the ubiquitous bamboo *Bum-busa arundinacea* (Retz) call by the people "mūngil," "mula," and "illi." Its uses are too well known to need description. It dies down at intervals of from 25 to 30 years, and 8 or 10 years must elapse before full sized culms can be obtained. A general seeding occurred in South Travancore when Lieut. Ward was travelling through the country in 1817, and another in 1870. This last was confined to the area south of the Acchankōvil river. North of this river, the seeding occurred about 1879–80, and it has been impossible to get full sized bamboos in North Travancore ever since. The male bamboo or "kal mūngil" *Dendrocalamus strictus* (Nees) is, to my knowledge, only found in Travancore in the Anjināda valley at an elevation of about 3,000 feet. Its almost solid culms are used for carriage shafts, spear handles and other purposes. A species of thornless bamboo which Mr. Gamble has called *Oxytenanthera Bourdilloni* is found only on the hills at elevations over 3,000 feet, and there only growing on rocky cliffs. It is known to the Hillmen as "pon mūngil," "kāmbu" and "arambu." This bamboo attains a diameter of only about 4 inches, but the internodes are unusually long, and the walls of the culms are very tough. The Hillmen use it for making combs and other household implements. It is said to flower only at long intervals and did so in 1888–1890. The "eetta" or "irūl" reed *Beesha Travancorica* (Bedd) forms the undergrowth of our forests over immense areas in different parts of the country, and near the crest of the hills it often occupies the whole of the ground, covering the slopes with dense and almost impenetrable thickets. Its presence is generally indicative of free but poor soil. Land that has been cleared for hill cultivation and has been abandoned often becomes covered with thickets of this reed, though, before the clearing was made, there may have been only a few clumps of it growing in the forest. The Hillmen use this *Beesha* largely for their temporary huts, the reeds themselves being employed for the frame work, and the leaves to thatch them. The reeds are also much used for fencing, basket making, mats &c. An excellent paper is made out of the fibre. The Hillmen recognise two species of this reed; they have not been scientifically distinguished, but further investigation will probably prove this to be the case. This *Beesha* seeds once every seven years. The "amma" reed, *Beesha Rheedii* (Kunth), which is found along the river banks in the low country, is used for basket making. Another reed, *Teinostachyum Wightii* (Munro) is found on the hills. It is supposed to flower at long intervals, and did so on Peer-merd in 1887–1889. Other reeds, not yet identified, are indigenous on the higher hills. Of grasses, the most important is the lemon grass *Andropogon schœnanthus*, from which an oil is extracted. As time goes on, other kinds will probably be found, useful for the manufacture of paper, and for various purposes.

Bamboos reeds and grasses.

1126. The "vakka" or "murutthan" năr, which is exclusively used for the dragging of timber by elephants, is made of the bark of *Sterculia villosa* (Roxb.), a small tree of rapid growth, which is becoming scarce in many places on account of the great demand for it. The fibre is unusually strong, as the strands not only run lengthwise, but are formed into a network by other strands crossing them diagonally. For common purposes, a very useful fibre is obtained from the bark of *Helicteres isora* (Linn), a small shrub growing to a height of 10 feet, and very abundant in the low country. Other trees and shrubs like *Tremá orientalis* (Blume) and *Debregeasia velutina* (Gaud) produce good fibres, and numberless creepers may be put to the same use.

Some indigenous fibres.

1127. This completes the list of the more important indigenous products of Travancore. A full list of the trees, as full as I have been able to make it, and brought up to date, composing about 400 species, will be given in the Appendix.

List of trees in the Appendix.

CHAPTER III.

Past and present management of the Forests.

Early history of Travancore. Up to the beginning of this century, no timber was exported, though the forests were much richer than they are now.

1128. Travancore has been a country of forests from the earliest times. Nature has fitted it for the production of vegetation of all kinds by the character of its climate, by the warmth of its atmosphere, and by the almost perennial moisture which prevails. As I have explained in the first chapter, the earliest inhabitants occupied merely a fringe along the coast, and their villages were scattered along the banks of the chief rivers and the few foot-paths that intersected the land, while the rest of the country was covered with jungle. Up to the middle of the 18th century, the country was broken up into small principalities, the chiefs of which were constantly at war with each other, and the extent of the forests probably varied little from one generation to another. The land that was cleared of trees and brought under cultivation during a time of peace, would be abandoned and would lapse into jungle when war again broke out. Externally too, there was very little trade, and the only timber that was felled was for private use. It was not till the commencement of the 19th century that even teak was exported. But, with the completion of the 18th century and the death of Tippu, tranquility was permanently restored, and though an insurrection broke out in 1808, which had to be quelled by British troops, it was soon brought to a close, and the 19th century has been, with this exception, a period of complete rest and peace. This restoration of tranquillity permitted the people to make some progress in arts and civilization, and a demand for timber soon sprung up. It was not long too before the notice of the British was turned towards the timber wealth of the country, and we find both Lieut. Arthur and Lieuts. Ward and Conner drawing attention to it * in their memoirs of Travancore. Some time early in the century, a Mr. Edye was sent out by the Admiralty to see if a regular supply of timber could be § obtained for the dockyards; he visited Ceylon, Malabar, Canara and many parts of this country, and, among others, the Idiyara valley then famous for its large teak, but no results seem to have followed his visit. At later times, similar enquiries have been made; in 1858 it was proposed to arrange for a regular supply of ànjili from Travancore for the Bombay dockyards, but the quantity forthcoming was not sufficient, and the idea was given up. Some twenty years ago, a Bombay syndicate suggested to Mr. J. D. Munro to take up for them a large tract of land in Travancore in order to plant teak, but the Government did not approve of the proposal, and the scheme was abandoned.

At the beginning of the century, the forests under the charge of the Commercial Agent of Alleppey.

1129. The first reliable information as to the working of the forests comes from Messrs. Ward and Conner, who were employed on the survey of the country from 1817 to 1820. They remark that, shortly before their visit, it had been customary to rent each river for its timber (apparently only teak), and that this had nearly ruined † the forests, as it was becoming increasingly difficult to get large teak anywhere but in the far interior, but that at the time of their visit a new system had recently been introduced, and the Government had begun to work down timber on its own account. These operations were entrusted to Captain Robert Gordon of the Bombay Engineers who held the post of Commercial Agent at Alleppey. The work of the Commercial Agent was at that time much more varied and necessitated much more travelling than it does now. Besides attending to his duties at Alleppey, Captain Gordon was Conservator of Forests, and had charge also of the collection of

---

* Seventy years ago many parts of the country now completely destitute of trees were then covered with forest. In the Survey of Travancore, mention is made of thick wood at Pallipuram, a little north of Trivandrum, and also near Quilon. The neighbourhood of Changanachéri was "densely wooded." Elephants were caught near Kottárakara, now a bare plain country without a tree, while the forests were said to be rapidly approaching the sea board, and villages were being abandoned through the incursions of wild elephants, with which the inhabitants were unable to contend in consequence of the then recent disarmament. See Memoir of the Survey of Travancore pp. 39.

§ An essay on this subject was read by him before the Royal Asiatic Society, May 16th 1835, and was printed in vol. 2 of their Journal.

† In North Travancore, the name of Mátthu Tháragan is mentioned as a great teak contractor. I have been unable to find exactly when he lived, and have only ascertained that he flourished in the first half of the century, was in all probability the man, or one of the men, who rented the rivers for their timber. So much is certain, that he was a man of great enterprise, that he had numerous elephants, and that he travelled into the most distant parts of the forests of the Idiyara valley and just outside of it to get teak of large dimensions.

fores'
f. the cardamoms from the gardens on the hills, as well as the arrangements connected
with their sale and that of the timber collected. During the wet season, from June
to November, he used to move his office out to Kōthamangalam or Thodupura,
whence he could supervise the felling of the teak on the hills, and its removal to the
riversides, as well as the collection and transport of the cardamoms to Alleppey.
On the Rāni river was stationed his assistant Mr. Walcot, who superintended similar
operations in that neighbourhood. From December to May, the Commercial Agent's
office was removed to Alleppey, where he occupied himself with the sale of the produce that had been collected. At this time, teak was the only timber monopoly
claimed by the State, and all other * timber was free to the people, except that it was
subject to a light river duty when conveyed by water.

1130. Shortly after this, the duties of Commercial Agent and Conservator of
Forests were separated, and Mr. U. V. Munro was appointed
Mr. U. V. Munro the   the first Conservator. Besides the charge of the forests, he
first Conservator of For-
ests. His duties and staff.   had also the duty of collecting the cardamoms from the hills,
and delivering them at Alleppey together with the timber that
had been felled. The timber was cut in the forests, dried, dragged by elephants to the
river sides, and then floated down the rivers in rafts to Alleppey. The cardamoms
were collected on the Cardamom Hills, despatched by bullocks to Nēriyamangalam on
the Periyār, and thence either direct to Alleppey, or to Thodupura where they were
stored, and whence they were sent on later. The § establishment under Mr. Munro in
1844 was as follows :—

For Forest work.	For Cardamoms.
2. Aminadars.	2. Aminadars.
14. Pillays.	1. Sumprethy.
7. Vijārippukārar.	1. Writer.
6. Accountants.	5. Proverthikārar.
74. Peons and watchers &c.	6. Vijārippukārar.
——	14. Pillays &c.
103. costing 420 Rs. a month.	97. Peons and watchers
	——
	126. costing 750 Rs. a month.

The Forest Department at that time confined its attention to working down teak on
the Periyār and Acchankōvil rivers, the total number of logs delivered averaging
1,500 a year between 1833 and 1842. The Southern portion of Travancore was not
worked at all for timber. In addition to teak, " blackwood and ānjili &c. " are
mentioned as being monopolies in 1844, but, whether these woods were worked
down by Government agency or were merely subject to a seigniorage, does not appear. Wax as well as cardamoms seem also to have been regarded as Government
monopolies, and we find the collection of wax regulated by Proclamation No. 1929 of
1st January 1838.
18th Dhanu 1013.

1131. On the death of Mr. Munro in 1844, Mr. West was appointed, and held
the post for about 8 years, when he was transferred to the
Mr. West and then Mr.   Judicial Department, and Mr. Kohlhoff succeeded him. This
Kohlhoff appointed Con-
servator of Forests. Pro-   gentleman held the office for about a dozen years when he too
clamation regarding kōl-
teak.   was transferred to the Sadr Court, and handed over the duties
                                                           15th August 1864
of Conservator to Mr. Vernède on 1st Thoolam 1040 · It was during the time of Mr. Kohlhoff that kōl-teak was made a closer monopoly. Before
1853
1028 it had been the custom for any one wanting kōl-teak for " private pagodas, Nam-
boodries' illams, and Christian churches " to ask permission, and he was allowed to
fell what he wanted from private lands without charge. But a custom arose for such
persons to apply for much more than they required, or, when they had obtained permission, they sold the teak and used jungle woods. Accordingly, Proclamation
             16th February 1853
No. 143 of 6th Kumbam 1028 was issued, by which a royalty of 2 rupees per candy was
levied on kōl-teak, and private persons were no longer allowed to fell the tree, but
application in writing had to be made to the Revenue authorities for the quantity
required. Sanction had then to be obtained from the Huzur. This Proclamation

---
* Memoir of Travancore Survey pp. 39.
§ Selections from Travancore Records pp. 69.

expressly fixes the size of köl-teak as under 10 virals (12½ inches) quarter girth a, the middle of the log, and it prohibits its sale for mercantile purposes. The Forest Department at that time felled nothing for sale that was less than 12 virals (15 ins.) quarter girth at the middle, and the logs averaged 3 candies each, and were often double that size. Other trees were felled on seigniorage, varying from time to time, at one time 3 fanams a candy, at another 1 rupee per log, and so on.

1132. The transfer of Mr. Kohlhoff from the Forest Department to the Judicial line marks an epoch in the history of our forests. At the end of 1864/1039 by for the greater part of the country was still clothed in * forest. The population in 1854 was 12½ lacs, but in 1875 it was 24 lacs, a most extraordinary increase if the Census of 1854 was correct.

Condition of the country as regards its forests at the time of Mr. Kohlhoff's transfer to the Sadr Court in 1857.

This increase was, no doubt, due to the wise administration of the late Sir Madhava Row who instituted numerous reforms, improved the administration of justice, regulated the collection of revenue, and introduced system where before there had been chaos. Consequently, not only did the indigenous population increase, but men of ability began to be attracted from other parts of India. This rapid increase of the indigenous population began to tell upon the area of forests, for it must be carefully noted that almost the whole of this extra population, dependent as it must have been on agriculture for its subsistence, must have spread over the country. It has not, as would be the case in most European countries converged to the large cities. † Alleppey and other chief towns contain to-day no larger a population than they did 70 years ago, area for area. Places like Trivandrum have, indeed, increased by the addition of suburbs, but there is no large congestion in certain parts of it. In the town itself, the people probably have just as much room as they had at the beginning of the century. In fact, of the 11¼ lacs increase in population between the years 1854 and 1875, probably 10 lacs scattered themselves over the country and settled there. What an effect this must have had on the character and extent of the forests may be easily imagined. Whereas the estimate of hill cultivation is at the present time 50,000 acres a year, in 1854, probably not more than 10,000 acres a year were cleared. Again, in 1864 the cultivation of coffee had not commenced, and the 40,000 acres since sold for that purpose were clothed in heavy forest. Mr. J. D. Munro, Mr. Richardson and others of the early pioneers of coffee planting have told me how rapidly and thoroughly the forests have disappeared from many parts of North Travancore in the last 20 years. In fact it cannot be doubted that the area of forests now is very much smaller than it was in 1864. It was not to be expected that a minister of Sir Madhava Row's ability would overlook the destruction of the forests, or would undervalue the importance of their preservation. Consequently, we find more attention paid to their management from 1864 forward, a constant increase in the staff of Forest officials, and the publication of rules from time to time on subjects connected with the forests.

1133. Mr. Vernède, then, was appointed Conservator at the commencement of the Malabar year 1040 (Aug. 1864), and, a week later, his nephew Mr. J. S. Vernède was nominated Assistant Conservator and stationed at Malayáttur. In June 1865 the duty on timber was adjusted, and on the 20th of the same month were published the rules for the sale of waste lands both in the low country and on the hills, which permitted capitalists and others to take up land for the cultivation of coffee and other products. These rules declare ebony and sandalwood to be monopolies as well as teak and blackwood. In the following month the kudi-vila, or amount paid to the cardamom ryots for the collection of the spice, was adjusted. At the time of Lieut. Ward's visit in 1817-1820 the cardamoms grown in the gardens on the Travancore Hills were regarded as a monopoly as at present, and the owners of the gardens were obliged to deliver them to the Government at a fixed rate, but wild cardamoms were ‡ allowed to be removed on payment of light transit dues, amounting to about 2¼ rupees per thulám of 20 English lbs. Up to

Mr. Vernède appointed Conservator in 1864. An Assistant Conservator also appointed. Various rules passed in 1040 (1864-5) about the sale of waste lands, rates paid for cardamoms, and the felling of trees on chôrikkals &c.

---

* Sir Madhava Row was appointed Dewan in 1858, and the effects of his reforms cannot have shown themselves till two or three years later, so the large increase of population could only have commenced about 1860 or 1861.

† Col. Welsh in his Reminiscences of Travancore, 1819, says that Alleppey contained at that time 26,000 inhabitants, exactly the present figure.

‡ Ward's Diary Nov. 14 1818.

f. 1822, the price paid to the cultivators was 48¼ fanams (nearly 7 rupees) per thulám. From that date to 1829, it was 60 fanams. The price was then lowered to 55 fanams, and from this year it doubtless fluctuated from time to time as the selling price varied, but in 1865 the price paid was 240 rupees a candy or 16 rupees a thulám. By the Proclamation of July 1865, one third of the selling price was ordered to be paid to the ryots. In this same month (July 1865), the felling of teak and blackwood and of all jungle trees over 10 virals (12¼ inches) quarter girth was prohibited. In the following month (August 1865), the method of charging duty on forest timber was slightly altered. Hitherto, it had been customary for the duty to be charged on the cubical contents of the logs, but this led to great delay in measuring the timber in the forests and at every watch station it passed, accordingly, an order was issued directing that the duty should be levied per log.

1134. In the year 1041, several noticeable proclamations were issued. In
October, rewards were offered for the first time for information
Various changes and im- regarding the illicit felling of teak and blackwood &c. In
provements in M. E. 1041 November, the forced employment of watchers at watch stations
(1865-6). First attempts November, the forced employment of watchers at watch stations
at propagating teak by without pay was abolished. At present the only remnant of
sowing. Royal trees on this modern socage, or " úliyam " as it is called, existing under
private lands and avenue either the Forest or Cardamom Departments, is that the Hill-
trees protected. men are expected to work for us at a reduced rate, in consideration of their being allowed to clear land in the forests for their crops. On
the $\frac{\text{30th January 1866.}}{\text{9th Magarom 1041.}}$ an important Proclamation was issued prohibiting the felling
of teak, blackwood, ebony, or any trees planted on the banks of rivers for their sup-
port, or any avenue trees. It further directed that jack, palmyra, tamarind and
other tax-paying trees should not be felled without permission. A few days later,
the price of kōl-teak was raised from 4 to 6 rupees per candy. The year 1041 is
remarkable because it was then that the first attempts to artificially propagate the
teak tree were made. In this year the Vembūram island above Malayáttūr was
sown with teak seeds, but the result was not satisfactory, and it at once became apparent that if teak was to be successfully grown it must be carefully put out in regular
plantations.

1135. In 1042 (1866-7) the planting of teak was commenced on a small scale
both at Malayáttūr and Kōnniyūr, some 85 acres being com-
Progress during 1042 pleted this year, and, from this time forward, extensions have
and 1043 (1866-7 and been made, till the total area has reached more than 1,000 acres,
1867-8). Teak plant- the details of which will be given in the Appendix. On the
ations commenced. An- 25th December 1866 a Proclamation was issued forbidding
other Assistant Conser- the felling of forest by Hillmen or others for cultivation, if
vator appointed &c. the felling of forest by Hillmen or others for cultivation, if
suitable for coffee plantations. In the following Malabar year 1043, the staff of
Forest Officers was increased by the selection of 8 Amindars and 24 peons who
were appointed to "conserve the forest and catch elephants". Early in this year
(Nov. 1867) another Assistant Conservator, Mr. Thomas, who had previously had
some experience of forest work in the British service, was nominated, and put in
charge of the teak plantations and forests near Kōnniyūr. About this time, an
attempt was made to induce the owners of cardamom gardens on the Cardamom
Hills to settle there, instead of living constantly in the Kambam valley, and only
coming up to the hills to pick the cardamom crop. Many families migrated to the
neighbourhood of Udambanchōla, took up land and erected huts there, but the
attempt failed, because the locality was too unhealthy, and it is now generally
admitted that these hills are too feverish during the hot weather for any one to
remain there who has not been thoroughly acclimatised by long residence on them,
like the Mannáns.

1136. In the month of Dhanu 1044 (Jan. 1869) the charge of the cardamom
gardens on the Peermerd Hills was taken from the Conservator
Progress during 1044 and placed under a separate officer, Mr. J. D. Munro being
(1868-9). Charge of the and placed under a separate officer, Mr. J. D. Munro being
Cardamom Hills placed selected as the first Superintendent and Magistrate of the
in the hands of a special Cardamom Hills. Hitherto, it had been customary for the
officer. Conservator of Forests to move up to Udambanchōla in the
cold season when the cardamom crop was ripening, and to remain there till all the

crop had been gathered and weighed for transportation to Alleppey. This arrar ment occupied a great deal of time, and the Conservator was therefore preven from attending to his other duties in connection with the collection and sale timber. As the cardamom monopoly was paying very handsome profits at the tin and as there was much work to be done in cutting roads, inducing persons to op new gardens, and in arranging to change from Nēriyamangalam to Peermerd t route by which the spice should be conveyed to Alleppey, a Superintendent w appointed specially to superintend these works, and to exercise Magisterial pow on Peermerd where coffee estates were then being opened. In this year the fir cardamom gardens were started in the neighbourhood of Kōnniyūr, under the supe vision of the Forest Department.

1137. In the year 1045, two important Proclamations were issued, one date

Progress in 1045 (1869-70). First attempts to restrict Hill-cultivation which have not been successful. 16 October 1869. 3 Thulam 1045. prohibiting the shooting of elephants, and the othe 13 April 1870. 23 Meenam 1045. laying down rules for Hill-cultivation. By thi last named notification, the people were allowed to cultivate without permission "grass lands, eetta jungles, and marshy places" belonging to Government, but they were prohibited from clearing land on the "hills where clouds rest, or in jungles at the lower elevations where there are Sirkar trees such as teak, blackwood &c." Further, they were not allowed to clear land "where there are trees of more than 10 virals quarter girth." People desirous of taking up forest land where there were large trees were directed to apply to the Conservator, who would order an examination of the land required, and if he saw no objection, he would give a certificate to this effect to the applicants to present to the Tahsildar who would then charge tax upon the acreage, and the applicants would be allowed to clear. This was the first attempt made to restrict Hill-cultivation, and it has not been very successful. Eetta or reed jungles are usually accepted as meaning any jungle where there are reeds, however many large trees there may be growing among them, though it was evidently the intention of the Dewan to include only pure reed forests. Again, the rule regarding the land on which are growing trees above 10 virals quarter girth is usually construed to mean that any land may be cleared, provided trees above this limit are not felled. And lastly, the Hillmen constantly clear forest land without any permission from the Conservator and without paying any tax. Moreover, this Proclamation contains no penal sections, so that no punishment is fixed for infringement of its clauses.

1138. During the next three years 1046-7 and 8, no changes of importance

Progress in 1046-7-8. Shencottah depōt transferred to the Forest Dept. Changes in the method of working down teak. seem to have been made, except that in 1047 (1871-2) the charge of the timber depōt at Shencottah was transferred from the Revenue to the Forest Department. In the previous year, the returns had been only 9,521 rupees, but in this year, § they rose to 16,222. Apparently, the timber was sold by auction at irregular intervals, and not, as at present, by daily sales. In spite of the rules against Hill-cultivation, the destruction of forest continued, and, in all his Annual Reports from this time forward, the Conservator drew attention to the damage caused, which he was less and less able to prevent as the culprits waxed bolder from impunity, in the absence of a proper Forest Law and a sufficient staff to enforce his orders. Some dissatisfaction was also expressed by the Conservator at the manner in which the sales of timber were carried out at Alleppey. The Commercial Agent at that place, occupied with other work, did not sufficiently consult the interests of the timber merchants, and the teak delivered there by the Forest Department was often held over for too long a time, or was sold at a low rate in § order to clear it off. The Conservator therefore proposed that private persons should be allowed to fell timber under supervision as they required it. This suggestion was not approved at the time, but other changes were made soon after. For some time previous to the year 1870, teak was felled on contract in the neighbourhood of Kōnniyūr, and was delivered at the places named by the Conservator, but, about this time, the contract system was given up, and the Govt. worked down its timber all over Travancore on its own account for about 9 or 10 years. All this

* Conservator's Report 1043.
§ Conservator's Report for 1048.

time, nothing under 12 virals (15 inches) quarter girth at the middle of the log was permitted to be felled, though a limited quantity of stunted, or kôl-teak, was felled by the Taluq authorities for the use of churches or pagodas, on requisition being made to them.

1139. In the year 1049 (1873–4) the attention of the Conservator was drawn to the great mortality among the elephants which fell into pits, *Changes in the manner of capturing elephants. "Inâm" pits. The keddah near Kônniyûr.* and he suggested that some other means should be adopted for capturing them. Lieut. Arthur in his Memoir of Travancore, written in 1810, mentions that, at that time, the Government used to allow people to cut pits for these animals on payment of a tax per pit. The elephants thus taken apparently became the property of the persons who dug the pits. This custom accounts for the extraordinary number of old pits still to be seen in the forests in the most unlikely and distant places. Doubtless, many of these paid no tax whatever, for, when a party of hunters, timber men, or others went into the forests, they were in the habit of digging pits round their camps with the double object of protecting themselves and of capturing elephants or other wild animals fit for food, and we may imagine that these pits were often cut without the knowledge of any Sirkar official. At some period subsequent to 1810, the Government commenced digging pits on its own account, but it was still allowable for private persons to have "inâm" pits as they were called. The elephants thus captured by private persons had to be sold to the Government at 150 Rs. a head. It was not till Nov. 9. 1875 that inâm pits were prohibited, and even at the present time, at least one person (the karthâvu of Râni) has the right to dig pits and capture elephants. Under the inâm pit system a large number of elephants were uselessly destroyed every year. The secret of success in saving the lives of elephants falling into pits is to maintain an efficient and vigilant watch. Unless immediate information is obtained of such an occurence, and unless the elephant is at once supplied with fodder and water, it either succumbs from starvation, or damages itself so much in its efforts to escape that it dies. Few private persons were in a position to conduct a proper watch, and instances are not unknown where elephants have fallen into pits and no one has known anything about it till 20 or 30 years afterwards, when, in reopening the pits, the skeletons have been found *with the tusks unremoved.* Even in the case of Government operations there is great mortality unless the pits are properly watched, and unless there are tame elephants ready to effect the capture.* So, the Conservator recommended the formation of a keddah on the plan of Mr. Sanderson's keddah in Mysore, and he was deputed to go and visit that place. On Mr. Vernède's return, a site was selected on the Kalâr river near Kônniyûr, and the work was commenced in 1050 (1874–5). The keddah was opened in 1052 and was used till 1060 or for nine years in which time 100 animals were caught, but though it was very successful at first, its existence became known to the elephants, and, latterly, there were few captures, so it was closed temporarily, but I hope we shall be able to reopen it next year. While the keddah was in working at Kônniyûr, pits were still used in North Travancore, for it was § intended to construct 4 more keddahs before abandoning the pit system altogether, but, on account of the only partial success of the keddah at Kônniyûr, this scheme was abandoned, and, at the present date, all our elephants are caught in pits.

1140. Although the control of the forests near Shencottah had been handed over to the Forest Department in 1047 with the result that the *Various changes in the management of the Shencottah (Puliyara) depôt from the time it was taken over by the Forest Department.* revenue began at once to rapidly increase, many changes have since been made from time to time with the view of seeing what was the most profitable way of working these forests. Thus, in 1047 the timber was delivered at the depôt and sold by auction, as already stated ; in 1049 the people were allowed, for one year, to fell the timber for themselves, on permit. The time was subsequently extended for 2 or 3 years longer, but, though the returns were good, the forests suffered severely, as the people cut down the timber in the most wanton way,

---

* I have in my mind the case of a very fine young tusker which was caught near Thodupura two years ago, but we had no tame elephant within 100 miles large enough to control him, and he was 11 days in the pit before he could be taken out, with the result that he has been under treatment ever since and I fear he will have to be killed. Since writing this I hear that he has died.

§ Administration Report for 1053.

and much of it was smuggled into Tinnevelly without payment of our dues, *in* 1054, we find the old system again in force, and timber sold by Government *from a depôt.* In this year, an officer was appointed on a good salary to superin*tend the* work, but abuses were commoner than ever, and the Government was defra*uded in* the most barefaced manner. This officer had therefore to be removed after *a very* short probation. The depôt system was continued, an Aminadar being in *charge,* but in 1058 (1882-3) we find a return to the old seigniorage system, merchan*ts intensively* being allowed to fell timber on permits, but after a year or so, a Superintendent was appointed (in 1060) on 100 Rs. a month, and the depôt system was introduced again, and from that date this system of daily sales from the depôt at fixed rates and occasional auctions has been maintained until the present time. In 1063, four sub-depôts were opened near Shencottah, but though the returns have increased steadily, this arrangement has led to the most dreadful smuggling, and tens of thousands of rupees worth of timber have been taken out of the country without payment, through want of sufficient supervision and proper rules, and through the dishonesty of the officers of the depôts. These sub-depôts are now about to be closed, for they have increased the facilities for smuggling instead of preventing it. Until a thoroughly reliable officer is put in charge of the Puliyara depôt and the forests which supply it with timber, who will spend a good deal of his time in travelling through the forests and seeing what the contractors are doing there, these malpractices are likely to continue, for it is impossible to control the felling of the timber so long as the officer remains permanently at the depôt 20 or 30 miles away from the forests, more especially as the people of Tinnevelly are very bold and enterprising, and timber is very valuable there.

1141. Up to the year 1051 (1875-6) the teak and blackwood felled in the for-

Teak and blackwood contract with Messrs. Wallibhoyi and Co., and the method followed for collecting the timber.

ests, either departmentally or by contract, had all been delivered at Alleppey, and was sold there by auction, at prices ranging from 14 to 18 rupees per candy. This was all first class timber, above 15 inches quarter girth. But there were great complaints about the arrangements, and great confusion in the accounts, for, while the Forest Department credited itself with the value of the timber as soon as it was delivered at Alleppey, it only received credit for it after it had been sold, which was perhaps a year or two later. In this year, an enterprising Bombay firm, Messrs. Wallibhoyi, Kaderbhoyi & Co. contracted with the Travancore Government to purchase * 14,000 candies of teak at 17, 13 and 10 rupees per candy, according to class, and 2,000 candies of blackwood at from 20 to 12 rupees. In 1053 and 1056 other contracts were made with Messrs. Wallibhoyi & Co. for 30,000 candies of teak at the same rates, and 100,000 sleepers at 1¼ rupees each. In 1058 the same firm took a contract for 10 years at the same rates, not for any fixed quantity of timber, but for as much teak and blackwood as we could deliver to them within that time. This contract expires on $\frac{1 \text{ November } 1892.}{17 \text{ Thoolam } 1068.}$ Finally, they agreed in 1061 to take another lac of sleepers at 2 rupees each. At first, this timber was brought down departmentally, but, in 1054, contracts were given out for the delivery at our depôts of 1st and 2nd class timber, and in 1057 further contracts were made for the delivery of 3rd class timber, and departmental work was practically stopped. At first, the number of contractors was limited and timber was abundant, but more and more contracts have been made until there are at present on our books the names of no less than 135 men who are either working for us, or who have not had their accounts settled up. When the contract was made with Wallibhoyi & Co. in 1058, it was agreed that delivery should be taken from the Forest Department direct, and not from the Commercial Agent, and, from that date, the Forest Department has had no dealings with Alleppey except in respect to the cardamoms and minor produce sent there, and this has saved very much trouble. About this time, the felling of kôl-teak by the Taluq authorities was stopped, and the work was transferred to the Forest Department, the Sirkar demand per candy being raised to 8 Rs. instead of 6 Rs., exclusive of the cost of cutting and delivery.

* 1st class teak is above 12 virals (15 inches) quarter girth.
2nd class teak is above 10 virals (12½ inches) quarter girth.
3rd class or kôl-teak—below this.

1142. On $\frac{18\text{th March 1881}}{6\text{th Meenam 1056}}$ an important Proclamation was issued defining the limits of Hill-cultivation. This in part re-affirmed the Proclamation of 1045, but gave permission to the people to clear land " within four miles of inhabited places, " a most vague limit so long as the definition of an inhabited place is not given, for, by this clause, any man could build a hut in the jungle and claim to clear land within four miles of it. This Proclamation assigns no punishment for the infringement of its clauses, though it makes provision for the confiscation of the produce raised upon the land. In the following year $\frac{26\text{th April 1882}}{15\text{th Chittray 1057}}$ a reward of one quarter of the value of the article was offered for any information which should lead to the conviction of any smuggler of teak. The year 1057 is chiefly remarkable for the very disastrous flood which occurred in June 1882. In South Travancore it was hardly felt, but in North Travancore much damage was done. In addition to the destruction of houses and property, the Forest Department and the contractors suffered by the sweeping away of thousands of logs, some of which were recovered, but a great many were carried out to sea or otherwise disappeared. A large area of land at Könniyûr had been planted with teak, and Mr. Vernède estimated that 750,000 out of 900,000 plants put out that year were destroyed or swept away by the flood. It was in this year that the Conservator resigned the duties of a Magistrate to try forest offences, which he and his predecessors had held for a long time. As the High Court would not uphold his convictions because they were not passed under any special law, Mr. Vernède considered it useless to retain his powers any longer.

(Side note: Proclamation about Hill-cultivation in 1046. Disastrous flood in 1057 (1882). Mr. Vernède resigns his magisterial powers.)

1143. In the year 1058 § (1882-3) the forests on the eastern slopes of the Māhinthragiri hills (known as the " kāda-kāval " lands) were handed over to the Forest Department, in consequence of complaints from the people of Tinnevelly that the reckless destruction of trees there had reduced their water supply, and cultivation within this area was prohibited, while an Aminadar and staff were appointed to protect it. In the same year, an attempt was made to open a sandalwood plantation, 50 acres in extent, a few miles out of Quilon, but it proved a failure. On $\frac{3\text{rd April 1883}}{22\text{nd Meenam 1058}}$ an order was published forbidding the burning of grass in forests adjoining Tinnevelly, for fear of its spreading into that District, but as no one was appointed to enforce this rule, it has never been observed. A month or two later, we find notices issued by the Conservator directing that all timber sold by the Department should be stamped, and that receipts given by the Department should all be on printed forms. In this year Mr. J. S. Vernède, Assistant Conservator at Malayāttûr, died, and was succeeded by Mr. R. Stevenage who was appointed on $\frac{24\text{th April 1883}}{13\text{th Chittray 1058}}$.

(Side note: Progress in 1058 (1882-3). Transfer of " kāda-kāval " lands to the Forest Department &c.)

1144. At the beginning of 1059 (1883-4) the forests of South Travancore near Vīrappuli were transferred from the Revenue to the Forest Department, and Mr. D. Grant was appointed Special Assistant Conservator with control over them, in addition to his Magisterial powers. The year 1059 is chiefly remarkable for the appointment of a special Commission to discuss the management of the forests, and to draw up suggestions for their better administration. The Commission sat for about a month, examined witnesses and collected evidence, and then drew up a report, which, among other things, recommended the appointment of two Deputy Conservators, the abolition of the seigniorage system, and the substitution in its stead of a depôt system for the sale of timber other than the " royalties. " It also prepared the draft of an Act very much on the lines of the Madras Forest Act of 1882. On February 12, 1884 a notice was issued placing an assessment of 2 fanams a parra on all Government land cleared for paddy. Before this, Hill-cultivation had been permitted free except in special cases. (See para 1137.)

(Side note: Progress in 1059 (1883-4). Appointment of Mr. Grant, Special Assistant Conservator, and of a Commission to discuss forest affairs &c.)

1145. At the beginning of 1060, Aug. 26 1884, the depôt system, as suggested by the Forest Commission, was introduced, and a reward of one-third the value of any timber, other than teak and blackwood, which might be confiscated for smuggling was promised to the informer. For teak and blackwood the share was and is one-

(Side note: Introduction of the depôt system of selling timber in 1060 in place of the old seigniorage system. The change not such)

§ In the year 1058 an important change was made with regard to the method of selling timber. Before that year, all the teak collected by the Forest Department was sent to Alleppey for sale, but in that year it was arranged that the Forest Department should sell the timber that had been collected, and it was thus made quite independent of the Commercial Office.

42

an improvement as was
expected, for various rea-
sons. fourth. Previously, any one who wanted timber other than the royalties, and anywhere but near Shencottah, had been able to obtain permits at the rate of one rupee per log, under which they were allowed to fell and remove the timber at their pleasure. Very little supervision was exercised over these people, the only check on smuggling being the watch stations which were dotted along the river and road sides, but if the permit holders could escape these they could smuggle as they pleased. Many of the watch pillays and watchers made a good income by passing timber without entering it in their books, and the post of pillay was eagerly sought after, the fortunate holder of such a post, whose pay was 3 rupees a month, being able to keep a servant and to live at the rate of 30 rupees on the fees he levied. But a very small proportion of the trees felled in the country ever paid our light charges, and the same permits were used over and over again for different lots of timber. These malpractices were so notorious and so flagrant that I, in common with the other members of the Commission, recommended that the depôt system should be introduced, and that all timber of whatever kind should be brought to the depôts for sale. This change did not in itself prove much of an improvement on the old seigniorage system, who agreed to deliver timber at our depôts, engaged hosts of sub-contractors who did almost as much mischief as the old permit holders, and they were found just as difficult to control, but attention was drawn to the existing abuses, and arrangements were accordingly introduced for checking the timber in the forests. Another disadvantage of this depôt system was that poor people, who in the old days could take out permits for small quantities of timber, were, under the new system, obliged to go to the depôts, often far distant from where they wanted the timber, and to buy wood which was perhaps not at all what they required. Thus they were often driven to fell trees clandestinely on account of the difficulties attending the honest purchase of their requirements. The contractors held a sort of monopoly and brought to the depôts only such timber as they could conveniently procure, and as a part of their agreement was that they should themselves purchase what could not be sold, they took care to deliver only what they could easily sell, and thus the new system very closely resembled the old, except in name, and except that the right to fell was given to only a few persons instead of to many. This depôt system was abolished in 1063, and the old seigniorage system was re-introduced.

1146. In the year 1060 (1884-5), a Proclamation was issued increasing the
Increase of the ryots'
share of the value of cardamoms. Successive Super-
intendents of the Carda-
mom Hills. share given to cardamom ryots from one-third to two-fifths the selling price of the spice. In para 1133 I have shown that in $\frac{1865}{1040}$ the price paid to the ryots was fixed at one-third the selling price instead of being an arbitrary sum selected at the discretion of the Government and constantly changed. At that date, cardamoms were selling at a very high figure, and a third of their value often amounted to 30 Rs. a thulam, § but the planters in Ceylon began to introduce this product there about 1880, with the result that the prices had in $\frac{1885}{1060}$ been forced down to a very low level, in fact to about one-fifth of the price in 1865, or to 6 Rs. a thulam. This could not pay the ryots who had to give higher wages than formerly to their weeding and picking coolies, and, on perceiving the truth of their complaints, the Government conceded to them two-fifths of the value instead of one-third as formerly. In para 1136 I have shown how Mr. Munro was appointed the first Superintendent and Magistrate of the Cardamom Hills in 1044. This office he held until the year 1051 when he resigned, after opening up the area of his charge, cutting roads, commencing new gardens, erecting buildings and putting the Department into a thoroughly efficient state. Mr. Munro also caught elephants and opened a depôt at Râmakkal for the sale of timber, but it was subsequently closed by one of his successors, and has been recently re-opened by the Forest Department. The Cardamom Department still catch a few elephants each year. Mr. Robert Baker succeeded Mr. Munro but died in $\frac{1878}{1063}$ and Mr. Bensley took his place. On the appointment of this gentleman in 1881 as Superintendent of Police, Mr. Maltby, who now holds the post, was nominated Superintendent and Magistrate of the Cardamom Hills. At the beginning of 1063 Mr. D. Grant was sent up as Assistant Superintendent but he resigned at the end of the year, and, after an interval of twelve months, Mr. J. S. Sealy was taken from the

§ Equal to about 20 English lbs.

Forest Department where he had been acting as Deputy Conservator for a short time, and was attached to the Cardamom Department in Mr. Grant's place.

1147. In the year 1061 I was deputed to make an exploration of the forests of the State and to write a report on them. Early the following year, Mr. D. Grant, who had been in charge of the Southern Forests, went on leave, and on his return to the country after nearly two years, he was sent to the Cardamom Hills, as I have already shown. In 1063, the first real Forest Act was passed as Regulation IV of 1063 on December 6 1887. So far as it goes it is excellent, but it only comprises a portion of the draft proposed by the Forest Commission in 1884. In the same year, the depôt system was abolished except near Shencottah, at Trevandrum, and in the southern portion of the State, and the seigniorage system was reintroduced with much higher rates, the charge being levied per candy, and not per log as formerly, January 24 1888. At about the same time, I was appointed Deputy Conservator and stationed at Malayáttûr. A new depôt was opened at Kunili on the Peermerd plateau this year, and four new sub-depôts near Shencottah. I have already alluded to them in para 1140. Towards the end of this year, the Conservator issued a notice regarding the particulars required from applicants for free permits to remove timber grown on private taxpaying land (July 10. 1888). Hitherto a great deal of timber had been smuggled out of the forests, on the plea that it had been grown on private and taxpaying lands and was therefore exempt from duty, but from this date it was ordered that before a free permit could be granted a certificate from the Proverthik-káran must be produced, showing that the timber in question had really been felled on private land, as stated. The year 1063 saw the completion of the demarcation of the boundary between Travancore, and Tinnevelly, Madura, Coimbatore and Cochin respectively. This work had been commenced upwards of a dozen years previously, the watershed being accepted as the boundary except in a very few instances. At the present time, there still remain two or three small portions of boundary, about which there is some dispute, but these are in the low country near Shencottah and Panagudy, where the watershed is not taken as the guide, and the line has to be decided by obtaining evidence of occupation.

Progress in 1061–3. (1885–1888). Act IV of 1063. Appointment of a Deputy Conservator. Completion of the demarcation of the boundary between Travancore and British territory or Cochin.

1148. The first Forest Reserve in accordance with the Forest Act was proclaimed in the year 1064. It was selected by me near Kônni-yûr, and is estimated to contain 300 square miles (October 9. 1888). The following month, the Conservator issued a notice prohibiting the felling of trees, cut on permits, which contained less than 2½ candies each. (November 13, 1888). Another Reserve in the neighbourhood of Ariyankávu and Kulatthuppura extending over 121 square miles was declared on February 5, 1889, and a third of 116 acres in area at § Váli near Trevandrum was gazetted on March 26, 1869. The following month, the Conservator issued a notice to the effect that any one wishing to obtain a free permit to move timber growing on taxpaying lands should get the necessary certificate from a Forest officer and not from the Revenue Department. (April 30, 1889). A fourth Reserve, comprising 345 square miles, and situated near Malayáttûr in the extreme north of Travancore was proclaimed on November 4, 1889, and a fifth, 5 square miles in extent, at Mukkana mala near Trevandrum was declared on November 19, 1889. In September 1890 the supervising staff was strengthened by the appointment of Mr. A. C. Watts as Assistant Conservator, who had attended the course of lectures at the Dehra Dun Forest school. In May of the present year, a student was nominated by Government to attend the forest course at Government expense at the same institution, and he will be available for employment two years hence. Very recently, Mr. Vernède, who had held the office of Conservator of Forests for nearly 27 years was obliged to resign on account of ill health, and I was appointed to succeed him

Progress from 1061–6 (1888–91). Formation of 5 Reserves. Appointment of Mr. Watts. Resignation of Mr. Vernède.

14, June 1891
1, Mithunam 1066. This brings us up to the present time, and in considering the various branches of the management of the forests it will be more convenient to discuss the administration under different heads, and I will first describe, as the most import-ant branch of forest work, the arrangements now in force for

§ The Váli Reserve was, properly speaking, not noticed under the Act until lately, but it was closed at the date mentioned.

(1) *The conservancy and improvement of the Forests.*

1149. As stated in the last para, five Reserves, aggregating 771 square miles, have been proclaimed, but, as yet, little more has been done than to gazette them. § Some demarcation has been carried out in the Kŏnniyŭr Reserve, but a considerable portion of the line has yet to be cut. This is a most important work requiring close attention, for, until the Reserve has been actually marked out, people who trespass inside the prescribed area and clear land for cultivation can always plead ignorance of the position of the boundary, and they escape with a very light punishment, or without any at all. Recently, an officer of the Revenue Survey Department 'registered land within the Kŏnniyŭr Reserve in the names of several applicants who represented to him that the land they required was outside the gazetted area. The case is still "sub judice" so I will not say more about it here, but it seems pretty clear that the Revenue officer, without ever going to the spot, accepted the statements made to him that the land in question lay in the middle of compounds and cultivated land, whereas it is really fine forest land, four or five miles inside the boundary of the Reserve, and far from any habitation. This work of demarcation, therefore, is one of the greatest importance, and must be carried out without delay wherever it is necessary, for in many places a river or a high range of hills forms the boundary, and no cutting is required. To protect these Reserves, a staff of guards has in most cases been appointed, and so long as their work is merely to prevent hill-cultivation their number is sufficient, but when we come to work these forests in rotation, and to thoroughly supervise the fellings their number will have to be increased. For instance, in the large Malayättŭr Reserve, extending over 345 square miles, we have at present only one Inspector and six guards: fortunately, the people are not aggressive, and when the necessary portion of the boundary has been demarcated (the Periyär forms the greater part of one side) I trust that even this small number will be able to prevent encroachment. The object of forming Reserves is not merely to prevent the destruction of the forest by hill-cultivators, though this is very necessary. The forests can never improve so long as the timber in them is felled at random and without any system, whether by Government or by the holders of permits, and the object of working plans is to arrange that timber should be felled over a certain area each year or couple of years, while the rest of the forest is left undisturbed till the time comes round for the trees in each portion of it to be cut down. I will return to this subject later on, but for the present it is sufficient to say that no working plans have been framed or even thought of in Travancore. Neither have any surveys been carried out, to determine the areas to be cut over during each period. We are fortunate in having a very excellent map on the scale of one inch to the mile, and this will assist us greatly in making a commencement, but the services of surveyors will certainly be required when we take in hand the development and improvement of the forests. The area set aside for Reserves is not sufficient for the requirements of the country, so that much remains to be done in this direction. The settlement of the Reserves already proclaimed has not been thoroughly completed, and especially have we to deal with the Hill-men who, from time immemorial, have been accustomed to make their clearings in these forests wherever they pleased. Outside the Reserves, there is a very wide extent of forest land, at present quite unprotected, and much of it seldom visited by the officers of the Forest Department. Here timber is often felled surreptitiously, and we know nothing of it. The improvement of the forests outside the Reserves will be a work of much greater difficulty than the protection of the Reserves themselves.

1150. In order to thoroughly supervise our forests, whether with a view to their conservancy and improvement, or to the felling of timber therein, it is necessary that we should have lines of communication running through them in all directions. At present, we are very badly off for roads or paths of any kind, and many parts of the interior have never been visited by any one. Throughout the length and breadth of the land, no money was ever spent by the Forest Department on roads till two years ago, with the exception of a small sum expended on a foot-path which runs for about 30 miles from the village of

Side notes:
Our five Reserves only just proclaimed. They have not yet been demarcated, nor has any work been commenced on them. The area is too small, and more Reserves will have to be selected.

Roads absolutely necessary for supervision. Our forests almost destitute of them. Much work to be done in improving waterways and making slides &c.

§ The demarcation to which I allude will be only preliminary until the Reserves are finally settled, but demarcation is necessary to allow where the area closed to cultivation commences.

Könniyūr through the teak plantations to the Travancore-Tinnevelly boundary east of Acchankōvil. This was opened about a dozen years ago, or rather, the old track was cleared out at a trifling expenditure, and it is now kept open, and a ferry-boat is maintained on the river at Thora. This path is not suited for cart-traffic. With this exception, I do not know of a single road or foot-path having been cut by the Forest Department before the year 1065. Some cart roads and foot-paths have been cut at Government expense by, or at the request of, the planters, and these have often been of great assistance to us in getting out timber, but they would never have been made but for the exertions of the planters in applying for them, and in carrying out the work themselves in most instances. In those forests where there are no estates the means of travelling are either by boat, or by narrow tracks impassable for ponies and leading steeply up and down hill as the wild animals originally made them. They are usually much overgrown. Sometimes, advantage may be taken of the intricate foot-paths cut by the hillmen from one clearing to another, but these cannot be depended on, for, on a second visit, all traces of them will have disappeared, as the clearings which they connected have been abandoned. Thus, travelling in the interior of our forests is most fatiguing, occupies very much time, and is often very unsatisfactory, as the paths often do not lead in the required direction. During the last two years, about 4,000 rupees have been expended by us in widening the bridle path from Ariyankāvu to the estates near Arundel for a distance of about 4 miles for the removal of timber, and also in making three of the sub-depôts outside the forests and near Puliyara accessible for carts. From a consideration of the large areas of forest that we possess, and the remoteness of many parts of them, some idea may be formed of the large amount of work that lies before us in opening up lines of communication everywhere. Besides making roads to visit the forests, we shall find as time goes on, that there is much to be done in the way of improving the lines of exit to reduce the cost of removing our timber. On some rivers a few hundred rupees judiciously spent in blasting rocks will make forests accessible by water, where previously the only means of working down the timber was by land. In other places, the cutting of slides or the removal of obstructions will cheapen the cost of removal. So obvious is this, that one or two enterprising merchants have gone to the expense of improving the timber-ways, in order to reduce the cost of transport, and in every instance where this has been brought to my notice, I have endeavoured to give them the full benefit of their improvements, by preventing other people from using their slides. Ordinarily, the holder of a permit to fell trees goes into the forest, and, choosing the nearest line from the tree to be felled to the river or road by which he means to transport the timber, clears all the brushwood off it, and cuts down numbers of young trees to form rollers. After removing the one or two trees he requires, the slide is abandoned, and another is cut. The expense is much greater and there is much more waste of timber than if permanent slides were made to serve each valley or block of forest. None of these works of improvement have yet been commenced.

1151. In order to examine and control the forests properly camping sheds are necessary, for, in a country of such heavy rain as Travancore, tents are almost useless. Of the buildings that belong to the Forest Department there are three good bungalows, one at Könniyūr, one at Puliyara, and one at Malayāttūr, besides a smaller bungalow at Könniyūr which is at present uninhabitable, and will have to be replaced by a new one. Then, there are the usual offices at Quilon, Malayāttūr, Könniyūr, Nagercoil, Puliyara &c., smaller offices at the depôts for the sale of timber, and various Amindars' cutcherries and watch stations scattered about the country. All these lie outside the forest area, are in the open country, and are used for the transaction of business, while inside the forest area we have but few buildings. There is a small bungalow near Panagudi in South Travancore, but the site was badly selected, as the wind, blowing through the Shūravali gap, strikes the building with great force, and it is now falling into ruins. Leaving this, we have to travel more than 50 miles by the map and over 100 in reality before the next building of any kind within the forest area belonging to the Forest Department is reached. This is at a place called Pilmotta, at an elevation of about 3,500 ft., near the Travancore-Tinnevelly boundary, and not far from the Chimmūnji peak. A combined bungalow and watch house, surrounded by an elephant trench, has been lately built here for us by the D. P. W. and a couple of peons are

The buildings belonging to the Forest Department are almost all outside the forest area. A great want of camping sheds inside the forests.

43

stationed here to put a stop to the smuggling of cardamoms. An Aminsdar with a staff of guards to protect the Reserve has been stationed at Kulatthupura, but no building has yet been erected for them. Near Ariyankāvu there is a rough building on the teak plantation for the accommodation of coolies. At the Kōuniyūr teak plantations there is a small wooden bungalow and a permanent camp surrounded by an elephant trench : a small bungalow and camp with an elephant trench are located near the keddah on the Kalār branch of the Rāni river ; and, on the hills above, another similar camp is situated at Pongampāra, where there are cardamom gardens. This is occupied only during the cardamom season. North of this place there is not a single permanent camp or building occupied by the Forest Department within the forest area, excepting on Peermerd, and the whole of these forests, which form by far the greater part of the forest area of the State, have hitherto been only occasionally visited by the officers of the Forest Department. During the elephant capturing season, indeed, camps are formed at different places, and these are occupied for about 3 months in each year, but they are merely used for this purpose and are abandoned as soon as the season is over. On Peermerd there are depôts at Kumili and Rāmakkal where timber is sold, and a permanent staff is maintained at each of these places, but the buildings are of a very temporary character. On the whole then our forests are almost destitute of any camps where a halt can be made, and whence visits of inspection in different directions can be arranged. Except in the neighbourhood of Kōnniyūr such a thing is impossible. Elsewhere, wherever the Forest officer goes, he must, unless he has time to get huts put up for him by the Hillmen, either carry a tent (if in the dry weather) or run up a grass hut for his accommodation. He will find nothing ready for him.

1152. In paras 889 to 894 I have shown the great damage done by forest fires, and that about 1,500 square miles of forest land are annually burnt over to the great detriment of the timber, young and old, growing thereon. No attempt has yet been made to commence fire-protection, and it is a subject of no small difficulty, especially in the case of land outside the Reserves where the grass and timber forests chiefly lie, but it is a difficulty that must be faced ere long.

No fire-protection.

1153. Similarly, in paras 867 to 888 I have pointed out the great damage done by hill-cultivation, carried on chiefly by low country people, and I have suggested some remedies to check it. When forest land has been formed into a Reserve this destruction of forest can be prevented, but it is different with the land outside the Reserves where such cultivation is permitted. There are certain rules in force against felling large trees, but when the culprits are caught, which is not always easy, it is necessary to prove against them an intention of "mischief" in order to obtain a conviction, and this is difficult, besides which, many Magistrates refuse to entertain such a charge. Many of the cultivators also evade the rules by leaving the big trees standing and clearing only the smaller ones, and, if the large trees are burnt, it is of course not their fault. The most effectual remedy seems to be to immediately increase the area of the Reserves, yet there are reasons why this work should not be hurried over.

Difficulties connected with the prevention of hill-cultivation outside the Reserves.

1154. Turning from the protection and improvement of the natural forests to artificial reproduction, we find that the area planted with teak was 149 acres at Malayāttūr, 954 at Kōnniyūr, and nominally 150 at Ariyankāvu, making a total of 1,253 acres. At Malayāttūr the first plantations were begun in 1042 (1866–67), and further plantations were opened till 1049 (1873–74) when work there was stopped, except as regards the maintenance of the existing plantations. The work has suffered from several causes, which have interfered with the uniform growth of the trees. First of all, the land selected was not all equally good. The plantations run in a narrow strip along the banks of the Periyār about four miles above Malayāttūr, the greater portion of them lying on the southern bank, and though the soil is mostly alluvial, where the trees grow remarkably well, there is some of it in which laterite comes to the surface, and here the teak trees have made very poor growth. Their size is therefore very unequal, for while some trees are unusually fine, others are very much below the dimensions they should have attained in the time. Another enemy from which

Teak plantations. acres at Malayāttūr. 149

these plantations have suffered is fire, which has, almost every year, swept through them. The consequence is that about half the area has been burnt out, and only about 80 acres remain. The weeds too have not been regularly and systematically kept down, and though the trees are now so large that they do not appreciably suffer from them, yet the effects of former inattention are shown in the unequal growth of the timber and the blanks and open spaces in the plantations. The Malayāttūr plantations were at the commencement very unhealthy, and coolies could never remain on them without getting fever, so that the cost of opening was higher than it should have been. Wild animals, bison as well as elephants, did much damage to the plantations, and, for these and various reasons, operations were discontinued after 1049, and all further extensions have been made near Kōnniyūr. The position of these plantations is excellent, and when the timber arrives at maturity it can be felled into the Periyār, and be floated down to Virappura at a trifling cost. Being now from 18 to 25 years old, we shall be getting thinnings soon that should more than pay for the upkeep, but the trees will not attain a large size sufficient for a general felling for about another 40 years.

1155. At Kōnniyūr also a commencement was made in 1042 with 38 acres,
The Kōnniyūr plantations. 954 acres planted. Well situated.
and by the end of 1049, 200 acres had been planted : 130 more were added in 1054 and 1056, and 394 in 1057. Finally, 230 acres were opened in 1059 making a total of 954. These are not all in one place. The first plantation was made a few miles to the south-east of Kōnniyūr on a stream running into the river, and at a place called Aravi pālam, and in 1057 the Vālapāra plantation, of the area of 70 acres, was opened near Kālanyūr about 6 miles south of Kōnniyūr. The other 846 acres are situated on the two banks of the Acchankōvil river, and the plantations form two narrow strips running along the river sides for several miles, the nearest plantations to Kōnniyūr being only some 4 miles distant. The site selected and the character of the soil all over this large area are admirable. The trees can be felled directly into the river, and can be floated down to the depôts at a trifling cost, when the time comes for their removal. It might have been better, perhaps, if the plantations had been made a little wider so as to concentrate the work, but to do this it would have been necessary to take in land whose soil was less fertile, and where the growth would have been less rapid. The two smaller plantations, though well situated as regards the facilities for the removal of their timber, and though they possess good soil, are isolated, and the cost of working them will therefore be greater, for large areas can be worked at a much cheaper rate than small ones.

1156. In spite of these advantages of climate, soil, and position, the Kōnniyūr
These plantations have not done so well as they should becatse the money allowed for them was not sufficient.
plantations have not been so successful as they should have been, nor has the growth of the trees on them been equally good, for several reasons. The chief mistake made was in attempting to open these plantations at too low a cost, and especially in endeavouring to dispense with weeding. The usual estimate for opening land including weeding and up-keep for a year has been 18 rupees an acre, just half the sum allowed on the Nelambūr plantations, as shown below, and though labour may be slightly cheaper in Travancore, any one who thinks over the subject will see that one weeding a year, when the plants are young, is not nearly sufficient in such a forcing climate as we have, to keep the teak from

Memorandum on the planting of teak by Mr. J. Ferguson of Nelambur, taken from Logan's Malabar.

1st year.				3rd weeding...	...	1—4.
Felling and firing	...	20—12.		4th weeding .	...	1—4.
Pitting and planting	...	3—6.		Nurseries and seed...	...	3—8.
1st weeding	...	...	2—8.	Contingencies	...	2—0.
2nd weeding	...	...	1—4.			
					Rs.	36—0.

2nd year.		4 weedings and 1 pruning	..10— 4.
3rd year.	3 do. and 1 do.	...	7— 4.
4th year.	3 do. and 1 do.	...	1— 8.
5th year.	3 do. and 1 do.	...	3—12.
6th year.	2 do. and 1 do.	...	2—12.

Rs. 27— 8.
Rs. 36— 0.

Total expenditure for 6 years Rs. 63— 8 per acre.

being choked. The sum mentioned, too, does not make any allowance for filling up the vacancies caused by plants dying, nor, for the money, can proper pits be cut for the reception of the plants. In order to do the work within the sum allowed, the land selected was not such as was covered with heavy forest, but was, usually, such as had been previously cultivated, and upon which there were grass or low shrubs growing. The cost of clearing this was very light, but the weeds, and especially the grass, have been most troublesome ever since, and the latter has in some places completely killed out the teak. When I visited these plantations in 1887 I found that they had not been weeded for 10 months, and that on some of the clearings the teak trees, which were then 3 years old and should have been more than 20 ft. high, were completely buried in the long grass, and could not be seen. To remove this grass would have cost at least 5 rupees an acre, and the sum allowed, when sanction was obtained to clean the plantations, was nothing like this, so that all that could be done was to beat down the grass and give the teak plants a little fresh air for a time. In June 1891 I visited these plantations again with the intention of abandoning the worst parts, but I found that owing to a more liberal expenditure on weeding the teak trees had succeeded in getting the better of the grass, and, though they are small for their age, they are healthy, and are deserving of further expenditure on them. The older plantations were more carefully opened, and have suffered less than the later ones. I am now about to have a new survey made of the plantations that are still kept up, and I expect to find that about one-quarter of the area has been abandoned, while of the rest, another quarter or more of the whole has been very much retarded for want of weeding, and the trees are small for their age, while the remaining half is good. The same amount of money spent on a smaller area would have given much better results. †

1157. Another mistake made on the Könniyür plantations was to over-lop the
A great mistake made in over-lopping the trees in these plantations, and especially in lopping during the monsoon.
trees. When Forest operations were first commenced in India it used to be the custom to remove all the side branches as the trees grew, leaving only a small tuft of leaves at the top of each, the object being to get clean and straight boles. But it was soon found that the removal in this manner of so much foliage seriously retarded the growth of the trees, inasmuch as the annual increment depends entirely on the extent of leaf surface during the growing season. This system of lopping has been completely abandoned * in European countries and in other parts of India, the trees in plantations being purposely put in very close together that they may draw up and protect each other, and such side branches as do grow are very weak and soon die back when the sun no longer reaches them. At Könniyür the lopping, besides being very heavy, was usually carried on during the monsoon or period of growth, when the weeds were being removed. Now, it is well known that wounds inflicted on trees and shrubs during their period of growth do much more harm than those given when the trees are at rest. Nature tries to replace what has been removed. The lopping of trees during the growing season is like damming a stream in the monsoon. You close up one channel, and the water bursts out in tiny streamlets at innumerable points, so the result of removing the branches of the teak trees § during the growing weather is that great numbers of small shoots spring out all up the stems of the trees, and these have again to be removed, causing knots in the wood. In any case, the teak is not a tree that requires, or will stand, much lopping, for it grows with such rapidity at first, that trees which have never been touched are often seen broken off or blown over by the strong wind. I have already alluded to the damage done by the floods in 1057.

1158. The Malayāttūr and Könniyür plantations mentioned above were
Value of the plantations.
valued in 1886 (see Administration Report for 1061 M. E.) at 3¼ lacs of rupees. After the survey of them has been completed I propose to make a valuation based upon actual measurements and enumeration.

---

† Since writing the above, the plantations have been surveyed and turn out to be only about 500 acres, instead of 954. This too includes much land where teak has not thriven.
* Col. Campbell Walker's Report on Forest management, p.p. 164.
§ There is also a great risk in lopping teak in the wet weather on account of the ravages of the borer described in para. 961 of this Report. It this insect once got into our plantations it might do immense damage.

1159. In the monsoon of $\frac{1890}{106}$ a commencement was made on a plantation of

Ariyankāvu teak plantation.

teak at Ariyankāvu, and about 150 acres of land were cleared, but the monsoon was unusually light, and there was some hitch about the regular supply of money for the work, so that the planting was not well done. I have this year been endeavouring to fill up the vacancies, and to plant up more of the land which was cleared, for only a portion of the area cleared was planted. At present, it is impossible to say what is the actual area under teak, but I will get it surveyed. There is some little doubt as to whether the rains are sufficiently heavy for the successful growth of teak here, but trees of large size in the natural forest are found in the neighbourhood, so that there should be no fear that the planted teak will not succeed. The soil here is excellent, and the site favourable as regards the chances of selling the timber, for there is a very great demand for teak at the Puliyara depôt, only 6 miles distant, and if the railway is made, the facilities for selling the timber will be still greater, for it will pass almost through the plantations. The only objection to the formation of a plantation here is that small areas always cost more in proportion than large, but this objection is overbalanced by the high price we shall obtain for our timber in Tinnevelly, and, after a year or two's trial of the locality, we shall, I hope, be able to extend the plantation, and so reduce the proportionate cost.

1160. An attempt to open a sandalwood plantation was made many years ago

Attempts at sandal-wood planting have all failed. The trees found at Ariyankāvu. Mahogany.

at Kōnniyūr, but apparently not on a large scale, and it proved a failure. In $\frac{1883}{1068}$ about 50 acres of land near Quilon were planted with sandalwood, but the enterprise was doomed from the first, the soil being poor, the site exposed, and the elevation much too low. The land was abandoned very soon. As stated in para 989, there are not many places in Travancore where sandalwood will grow and yield a wood that is scented, though there is no difficulty in obtaining scentless wood. About three years ago, a few hundred trees of good wood were found growing above the Ariyankāvu Pass. They had been sown through the agency of birds, who carried the seeds from an old tree which had been planted in the Pass itself. These trees can take care of themselves, with a little assistance from us in the way of clearing, and no great expenditure in the way of planting up or extending the area is recommended. A hundred or so of mahogany trees have been planted in various places and they have done well, but no attempt has been made to grow them on a large scale. At Malayāttūr about an equal number of Ceara rubber trees were planted some 10 years ago and have thriven well, but the yield of rubber is trifling.

1161. I must not omit to mention, as a measure of conservancy, the closing of

The conservancy of the "Kāda-kāval" lands.

the "Kāda-kāval" lands overhanging the British village of Panagudi, and situated in South Travancore. These lands were handed over to the Forest Department in $\frac{1882-3}{105s}$, as explained in para 1143, because the timber on them was disappearing. All cultivation within the area was prohibited, and the felling of trees forbidden. The people are, however, allowed to collect firewood and to graze their cattle, upon payment of certain fees. The prohibition against felling is a step in the right direction, but it will be further necessary to arrange for the closing of portions of the area to all entry, and the opening of only small portions in rotation to firewood-collectors and cattle.

(2) *Arrangements for the felling and removal, or delivery of the timber.*

1162. As already explained, the following trees are royalties, and may not be

The royal timbers.

felled by any one who has not obtained permission or a license from the Conservator of Forests. They are teak and kōl-teak, blackwood, ebony and sandalwood. The last named is only indigenous in Travancore in the Anjināda valley, and no revenue accrues to Government from it. The Pūnyātta chief is said to fell a little every year, but he has no authority to do so. Ebony also brings no returns to speak of, because only a tree now and then is felled for some special purpose; but teak and blackwood yield a large revenue. As already explained in para 916, kōl-teak is, properly speaking, only stunted teak of poor appearance, which would never attain a large size however long it was left to

41

grow, but the term is more loosely applied to all teak under 10 vannams (12½ inches) quarter girth at the middle of the log, and I shall so use it in future. Any one felling a royalty growing on his own property, unless he has permission to do so, is liable to be charged under section 426 I. P. C. for mischief, or, after a departmental enquiry, he can be called on to pay a prohibitory assessment for the trees he has cut down. In the former case, if the trees can be found, they have to be handed over to the Forest Department, by whom they are either taken to the depôts, or are auctioned where they lie; in the latter, the owner of the property is allowed to keep them. An exception exists in favour of the managers of coffee estates, who are at liberty to use the royalties growing on the properties they supervise, on payment of certain rates, after they have obtained the permission to fell the trees.

1103. All the royalties in the country are, therefore, felled by Government agency or by contract. In South Travancore, and for the Puliyara depôt, teak and blackwood are felled by the depôt-contractors, who deliver them at the depôts like other timber, and they are sold to all comers. In Central and North Travancore, these woods are all brought down in log, and are collected at Quilon, Vīyapuram, Cottayam, Vettikātmukku, and Virappura. Contractors and Government officers are at liberty to enter upon any property, and to fell and remove the teak or blackwood growing thereon, and, as a rule, the owners are only too glad to get rid of them to complain of their removal, but sometimes they do object. If a proprietor wishes the trees on his property removed, he applies to a contractor, who generally agrees to take the timber, for delivery at one of the above depôts, the proprietor bearing the cost of felling.

They can only be felled by Government.

1164. Teak and blackwood are found in the open forests on the lower slopes of the hills, where the soil and other conditions are suitable, and also at the foot of the hills and in the low country, but neither tree ascends the hills to any height. The finest trees of both species are found in the interior, and those met with in the low country are kōl-teak, or blackwood of small size and poor colour. As I have shown in para 1141, these trees were felled almost entirely by Government agency up to the year 1054 (1878-9), and, till that date, all the teak felled was first-class, or above 12 vannams (15 inches) quarter girth. Very little blackwood was felled then. To obtain such large trees it was necessary to go well into the interior, to the Idiyara valley, Velliyāmattam and similar places, and the kōl-teak was not touched, except when a few trees were required for pagodas or churches &c. In 1054, contracts were made with private persons to deliver teak above 10 vannams quarter girth, in order to fulfil the agreement made with Messrs. Wallibhoyi and Co. of Bombay, but kōl-teak was left still untouched. At first, the number of contractors was small, and the supervision over them was fairly effectual. The Aminadars were directed to accompany the contractors' men to the forests, and to mark the trees to be felled by them, but, as the number of contractors was increased, this rule was either countermanded or fell out of usage, as it was found that the contractors were much delayed in their operations by having to wait for the arrival of the Aminadars. Since then, the contractors have been left entirely to themselves to fell when and where they pleased. In the year 1057, contracts were entered into with men to fell and deliver kōl-teak, and in 1059, as the timber felled in this way was found to be yielding a very handsome revenue, advertisements were published calling for more contractors. This attracted a large number of men, many of whom, unlike the earlier contractors, were persons of no property whatever, and took up the contracts with the intention from the first of defrauding the Government.

Teak and blackwood felled formerly by Government agency, but now almost entirely by contract.

1165. In order to understand this, it is necessary to explain that the teak trees are felled in the forests, and are allowed to dry for from one to three years, according to the size of the logs. They are then dragged by elephants to the sides of rivers or streams, whence they can be floated down to the depôts, and, when the water rises sufficiently high, they are sent down either singly or in rafts, touching at each watch station on the way and being checked there. Of the different portions of the work of delivery, the felling costs very little, the expense varying from one-quarter to half a rupee per candy according to the size of the trees

Many of the contractors received large advances which they can never work off. Consequent loss.

felled and their abundance in the locality. The cost of rafting is also comparatively small, but the great expense is on dragging the logs to the water-side. It may happen, indeed, that a number of trees can be found growing beside the river, in which case the contractor makes a very large profit, but, as a rule, the elephant-hire swallows up the greater part of the money allowed. Now, it was known that the supervision in North Travancore was very lax, and that the checking of the timber was left to the Aminadars and their subordinates, and the contractors first applied for advances to fell the trees, felling, usually, many in excess of the number contracted for. They then applied for money to bring the logs to the depôts, asserting that they had been dragged to the river-sides and were ready for floating. The Aminadars were deputed to certify to this, and, in very many cases, they told off their subordinates to attend to the work, or they were induced to give certificates for timber which was still lying where it had been felled, but which they asserted was on the river-sides and was ready for floating. It is universally believed in North Travancore, and I have no reason to doubt it, that many thousands of logs were entered in the lists which had never even been felled. On the strength of these certificates, advances were given, at the rate of 2 rupees a log, to float the timber down, but year after year has passed by and the logs have not come. In one case, over 14,000 rupees were in all advanced to a certain contractor, who died not long ago, leaving a debt of nearly 10,000 rupees. In another case, a man received 3,000 rupees to deliver 717 logs, and he brought to the depôt not a single one, nor can more than half of them be found. When the man's property was attached for the debt, it realised rather less than 40 rupees. Other contractors received 3,000, 4,000, or 5,000 rupees, delivered a few hundred logs and then stopped. Altogether, the debts due by contractors at the end of 1066 amounted to 74,540 rupees of which probably 40,000 will have to be written off if the money cannot be recovered by distraint.

1166. It is bad enough to lose a large sum of money thus, but, in addition to this, we now frequently receive complaints from sub-contractors, who say they have not been paid for their work. It was very usual for contractors to give out sub-contracts to men on the understanding that they were to fell and deliver logs at the depôts, and to get for them half the "kōlvila" or rate allowed by Government. Thus the contractor cleared half the "kōlvila" for lending his name. In most cases, the sub-contractors, after felling a large number of logs, had no money to work the timber to the depôts, and, as the contractors would give nothing, they applied to us. The Forest Department cannot of course recognize any but the original contractors, and so the sub-contractors have to lose the payment for their work, unless they can manage to dispose of the logs to some other contractor. If they fail to do this, the logs lie in the forests till we find them and bring them down departmentally, but in any case the original advance to the contractor is lost, for when the logs are worked down departmentally they cost about as much as the kōlvila, for such logs are generally those farthest from the river, all those easy of removal having been removed by the contractor, or by other people who appropriated them, and, secondly, because departmental work is always more expensive than that done by private persons.

In addition to the money-loss, a great waste through logs being left to rot.

1167. Had the number of contractors been kept down to a reasonable figure, it would have been possible to exercise some check on them, but what could 6 Aminadars, already burdened with other work, do with 135 contractors and numberless sub-contractors felling in every part of the country at once. Not only so, but no information was conveyed to them as to who had received contracts. The agreements were written out at Quilon, and, till lately, there the matter rested: the first intimation the Aminadars received of new contracts having been given was that felling men would appear and commence operations. No notice was sent to them, nor were the felling men obliged to show their licenses, so that, for a long time past, there has been no supervision over the felling of royalties in the forest. As a consequence of this, many irregularities have been going on. First of all, it is no uncommon thing to find one contractor stealing the timber belonging to another. One man will go into the forest to work down his logs, and, finding a pile belonging to another contractor, it is the easiest thing in the world for him to chip off the original names and to substitute his own. The owner will perhaps discover his loss,

The number of contractors employed far too large. Great irregularities in consequence.

when the logs are on the way to the depôt, or he may only do so after they have arrived there, and when the logs are examined it will be easy enough to see that there has been some chipping, but, as the thief takes very good care to remove every vestige of a name, it is impossible to say who is the rightful owner, if it is not the man who delivered the logs. It is no remedy to order that all logs with fresh chipping on them shall be credited to Government. This has been tried, and instances have occurred where a man has chipped off another's marks and recut them again, to prevent his getting any money for the delivery of his logs.

1168. Felling on contract without supervision has been productive of very great waste. Even those contractors who signed their agreements with no intention to defraud Government did so in the hope of making money, and they are not to be blamed if they consulted their own interests before those of the Government. Their contracts were to deliver logs at the depôts. Very frequently, the trees they felled would each have yielded a log and a piece at the top which would have sold as a sleeper. In almost every instance, these top-pieces have been left lying about, and, in the Mīnacchal taluq alone, there must have been 15,000 or 20,000 of them. Had they been brought down with the logs to which they belonged, the cost would have been trifling, and they would have fetched 2 or 3 rupees each, but it would not be worth while, now that the logs have been removed, to work down these top-pieces alone. Again, many logs are burnt in the forests through not being properly fire-traced. Again, another evil, due to want of supervision, has lately been brought to my notice. Many contractors, who have agreed to fell a certain number of candies each, have cut down trees far in excess of their contracts. One man, who had delivered all but 9¼ candies of the number allowed by his contract in last February, and has been delivering more since that date, gave out a sub-contract for 1,300 candies only last month. A second, who had a contract for 500 candies in February, and has been delivering since, admits that he has over 2,000 trees felled and ready on the hills, and that he has made a sub-contract for 600 more. A third, who held a contract for 500 candies in February, and has delivered many logs since, has just given out a sub-contract for 600 candies, and so on. Of course we can prevent these people getting any benefit from their excess fellings, by confiscating the timber, but the trees are felled and the mischief is done.

Great waste of timber, and excessive felling due to insufficient supervision.

1169. A great mistake was made in giving out contracts for kōl-teak, and in not laying down an inferior limit of size. The agreement made with Messrs. Wallibhoyi & Co. was to deliver to them sleepers of 6 vannams quarter-girth (7½ inches), in the expectation that these would be supplied by the top-pieces of the logs, or by trees of really stunted teak. The result of this has been that young and immature teak trees of all sizes from 6 vannams up to 10 have been indiscriminately felled, which, had they been left alone, would have grown into large trees in the course of time. Thus, for a small present revenue, the future supply of teak has been greatly imperilled. Another result of this is that it has given great scope for fraud, for, until contracts were given to fell kōl-teak, any one found cutting it down could be prosecuted with the certainty that he would be convicted, for no kōl-teak was ever felled except by Government. Now, a man can cut down a teak tree, either to remove it from his compound or to use it in his house, and, if he is caught, he says he was felling for such and such a contractor, and it is not difficult for him to get the contractor to connive at the fraud.

The kōl-teak contracts have been especially unfortunate.

1170. The contracts given specify that the timber shall be felled and delivered at such and such depôts, so the contractors have to bring it to one, two, or three of the depôts already mentioned. There it is received by the officer in charge of the depôt, and is hauled up on shore. The raftsmen obtain receipts, and the contractors are entitled to be paid as soon as the logs are measured and sold to any one. If not sold at the fixed rates, the timber is auctioned, and the contractor gets one-fourth of the selling price. The rates first paid to contractors were 4½, 3 and 2½ rupees per candy, respectively for the three classes. As timber is more difficult to procure now the rates have been raised to

The rates given to contractors for felling and delivering the logs at the depôts. The kōl-teak contracts very profitable to them.

5¾, 4¾, and 3½. The price given for logs of the first class hardly pays the contractor under any circumstances, because large elephants are required to drag these large logs; moreover, there is very little first class teak in the low country, and the cost of getting it from such places as the Idiyara valley is prohibitive. (See para 939). But, if the contractor is a man of judgment, he can and does make very large profits from the kōl-teak contracts. If he is careful to fell his trees along the river-sides so as to save elephant-labour, and if he keeps a sharp look-out that the logs are not stolen, he may make upwards of 1½ rupees profit per candy. Not long ago, I had some kōl-teak felled a little above the Vettikāt depôt, and although the contractors had been over the ground several times, selecting and rejecting, the trees (which averaged about ¼ candy each,) only cost 2 fanams each to cut and trim, and from ¾ to 1 rupee per candy delivered at the depôt, showing that, when the contracts for kōl-teak were first given, very handsome profits could be made.

1171. In his Administration Report for the year 1056 the Conservator of Forests remarked on the great gain resulting from the introduction

No comparison to be drawn between the old rates at which teak was worked down, and the much lower rates at which contracts were given.

of the contract system. Previously, the cost of felling and delivery by departmental agency had been 10 rupees per candy, and this had been subsequently reduced to 6 rupees, but, at the time my predecessor wrote, contracts had been taken up at 4½ and 3 rupees for 1st and 2nd class timber. The difference is considerable, and, doubtless, there seemed every chance that the work would be done equally well, but, as I have shown, these rates do not pay the contractors, and the quantity of 1st and 2nd class timber brought down now is only one and six per cent, respectively, of the whole supply, so that, had no kōl-teak contracts been given, the supply would have been inadequate, and there would have been a necessity to raise the rates to get a sufficiency of timber. The comparison therefore is hardly fair, because the 10 and 6 rupees above referred to were spent on 1st class teak which could only be obtained in quantity from the forests of the interior, whereas the 4½ and 3 rupees (now raised to 5¾ and 4¾) were given for isolated trees of 1st and 2nd class mixed with kōl-teak. These rates did not pay the early contractors who have nearly all failed, but those who obtained contracts for all 3 classes have made these rates pay because they have worked down the larger teak in company with the small. Had no kōl-teak contracts been given, the rates must have been raised long ago, to enable the contractors to go into the interior for their timber as we were doing. Nor can any contrast be fairly drawn between the average rate at which teak was worked down by departmental agency, say 10 rupees a candy, and the average rate paid now, say 3 rupees. The former was for first class teak fetching from 15 to 18 rupees a candy, and the latter is almost entirely for kōl-teak, and the average selling price is not much above 10 rupees. When we consider the inevitable waste that results from an unsupervised contract system, and the direct money loss that has resulted from the failure of contractors, I do not think we shall find the present system much cheaper than the old one, and, when we remember the uncertainty as to the quantity and time of arrival of the timber brought by contractors, we may well doubt if these apparently cheap rates have not been obtained at too great a cost.

1172. It is easy for us at the present time to be wise after the event and to point out the mistakes that have been committed. Ten years

Steps to be taken in future.

ago my predecessor was perfectly justified in drawing the conclusions he did. But certain unforeseen circumstances occurred which upset his calculations, and we have now to correct the errors which time has made apparent. All we can do is to endeavour to put things in order and to make a fresh start. Since I took charge of the Department I have not given out a single new contract, and I am endeavouring to close all the old ones. When this has been done, or when all but a few of the best contractors have been left it will be time enough to draw up new contracts, if they are to be given. Of the 135 contracts, more than ¾ have expired, but they have not been closed, as almost all of these contractors are in debt; of the other contracts several run for another two years, but the contractors have in some cases brought all the timber they agreed for, and their contracts may be cancelled. The logs of those contractors who have hopelessly failed are being worked down departmentally, so far as we can find them, but their number is much below what it should be, and the cost of delivery will quite amount to the contract

45

rate, so that the debts of the contractors are not likely to be reduced, but we get the timber and make the usual profit.

1173. To sum up, it may be said that the contract system now in force has failed because :—

Causes of failure of the contract system.

(1) The supervision was inadequate.
(2) The number of contractors was too large.
(3) Contracts for kōl-teak should not have been given.

1174. Besides what is brought down by contractors to the depôts of Central and North Travancore, and what is being worked down by us on behalf of contractors who have failed, a few hundred logs every year are felled and delivered by the Aminadars from places where the contractors have not been at work. All the teak and blackwood felled is thus delivered in the form of logs at the depôts mentioned, with two exceptions. The first exception includes the kōl-teak sold to applicants for religious purposes. It has always been the custom to allow kōl-teak to be thus sold in any part of the country to build or repair churches, mosques, pagodas or Nambudries " illams. " Application has to be made for a certain number of candies at a given place, and, on payment at a fixed rate (lately raised to 10 Rs. a candy, besides the cost of felling,) the required number of trees is cut down and the timber is handed over to the applicant. Requisitions from the D. P. W. are treated in the same way. The second exception refers to Shencottah, Peermerd and South Travancore. Here teak and blackwood are treated as ordinary woods and are brought to the depôts of Puliyara, Quilon, Trevandrum and Nagercoil, and to Kumili and Rāmakkal on Peermerd.

Besides what is delivered at the depôts some teak and blackwood sold to private persons &c

1175. Reference has already been made in para 1140 to the Puliyara depôt near Shencottah, the most profitable of its class in Travancore. All the timber delivered at this depôt, whether logs, sawn materials, or parts of bandy wheels, &c. is supplied by a single contractor who has the sole right to cut and prepare these materials for this depôt. The woods delivered are teak, blackwood, kongu, vēnga, ven-teak, mayila, chokkala and occasionally one or two others, and the rate paid to the contractor varies from 6½ to 7½ Rs. a candy. At Quilon there are two contractors who supply teak, kongu or thambagam, vēnga, and sometimes a little thēmbāvu. The rate paid to the contractors is 9½ Rs. a candy. At Trevandrum the woods supplied by the two contractors are thambagam, vēnga and thēmbavu and a little blackwood, but teak is not obtainable in the neighbourhood. The rates vary from 9 to 10 Rs. a candy. At Nagercoil the timber brought for sale includes teak,kongu, nāngu, vēnga, mayila, ānjili, ven-teak and other woods. The rate paid to the contractors varies from 5¾ to 12 Rs. a candy. The number of contractors has been 5 or 6, which is far too great ; there have been endless disputes about the ownership of the timber delivered, one contractor stealing from another again and again. The depôt has been crowded with timber far in excess of the demand, and prices have therefore fallen. I am now reducing the number of contractors to 2 or at the most 3, and am also arranging that the supply shall not in future be so excessive. At all these depôts, if the timber supplied is bad and has to be sold by auction, the contractor gets only half the selling price. At the Peermerd depôts of Kumili and Rāmakkal the arrangement is slightly different, the contractors engaging to bring their timber to the depôt to be measured, and then to remove it for sale themselves on payment of 5 or 5½ Rs. a candy. Bamboos are brought for sale in the same way as timber and in South Travancore can only be procured from the depôts.

Rates paid for the supply of timber at the different depôts of Puliyara and those in South Travancore.

1176. The agreements made with the contractors forbid the felling of any trees under 2½ candies in cubical contents, but as the greater part of the timber delivered is sawn or axe-squared this rule is often disobeyed, and, as the felling is carried on entirely without supervision, it is impossible to tell, when the timber is brought to the depôt, what was the size of the trees from which it was cut. After making their agreements with us the contractors send their men to the forest and the timber is prepared. A pass is given by them to the nearest watch station, specifying the number of pieces and the cubical contents of the logs or sawn materials despatched.

Method followed for the delivery of timber at the depôts above mentioned.

The clerk at the watch station gives a waybill, retaining the contractor's pass, and this waybill is checked at each station till the timber arrives at the depôt. Reports are sent in every five days from each watch station to the head-office at Quilon, specifying the quantity of timber that has passed the station, whether brought down on contract or under permits. On arrival at the depôt the timber is measured and entered in the stock book, all bad timber being put on one side as rejected, and the contractor is entitled to receive payment at once on all the good timber that he has delivered, and for which he receives printed receipts. Until quite lately, the contractors were allowed to stack their timber in the yards, but it was not measured and entered in the stock-book, nor was the contractor able to get payment for it until he could induce a purchaser to buy it. Thus, timber was often left for months in the depôts unmeasured, at the risk of the contractor, who could neither watch it himself nor claim a compensation if it was stolen, as receipts were not given till the timber was sold. This system saved the depôt-superintendents and the watchmen a great deal of trouble, as they were practically not responsible for any loss, but it gave room for much fraud and many complaints from the contractors, who took care to indemnify themselves in other ways for the losses they had suffered. The quantity of timber to be supplied at the depôts is fixed by agreement, but if the sales do not go off readily the officer in charge of the depôt has the power to order that no more be brought down for a time.

1177. It I have explained myself clearly, it will be seen that, with the exceptions mentioned, all the royal timbers have to be brought to the depôts and can only be obtained there. It is different with *The permit system.* *Permits how obtained* *and the rates charged &c.* the other woods. In South Travancore and in the forests which supply the 6 depôts, with timber those trees which are required for sale at the depôts may not be felled. and their timber can only be obtained at the depôts, but all the other woods can be procured on permit, and in Central and North Travancore every species of tree can be felled, and any kind of timber can be procured on permit, the royal timbers only excepted. Permits can be procured at Quilon on the applicant's stating how many candies of timber he requires, the kinds, where it is to be felled, how it is to be brought down, and what length of time he wants for its removal. On payment of the ordinary fees he then receives his permit, and if he wishes to fell in a Reserve he is also supplied with a pass to enable him to enter it, and he is then at liberty to begin felling when he likes. In the case of one Reserve we charge double rates for thambagam and ânjili boats, but with this exception no difference is at present made between Reserves and other forest. The rates charged run from 5 fanams (about 11 annas) for common wood up to 2 Rs. a candy .for ânjili, jack and thambagam in log, and nearly double for sawn materials. For mango planks only 3 fanams are charged, and for logs of nedunâr (Polyalthia fragrans) only ¼ rupee each, and for bamboos a rupee per hundred. The logs must be not less than 2½ candies each, except nedunâr. Dug-out boats are charged for according to the kind of timber used and their diameter, but the rates levied are much too low, being far below what would be charged if the same log was worked down as a log or sawn up into materials. Thus, a thambagam boat 30 virals broad and 12 koles long would be charged only 4 rupees, whereas the same quantity of timber in log would cost the permit-holder 23 rupees, and, probably, in cutting the boat, a piece almost as large would be rejected, as being too much for the one boat but not enough for a second. The rates for boats will therefore have to be adjusted.

1178. The permit obtained, the holder of it has to get his timber ready. if he has not already done so before applying, which is said to be *Procedure to be follow-* *ed in removing timber* *obtained under permit.* often the case. He then brings it to the side of the road or the river whence he can cart or float it to its destination, and applies to the Aminadar of the district to stamp it. The Aminadar measures the logs, stamps them, and enters the measurements on the back of the permit. The timber is then removed by the route specified in the permit, being stopped at each watch-station on the way and examined there. If all is correct, the watch-pillay notes on the back of the permit that he has examined and passed the timber, and it goes on its way to the next station. If there is anything wrong, the pillay detains the timber and reports. Not unfrequently, it is found that it has been stamped by the Aminadar within the time allowed, but that the last day has passed before it has reached the watch-station. In cases of this sort a fine of ¼ rupee a candy or one rupee per boat are usually levied, but, if the time has expired before the Aminádar has stamped it, the seigniorage has to be fined again, unless the permit-holder can show that the delay was caused by the Aminadar. If unstamped timber is caught going down the rivers or along the roads it is confiscated and sold by auction, unless it is being moved under a free-permit, as explained below. The informer or seizer gets ⅓ price of jungle wood confiscated and ¼ price of teak and blackwood. At present nearly all the watch-stations are provided with stamps, but I do not see the use of this, for the Aminadar is the only person authorised to stamp timber, and it is liable to lead to fraud. Permit-holders often endeavour to get an order that their timber may be stamped by a watch-pillay, alleging that it is not easy to find the Aminadar, but to allow this is very dangerous, as the Amina-dar and pillays hold a check on each other. Permission used sometimes to be granted, and I remember catching some unstamped timber at Erättapětta three years ago, which had, according to the permit, to be stamped at the lowest station of the two on that river, and had I not caught it, it would assuredly have been run past the station during the night and never have paid any seigniorage to Govern-ment, the permit being used for some other logs.

1179. The relative advantages of a depôt or a permit system have been much debated. Both systems are open to much abuse unless care-fully supervised. The depôt-system was tried for 3 years, as I have noted in para 1145, and it was abandoned in favour of the old seigniorage-system, which is certainly easier for the people and is better adapted to their wants. The arrangements are now superior to what existed in 1059 but there is still much waste and a good deal of smuggling, owing to the supervision being inadequate and not of a proper kind. Permit-holders are allowed to fell just where they please, and they frequently cut down several trees before they get one to suit them, or they find, after the tree has been cut, that it is inconveniently far from the road or river, so it is abandoned. In dragging their timber they use rollers cut from the forest around, thus destroying hundreds of nice young saplings which would grow into large trees in time. No one pays the least attention to them till the timber is ready for removal, and no assis-tance in improving the slides and road-ways is given them by Government. Evi-dently more supervision in the forest is required, and more concentration of oper-ations.

The permit system more convenient for the people than the depôt-system, but it wants careful super-vision.

1180. There are altogether 42 watch-stations scattered through the country on the road and river sides, with one pillay and two watchers at each. Their duty is to check the timber passing, to recover fees for bamboos and firewood, and to send in reports every five days. I am very doubtful of the efficacy of this sort of check on the passage of timber. In the first place it is a very easy matter to collect the timber for removal by river half a mile above a depôt, and then to run past at night. Even if the men are on the watch, the strong current hurries the logs away, and long before the boat can be got ready the logs are out of sight. We hear, too, of very few captures at night, and the pillay and his men usually prefer to sleep in their homes and let the watch-station take care of itself. Even in the day time, the station will be usually found unoccupied, unless a visit is expected. In the second place, the men are poorly paid, and a timber merchant will usually pay a good large sum to get his logs passed free of seigniorage. And lastly, the timber mer-chants are mostly men of wealth and influence, and the watch pillay can probably do very little to harm them or to stop their smuggling. The pillays are, indeed, in a very difficult position, with many temptations assailing them, and very little assist-ance to keep them straight. When there is perhaps only one watch station to be passed and a timber merchant has brought down a quantity of timber far in excess of his permit, is it any wonder if the pillay is induced to pass it as the proper quant-ity or quite free ?

The watch-station sys-tem a very slight check on smuggling.

1181. I have had several complaints from merchants of the inconvenience of the present rules. According to the old custom, the seignio-rage was levied on the log and not on the candy, and there was consequently no measuring. The logs were merely counted at each watch station and allowed to proceed. Now that each log has to be measured and compared with the measurements on the permits, there is great

Merchants complain about certain details of the permit system.

delay, and if the merchant is in a hurry he has to give the pillay a douceur. Another complaint is that the rule prohibiting the felling of trees of less than 2½ candies in cubical contents is hard. The object of this rule is to prevent the cutting of small trees, and if logs of less contents than these are brought down, they are charged as 2½ candies each. The merchants say that the same object could be secured by fixing a minimum of girth and leaving the length, and therefore the size of the log, at their discretion, and that, since for the transport of such large logs over rough ground very powerful elephants are required, they are thus heavily mulcted by having to pay high rates for the timber. Perhaps we may be able to adjust these matters later on.

1182. Hitherto I have been speaking of felling timber from Government land. Different rules prevail in the case of other lands. Petty chiefs,

Seigniorage has to be paid for timber grown on Jenmom land &c. when moved outside the area.

large land-holders or " Jenmies," and the owners of coffee estates are allowed to fell and use any timber, but royalties, growing on their land, free of any charge, but as soon as the timber is moved off their land, whether by land or by water, it becomes liable to our seigniorage. The Pūnyātta and Chenganūr chiefs charge their own rates within their own territories, but the timber cannot be moved beyond them without also paying our charges, and permits have to be obtained in just the same way as if the timber was brought from Government forests.

1183. Owners of tax-paying lands come under yet another rule. They are at liberty to fell and use any trees (royalties excepted) growing on

Timber grown on tax-paying land is free, so long as it is used on the land or conveyed by land, in which case a free permit has to be obtained. If taken by water seigniorage has to be paid.

their properties, and no charge is made on them, but if they wish to move the timber by land, or if they sell it and it has to be moved by land, they must apply to the head-office at Quilon enclosing a certificate from a Forest-officer or Provertthikāran saying the timber grew on their land, and giving the quantity to be removed and its destination. A free permit is then given and the timber can be moved, but if transported before this permit has been obtained it is liable to be confiscated. The reason that it was found necessary to pass this rule about obtaining a certificate was that a great deal of timber used formerly to be smuggled but of the forests and passed as timber from private lands. But, as regards timber that has to be conveyed by water, no certificate is of any use. It has always been the custom that such timber, even though growing on private lands has to pay seigniorage just as if it was cut out of Government forests. It does not matter if the wood is taken from a jack or anjili tree which the owner himself planted, the instant it touches the water it must pay seigniorage. It is not easy to grasp the theory of this charge. I was informed by one man that it was because Government got no return from all the area of the country under water, and that water-dues were accordingly levied in this way, but if so, every boat that piels ought to be licensed, or it ought to pay a toll every time it was taken on a journey, otherwise the charge is unequal. A charge is indeed made for boats whether cut out of the forest or brought from private lands, but that charge once paid, the boat can go on plying for 30 or 40 years and the owner never has to pay any more for it. The kind of timber which suffers most under this rule is mango-wood, which is purchased in large quantities from private lands in the low country, and is conveyed in the form of planks to the ports for shipment. The Forest Commission which sat in 1884 universally recommended the abolition of this water-due, but it is still levied. As a concession, however, to public feeling the charge on mango-planks has been reduced from 9 to 3 fanams a candy. This rule leads to a great deal of smuggling, for as the planks are cut in the low country they have probably only to pass one watch station, and the pillay is often induced to pass 50 or 60 candies as 20, for there is only a remote chance of his being caught. Not a few instances have occurred where this rule has operated with great severity on the people. I recollect an instance of a man who purchased some anjili for a church in the Mīnacchal Taluq. It was brought from private land, but, unfortunately, it was obtained on one side of the river and the church was on the other, and as he attempted to take it across the river without paying the seigniorage, the timber was seized and taken to the nearest depôt, fourteen miles down the river. After two or three months delay the timber was released, but the owner had to pay the seigniorage, and also the cost of taking it down to the depôt and bringing it up stream again.

46

1184. Thus, no timber of any kind, except firewood, can be moved without a permit. To sum up, it may be said that the permit system and the other rules in force for the felling and delivery of timber are adapted to the country, but

Remarks on the permit system.

(1) The supervision is inadequate.

(2) The system of watch stations is of doubtful efficacy.

(3) The charges on timber cut from private lands when it is conveyed by water should be abolished.

(3) *Arrangements for selling.*

1185. First, as regards "royalties." I have shown how they are felled and delivered at the depôts, generally by contractors. In Central and North Travancore logs only arrive from the middle of May to the middle of December, for, during the dry weather, there is not enough water to float them down. Until three years ago no regular stockbooks were kept at the depôts. The logs arrived, but they were not measured or numbered until they were sold, and they might lie at the depôts for several months before the measurements were taken, being, all the time, liable to be stolen or changed, or to have the names on them altered. The post of pillay in charge of the depôt was a very lucrative one, and was much sought after, though the pay was only 3½ rupees a month.

Sale of royalties to the Bombay firm or the D. P. W. &c.

By the present rules, receipts are given for the logs as soon as they arrive, and they are then measured and entered in the stockbook by the officer in charge of the depôt. When a good number has arrived, notice is sent to Messrs Wallibhoyi's agent to come and make his selection and the timber is measured over to him if he requires this to be done, or he accepts the measurements of the Vijâripukkâran. In any case, the Conservator, or the Assistant Conservator in whose division the depôt is situated, has to personally see measured not less than ten per cent of the logs measured by the Vijâripukkâran, and no logs can be sold until this has been done. This measurement to Messrs Wallibhoyi and Co. takes place at each of the 5 northern depôts about 3 times a year. When a batch of timber has been measured and the contents calculated, Messrs Wallibhoyi pay the value of it, and they then receive a "rahadâri" or pass to remove the timber, and the logs are stamped. This "rahadâri," which is only given in the case of teak and blackwood, clears the timber of duty at the Customs house, whereas an ordinary "nadachit" only certifies that the timber has been properly obtained, and the Customs-duty has to be paid. The D. P. W. and Marahmat department are allowed to select timber before Messrs Wallibhoyi and Co., but everyone else has to wait until they have made their choice. We make an exception at Quilon and Trivandrum, because, at the former place, the Bombay firm does not care to take teak, owing to its great distance from Cochin, and outsiders are therefore allowed to make their choice in the order of application. At Trivandrum and Quilon there is a good demand for blackwood on the part of private persons, so, although the Bombay firm generally takes this timber from these depôts, the public is allowed to purchase it at one rupee above the rate paid by them.

1186. One advantage gained by the new system of measuring on arrival by the Vijâripukkâran and checking by a superior officer is that it does not necessitate such frequent visits of the Conservator or Assistant Conservator to the depôt. Formerly, whenever even two or three logs had to be measured to any one, one of these officers had to make a journey to the depôt to measure them to him. Now although an officer of the superior staff is always present when the Bombay firm is taking timber, it is not always necessary for him to attend when small purchasers arrive, for all the logs up to a certain date have been examined and checked, and selection can be made from them. Moreover, as the timber was formerly not measured till it was sold there was room for fraud.

Present system simpler than the old and keeps a better check on the subordinates.

1187. Once a year, generally about April, when the former monsoon's supply of logs has ceased to come in, and before the arrival of the new supply, the surplus not taken by any one is sold at outright auction. When Messrs Wallibhoyi first took the contract, all the logs sold to them were measured without any allowance for

Rejected timber formerly sold with reductions or by auction. Change lately made.

flaws. They accordingly took only good timber, and there was a large surplus of inferior timber sold every year at auction. This led to smuggling, for the lots were eagerly bought up by people who lived above the depôts, merely for the purpose of getting the "rahadāri." They sold their purchases again to persons who engaged to take the timber away and return the "rahadāri" to them. This pass was then used by them to cover teak which they felled surreptitiously in the vicinity of their homes. Mr. Vernède accordingly devised the plan of selling all the timber at the depôts to the Bombay firm, allowing them reductions on account of defects, and for a time this worked well, but at last its agents came to expect reductions to be made on nearly every log. I have now decided to try the old plan of giving no reductions, and I believe that the Bombay firm will still take the greater part of the timber. What they refuse will have to be sold by auction. and I am arranging that if any one purchases timber for removal to places above the depôts, the logs will have to be again checked and examined for stamps on their arrival at their destination. This will at least ensure the timber going with the "rahadāri."

1188. The standard linear measure is the "kōl" which contains 24 virals, each 1¼ inch long, so that the kōl is 30 inches in length. The System of measurement in Travancore. standard cubic measure is the candy which is a cubic kōl : * it therefore contains 15⅝ cubic feet. The candy is divided into 24 thūvadas, and the thūvada into 24 perukkams ; the perukkam is further divided into 16 vīshams, and the vīsham into quarters. The contents of a log are calculated by measuring the length in kōls and quarter kōls, and multiplying this into the square of the quarter girth taken at the middle in virals and quarter virals. The result is a whole number (perukkams) and a fraction which may run to sixty-fourths of a perukkam. Up to the end of 1066 it was customary to make up all the accounts in candies, thūvadas, perukkams, vīshams and quarter vīshams, though a vīsham contains less than 3 cubic inches, and would be worth, at 10 Rupees a candy, about half a cash. § We now, in preparing the accounts, calculate fractions of half a perukkam and over as one perukkam, and neglect everything under half a perukkam. The errors tend to correct each other, and the accounts are very much simplified without any appreciable loss of accuracy.

1189. The prices at which teak is sold are 17, 13, and 10 rupees for the 3 classes (see note to para 1141). Formerly, the price was much Rates at which teak is sold. higher. Between 1866 and 1874 it varied from 13 up to 17 rupees though the candy was 20 per cent smaller than it is now. But in the Administration Report for 1053 M. E. it is stated that the timber sometimes lay at Alleppey for 15 years before it was sold, and in various ways there was heavy loss, so the Bombay firm received the contract at these low rates that they might be induced to take large quantities of teak and remove it at once, which they have done, and have made a handsome profit out of it. After the end of this year, 1067 M. E. the supply arriving at the depôts will probably be much less than it now is, and at the same time the demand is constantly increasing, so that next year we should be able to raise the rates as far as Travancore is concerned, but the rate that Messrs Wallibhoyi can afford to pay will depend on the selling price at Bombay. † I am told that in the State of Cochin the price of teak is 14 rupees a candy (considerably smaller than ours) irrespective of the size of the logs. With us the price of other timbers has risen much of late years, and in many cases this, our most useful wood, is employed where commoner woods would do just as well, because it is no more expensive than they are.

---

§ Half a cash= $\frac{1}{512}$ of a rupee.

* As a candy is 15⅝ cubic feet, prices may be compared with those ruling in British India by taking the number of rupees given as annas, and a candy as a cubic foot. Thus, 14 rupees a candy is almost the same as 14 annas a cubic foot.

† The following are the average selling prices of teak at Bombay since 1872. The quotations are per ton of 50 cubic feet, which is generally taken as equivalent to 4 candies, but is in reality rather less—

	Rs.		Rs.		Rs.		Rs.		Rs.
1872	82	1876	65	1880	70	1884	80	1888	74
1873	91	1877	70	1881	80	1885	72	1889	82
1874	93	1878	60	1882	103	1886	75	1890	73
1875	·90	1879	65	1883	105	1887	72	1891	63

Average of 20 years 78 Rs.

1190. Blackwood sells at 18 and 15 Rs. according to class, the classes being as follows. All logs of 12 virals quarter girth and above in the first, and all below in the second.

Rates for blackwood.

1191. As regards other woods, I have explained how they can be obtained on permit in Central and North Travancore, and that they must be purchased from the depôts at Puliyara and in South Travancore. At all these depôts the timber is offered for sale to the public at fixed rates, auctions being occasionally held to get rid of the inferior timber and to gauge the state of the market. The greater part of the timber sold at the depôts is exported to British India, principally to Tinnevelly, so the price in Tinnevelly regulates the price here. As a natural consequence, the depôts nearest Tinnevelly or those from which the timber can be most easily exported thither pay the best. At Puliyara, on the borders of that District, there is an almost unlimited demand for timber at a profit of 10 Rs. a candy for the best woods, and of $6\frac{1}{2}$ or 7 for the poorer kinds which are seldom sold there. The selling price running from 13 to 18 Rs. At Nagercoil, the next nearest to Tinnevelly, the selling price runs from 9 to 18 Rs., but the cost of procuring the timber is heavy and the profit varies from $1\frac{1}{2}$ to $7\frac{1}{2}$ Rs; a candy, and there is a ready sale. At Quilon, whence timber can be exported by ship to Tinnevelly, the selling price varies from 12 to 20 Rs., and the profit from $2\frac{1}{4}$ to 10 Rs., but these rates are rather high, and the timber does not go off readily. Lastly, at Trivandrum the most difficult of access from Tinnevelly, although the tariff rate is as high as at Quilon, the demand for export is not great and good timber has often to be auctioned at only half a rupee or so per candy profit to Government. I have already explained the rates which we obtain for timber at the Peermerd depôts (see para 1175). Further particulars will be found in the Appendix.

Prices realised for other timbers.

4. Miscellaneous works.

1192. Next to the control of the depôts and the management of timber operations, as already described, the most important work of the Forest Department is the capturing of elephants. These animals are wild in the forests, and are in some places particularly abundant. They do not always remain in the same spot, but move about over large areas, their movements being regulated by the quantity and condition of the food available, and by the state of the weather. Over the greater part of Travancore they descend from the hills as soon as the water begins to fail there, that is to say about January, and they are then to be found in the thickest and coolest parts of the lower forests in the vicinity of some river. As soon as the showers begin to fall in April, their instinct tells them that they can again obtain water on the hills, and that fresh grass has sprung up where the dry herbage was so lately burnt, and they immediately commence an upward movement to the higher ground. There they remain till about September when some, but not all of them, descend to the lower slopes of the hills and even to the low country, to see what they can get from the fields of hill-paddy then beginning to ripen, and they often destroy large quantities of grain. In November these migrants again ascend the hills and join their companions. Advantage is taken by us of the annual descent from the hills in the hot weather to catch these animals, but in November no attempt is made to capture them as the pits are then full of water. Towards the end of the year, new pits are dug and old pits re-opened in places where the elephants are sure to pass in the course of their migrations, and the pits are carefully covered with sticks and leaves, so that it is most difficult to see where they are. A temporary camp is usually formed in the neighbourhood, and watchers are appointed from the neighbouring villages or from the Hillmen's " kudies " to visit the pits every morning, and see if any elephant or other animal has fallen in, for, in addition to the elephants captured, we usually secure 5 or 6 bison, a dozen sambur, and often a bear, tiger or leopard each year.

The capturing of wild elephants is entrusted to the Forest Department.

1193. As soon as news is brought that an elephant has fallen into a pit the whole camp turns out and the men proceed to cut down trees and to place them over the mouth of the pit, or he would in a short time break down the sides of the pit and manage to escape, indeed a large tusker will often do this in a night, especially if the pit is not quite of the

Method of capture by pits described.

regulation depth. The next step is to divide the pit into two by a partition of poles, and, when this is sufficiently secure, men descend into the unoccupied space, and, putting their hands through the bars of the partition, manage to tie ropes round the neck and hind-legs of the animal. As soon as the tame elephants which have been kept in readiness arrive, the logs are removed from the top of the pit, and boughs and leaves are thrown into it, and in an incredibly short time the elephant treads them down, and manages to scramble out to firm ground again, but not to liberty, for with one tame elephant on each side to keep him in order, and a gang of men holding his heel-ropes he cannot do much harm, and before long he is safely caged. The animals have usually to be kept two or three days in the pits till the tame elephants arrive, and they often damage themselves very much in their efforts to escape. All this they have of course to be fed and watered. The work of taming and teaching the captured elephants is much simpler and occupies less time than might be supposed, from 2 to 6 months being usually sufficient to reduce them to submission, and to teach them to obey ordinary words of command.

1194. The pits are supposed to be 15 ft. deep (10 cubits), and the same in
diameter, but, as a rule, they are seldom more than 10 or 12 ft.
each way, and it is not desirable to insist on the full depth, as
the animals are much injured by falling from such a height.
The price paid for cutting these pits is 5 rupees each, and the
earth that is taken out is carefully removed to some distance that the elephants may not be alarmed at the sight of it. An expenditure of about 50 rupees is usually incurred when an elephant is captured and caged, in engaging extra labour &c., and the cost of watching, pit digging &c., distributed over the number of animals caught, generally averages about 100 Rs. each, so that every elephant when it has been safely caged may be said to have cost 150 Rs. Nearly all of the animals caught are young, ranging from 5 to 15 years in age, and to keep them till they are fit for work costs 500 or 600 Rs. at least. Most of the tuskers are given to pagodas, the cow-elephants being usually retained by the Forest Department as being of little use for processions. An auction of captured elephants was held last year at Trivandrum for the first time, at which the tuskers fetched most remunerative prices, the smallest of them, I am told, realising 2,000 Rs.

Cost of capturing elephants. Their destination afterwards.

1195. About 50 per cent of the animals that fall into the pits die or escape.
Some manage to scramble out, with probably some injury,
others are killed by the fall, and others again are so hurt that
they die in the course of a few months. This system of pit-
catching has therefore been much condemned, but the difficulty
is to find a better. A keddah was tried in Travancore (as stated
in para. 1139) but was closed after 10 years. The cost of capturing by this method was as great as that by the pit-method, and it was found quite impossible to train the older animals captured. I am very doubtful if keddahs can be worked with success in Travancore, because the herds of elephants are so small, very few of them numbering more than 20 individuals, and it is seldom that more than one herd could be captured in one season.* I am informed that the Forest officers in the Wynaad capture their animals in pits, and that they lose very few of them, as they adopt the simple expedient of putting a cushion of leaves or grass at the bottom of each pit to break the fall of the animals, and this I will introduce. Careless watching is also responsible for several deaths, and this can only be prevented by closer supervision.

Great mortality among the animals caught. Keddahs not suitable for this State

1196. The question has often been debated whether the number of elephants
in the country is increasing or decreasing. I believe that most
people would say that elephants are more numerous than formerly, but I am inclined to think that this impression is formed
from the increased damage done to cultivation of all sorts. If
we recollect that cultivation is yearly extending, we can well
understand that elephants are much more troublesome now than formerly, without there being any increase in their numbers, and if we could take a census of them we should probably find that their numbers are about stationary. I once attempted to

The number of elephants in the country probably neither increasing or decreasing. Occasional epidemics.

* Recently published reports show that keddahs are not a financial success in Mysore. Though the number of elephants captured was large, the expenses were so great that they were not covered by the money realised from the sales.

47

estimate how many there are in the State, and I came to the conclusion that there must be from 1,000 to 1,500, the greater number of them being found in North Travancore, especially the Cardamom Hills. Sometimes elephants die in large numbers, as in the year 1866, when a murrain attacked them in the forests near Malayáttūr, and 50 pairs of tusks were brought to the Forest offices at that place and Thodupurn in April and May of that year. Such epidemics would doubtless occur more frequently if the number of elephants increased unduly, and the supply of food fell short, and their rarity is a sign that the animals are not troubled for want of food, though their migrations show that it is not always to be obtained in the same place.

1197. For the purpose of capturing the wild animals, and for the moving of timber at our depôts and elsewhere a staff of tame elephants has to be kept. The number at present, capable of work, is 22, but of those only 3 are tuskers of any use to control the large elephants that may fall into the pits. It is on account of their insufficient number that I have decided not to re-open the keddah this year, for if a large herd entered it we should not be able to secure the members of it. The cow-elephants and small tuskers which we possess are sufficient for work at the depôts, but should we have to work down timber from the hills on our own account, as we shall probably have to do, owing to the failure of contractors, in fact we are doing this on a small scale now, we shall have to hire or purchase a good many more. In former days, when timber was worked down by Government agency a very large number of elephants was kept by the Forest Department, but the animals were sold or transferred to pagodas when the contract system was introduced, and they have not been replaced. The few animals we now have are fully employed, and we are much troubled by requisitions from pagodas for their services at processions. The pagoda festivals mostly occur in the hot weather, when the elephants should be kept without work on account of the heat, and when they are required to be in readiness to assist in the capture of animals that fall into the pits. At this time of year too it is usual to put them under treatment, and to feed them up against the next working season, and when they have to march 40 or 50 miles to attend a festival, and to be absent 8 or 10 days with insufficient food in the very hot sun, all the good that may have been done by resting and being fed up is lost. As most of the newly captured elephants are given to pagodas, and as a large number was handed over to them when the staff of elephants was reduced, we may well expect not to be called on any more to lend our animals whose lives must infallibly be shortened if they are subjected to this treatment.

The number of animals attached to the Department is small, and more are wanted Requisitions for pagoda-processions cause much trouble.

1198. The cost of keeping elephants is usually set down at 100 fanams a piece, or say 15 Rs. a month, that is about ¼ a rupee a day, but, as they can only be worked on an average through the year of once in two days, they cost one rupee a day each. When working, the elephants receive an extra allowance of rice and the keepers get batta in rice and money, and, in addition to this, interest and depreciation have to be charged on the original cost of capturing and training. On the whole, 3 Rs. is the lowest sum that can be charged as hire for each working day, but when we ourselves engage elephants we have to pay up to 10 Rs. a day for each of them The management of elephants is not an easy matter because they have often to be sent away to different places to work down timber, and it is not easy to properly supervise the work done in many localities at the same time. Elephant-keepers too are drawn from a very bad class of the people, and their custom is to malinger, and to starve their animals, or to hire them out to contractors or for processions without the knowledge of their superiors,

Cost of keeping elephants.

1199. The Forest Department has to collect any ivory that it can, and to forward it to the Commercial Office at Alleppey, where it is sold at the annual cardamom auction. Some tusks are picked up by the hillmen in different parts of the country, and some are taken from animals who have fallen into the pits, and have died there or after removal to the cages. A reward of 1 rupee per lb. is paid as "kudivila" for all ivory brought in. The weight of tusks delivered by us at Alleppey each year averages about 17 thulāms, but in years of murrain as in 1866 it is very much greater.

Collection of ivory.

1200. Cardamoms are, in the same way, collected by us and forwarded to Alleppey, and the average quantity for the last 5 years has been 500 thulāms [*] The greater part of these are obtained from the cardamom-gardens near Kōuuiyūr. There is a Cardamom Aminadar and a staff of clerks and peons under him, whose business it is to superintend the management of these gardens (supposed to aggregate 50 acres,) and to see the spice dried. In other parts of the country the wild cardamoms are collected and delivered to us, sometimes by the Aminadars who engage coolies to pick them. In South Travancore a man of the name Sangalakuricchi pillay obtained leave to take up 500 acres of forest for cardamom cultivation, but, instead of selecting it in one place, he has chosen a few acres here and a few acres there over many miles of country, and, has, in fact, done nothing to improve the natural growth of the plants. Cardamoms are a monopoly and must be delivered to the Government, but under the plea that there was no sunshine in Travancore when the spice was ripening this man has been in the habit of taking his collections to Tinnevelly to dry, and but a very small quantity of the cardamoms he collects is ever delivered to us. There is also a great deal of smuggling on the part of people from Tinnevelly, who can come over and carry off loads of cardamoms from the forests without our being in the least aware of it. In North Travancore there are in the forests supervised by us many gardens which are under the management of the Superintendent of the Cardamom Hills, as shown in para 746. These should be transferred to us. In other parts of the North the right to collect is put up to auction, and the different valleys or groups of valleys are allotted separately. The bidder who offers to deliver the greatest quantity of the spice is accepted, and he has to deliver the quantity agreed on, subject to a penalty, but anything in excess of the stipulated quantity is the bidder's own property, and can be used as he likes. In order to collect the spice it is a recognised custom that the bidder shall order the hillmen to collect for him without remuneration, he being a Government contractor, a certain quantity per head, the total being of course far in excess of his bid. The contractor then delivers to us the amount agreed on, and retains the rest as his profit, while the hill-men deliver their quota to the contractor, and exchange any further quantity they can collect for salt, knives, and cloths. This is a bad system, as it breaks the continuity of the monopoly, and permits a trade in the spice.

Cardamoms collected by the Forest Department chiefly from near Kōnniyur.

1201. The cardamoms of Travancore are divided into 3 classes (1) the "magara-ēlam" which are grown on the Cardamom Hills at an elevation of over 3,000 ft., and which, in the drier climate of that locality, ripen in the month of January. The scapes, which bear this variety, trail on the ground, for which reason the capsules, in which the spice is contained, are liable to be eaten by rats, unless the garden is carefully weeded. (2) the "kanni-ēlam" which are found on the slopes of the hills to the west of the Periyār, and below 3,000 ft. The scapes trail on the ground, but the capsules are smaller than those of the "magara-ēlam", and they ripen in October. (3) the "nila-ēlam" or long cardamoms of South Travancore. The plants of this variety are larger and the scapes stand erect. The capsules are longer and not so broad as those of the kanni-ēlam, but they ripen at the same time, (October). The value of the 3 classes run in the order I have named them.

Three varieties of cardamoms in Travancore.

1202. The money paid for cardamoms is nominally ¼ths of the selling price, but in reality this is never given to the collector The Aminadar first of all takes good care to secure good weight, as cardamoms vary much in weight, on account of their habit of absorbing, and parting rapidly with, moisture. When they reach Alleppey they are weighed in by Dutch, and sold by English, pounds, whereby the collector loses 10 per cent, and the storekeeper exacts extra weight, so that the collector really only gets ¼ or ⅓ the selling price. Then too, the account is not settled until the cardamoms have been sold, which may be months after the spice has reached Alleppey. An advance is given by the Aminadar as soon as the spice is delivered to him, but the balance is often delayed for years, if it ever reaches the collector.

Payments made for cardamoms collected.

1203. Beeswax is also a Government monopoly, and has to be delivered to the Forest or Cardamom Department by whom it is forwarded to Alleppey in the same way as cardamoms. The "kudivila" paid to the collector is 45 fanams (6½ rupees) per thulām of

Collection of beeswax. Chief species of bees.

20 English lbs. The hillmen recognise 4 different kinds of bees (1) "peranth-i" or large bee (Apis dorsata ?) which hangs its huge combs to the branches of "cotton" and "chīni" trees. (2) "thoda" or "nadaya-i" which makes its nest in hollow trees or in caves, often in large colonies, (3) "kossu" or "kotha-i" which suspends a small comb, consisting of a double row of cells, from the branches of bushes, and (4) "sirа-i" a very minute bee which builds its nests in the holes of trees, often taking possession of the tunnelings made by the borer (Zeuzera ?) in the teak trees. The honey of this species is very bitter. Of these four the greater part of the wax collected is obtained from the first, but the second and third kind· of the bees above mentioned contribute, the wax of the fourth species alone being useless. The hillmen collect the wax by ascending the enormous trees from the branches of which the combs are suspended, with the aid of pegs driven into the stem, or, when the nests are in caves, by letting themselves down over precipices by ladders made of rattans. About·8 candies of wax are sold each year at Alleppey, of which the Forest Department delivers one-half and the Cardamom Department the rest.

1204. The Government retain lac as a monopoly, but the quantity delivered at Alleppey is quite insignificant, and only reaches a few pounds annually. No attempt has ever been made to encourage the cultivation of lac, or the propagation of the coccus. Dammer, though a monopoly, is of such small value that only about 5 candies are delivered by us each year at Alleppey. The "kudivila" paid for it is 6 fanams a thulām. Honey is not a monopoly, but it is often collected by us for the pagodas at the rate of 17½ fanams a para.

Lac, dammer and honey.

1205. Over the greater part of the country no charge is levied on firewood, wherever it may be obtained, but, if it is taken by water in rafts, a charge of one rupee for 12 pieces is levied. Firewood taken in boats is free. There are, however, exceptions to this rule. At Quilon a charge is levied on fuel being brought into the town whether by land or by water, and in the Shencottah taluq fees are also charged. The reason why Quilon was singled out for special treatment was that a large cotton-mill was opened there some years ago, and, as the quantity of wood consumed was very large, it was felt that this could not be allowed without some return to Government. This charge, then, was aimed at the mill, but it really affects the whole town, though, as there is only one watch-station where the firewood is checked, the charge is often evaded.

Firewood free except in certain cases.

1206. Some revenue is obtained by the sale of minor produce in certain parts of the country, as, for instance, of tamarinds, firewood, and leaves for manure, and by cattle grazing fees from the eastern slopes of the Māhinthragiri hills. Again, by charges on charcoal, firewood, reeds &c. in the Shencottah taluq, and a trifling sum is collected from persons gathering " patthiri-pū " (the mace of a species of nutmeg) in North Travancore. The produce of His Highness the late Maha Rajah's estate at Pāl-arivi near Ariyankāvu is sold for 30 and odd Rs. a year, and under the name of "malam-pāttam," dues are levied on " pulinji-kā" (the pods of Acacia concinna), and ginger, and other products in the same locality. By a curious arrangement, persons willing to grow plantains were allowed to take up land in the neighbourhood of Ariyankāvu on ten-year leases at the rate of 4 rupees an acre a year,* the land to revert to the Forest Department after that time, but though the revenue should amount to 300 or 400 rupees a year, it can hardly be said that the forests will benefit by the clearing of land, often covered with virgin forest, for the purposes of this petty cultivation, when, as soon as it is abandoned, the area will be occupied by thorny scrub in place of the fine trees which grow there before.

Revenue from minor produce and other sources.

1207. About three years ago we were asked to replant the old bamboo hedge which formerly existed between Travancore and Cochin on the Mūvāttupura river, and which was very effectual in preventing the smuggling of tobacco into this State. The work is now finished and has been very well done, but the hedge will have to be watched for some time yet, as the villagers will undoubtedly attempt to make passages through it, should there be any relaxation of the supervision. It is now proposed that we should

Planting hedges along the boundary.

---

* According to the Revenue measurement, under which these leases were given, 8 acres of hill-land equal 1 acre in the plains, so we only get ½ rupee an acre instead of 4.

attempt a similar work in South Travancore between Arammula and Cape Comorin, where there was formerly a hedge (not of bamboos but of a species of *Acacia*) which has been neglected for many years, and has in many parts been cut down.

1208. Perhaps the most troublesome part of a Forest officer's work is the

Checking of hill-cultivation and prosecution of offenders most troublesome.

checking of hill-cultivation, and the holding of enquiries on charges of illicit felling. With reference to the first, it is often extremely difficult to find out who is the culprit, when people are discovered felling forest. The man who supplies the money and takes the profits keeps in the back ground, and, when he sees that it is really intended to prosecute some one, he puts forward some underling who is worth nothing, and who pleads before the Magistrate that he would starve if not allowed to clear land thus, whereas, really, this cultivation is usually undertaken as a speculation by wealthy Mahomedans. When, too, the case is proved, the Magistrate frequently lets the defendant off with a trifling fine of a few rupees, not amounting to one-tenth of the value of the timber felled, and covered, many times over, by the profit obtained from the cultivation, which he is then allowed to continue. Though the old proclamations declared that the crop in such cases would be confiscated, this is never done. Many Magistrates refuse to hear any cases of this sort. As regards the holding of enquiries, we are inundated with petitions often annoymous or pseudonymous, saying that such and such a person has felled trees without permission, or has stolen timber &c., and it requires days and days of patient enquiry, and often a journey of many miles, to arrive at the truth of the case. And the officers sent to enquire are not always themselves above suspicion.

1209. All the accounts of the Forest Department are kept in Malayalam, and in

System of accounts very cumbersome.

fanams, chuckroms and ca-h. A fanam is one-seventh of a rupee, a chuckrom is one-fourth of the fanam, and sixteen cash go to the chuckrom. Some idea may thus be formed of the long array of figures required to express a sum of money comparatively small. So long as the State accounts are kept in fanams, I presume that objections will be raised to the introduction of any other method in a Department which collects revenue, though I understand that both in the D. P. W. and the Police Department the accounts are kept in British money ; but, as the revenue of the State now exceeds 78 lacs of Rs., it is surely high time that, for the sake of simplicity, the accounts should be kept in rupees, and not in coin of such small value as the fanam. The accounts are kept by single entry, and accuracy is ensured by constant repetitions and checking. A better system would require less checking, and would thus save labour without any loss of accuracy.

1210. From all stations of any importance where there are Forest treasuries,

By the present system of accounts there is no check on the proper expenditure of the money, though provision is made to ensure the accuracy of the figures.

such as the various Assistant Conservator's offices and the chief depôts, a daily statement of the receipts and expenditure at each of them has to be sent to the Huzur, and bi-monthly accounts in brief, and monthly accounts in detail, are despatched from them to the head-office. On receipt of these returns the accounts of the whole Department for the month are made up and forwarded to the Huzur. At the end of the year, the accountants from all these offices, and the Aminadars and others who have received money for capturing elephants or for various purposes, are called on to attend the head-office to have their accounts adjusted, and, when this is done, accountants are sent to the Huzur from the head-office and the offices of the Assistant Conservators to explain the accounts submitted by them and to produce receipts, and when everything is settled, the annual report can be prepared. As the accounts have to be submitted at a very early date each month, the only way to arrange for this is for the head-accountant of each office, and every officer holding any money at all, to write direct to the head-office at Quilon, and not through the Deputy or Assistant Conservator in charge of his range. Thus these officers know very little of the expenditure in detail in their ranges, nor are the accounts in detail submitted to the Conservator in the head-office, and the checking of the whole of the accounts are practically entrusted to the Samprethy of the Quilon office and the accountants there. The accuracy of the accounts is thus indeed ensured, but there is no check on the proper expenditure of the money. The Conservator and his Assistants who travel about the country are the persons in the best position to judge if the money has

48

been properly applied or not, and not those who are stationed at head-quarters or the accountants in the Huzur, who think that a given work must always be done for a fixed sum, as, for instance, that teak must always cost 2 Rs. a candy to fell and deliver at a depôt, or that a cooly can always be obtained for 6 chuckrams a day, whether in the low country or in the most distant part of the forests. Similarly, the money paid to the Aminadars and others for carrying out works, as well as their pay each month, is disbursed from the head-office and not from the Assistant Conservators', and much hardship results from this. Thus, the Manjapra Aminadar whose head-quarters are only 4 miles from Malayāttūr, where the Assistant Conservator lives, has to send all the way to Quilon, some 80 and odd miles, to get his pay, and as the distance is so great he only gets it once in 6 or 7 months, much to the inconvenience of his staff. In France and in British India a great point is made that the subordinate officers have nothing to do with the spending of money, and even the Deputy and Assistant Conservators are not allowed to keep in their treasuries more than very small sums, but with us the Aminadars, Vijāripukārans and even the watch-station pillays are allowed sums often of large amount without any local check. True, their accounts are carefully examined at the head-office and in the Huzur, but the most important check is wanting, and so long as a man is clever enough to prepare his accounts correctly he is allowed great latitude, while another who is conscientious in his work but not good at figures runs the risk of dismissal for a few fanams, though he may have saved the Government many hundreds of rupees.

1211. Thus centralization is carried to extremes in the management of the accounts of the Forest Department. Money is paid to Aminadars and others for carrying out work, without any reference to the superior officer of the division or range, and without his even knowing that it has been applied for, and the accounts are sent in by the recipients of the money direct to the head-office. As the Samprethy of the Quilon office (himself not a very highly paid officer) and the accountants are held responsible for the accounts, they take very good care that the figures for receipts and expenditure are correct, and that all the vouchers for money expended are produced, but they do not trouble themselves with the adjustment of balances. An officer who has received money to execute a certain work dies or is transferred. What money he has actually spent is accounted for, but the balance is thrown into the limbo of un-adjusted accounts, from which it is only extricated after many years to be written off, whereas, if the matter was decided at the time, much of this money could be recovered. In any case, nothing is to be gained by postponing the settlement of the question. There are at the present time unsettled accounts of 10 years old and upwards, and, among others, the Malayāttūr accounts show a balance in hand of 70 odd parras of paddy, which disappeared over a dozen years ago when the Forest Department ceased to keep paddy for elephants. These unadjusted accounts are called "Thirrattu" (that which returns), and they are returned or carried on from year to year till they are finally written off, but as it is nobody's business to see that others do not accumulate, new debts are continually being added to the list while the older ones are being examined and written off.

As it seems to be nobody's business to see that advances are recovered, debts accumulate year after year instead of being written off or recovered.

1212. A third objection to the centralization of the accounts is that so many officers have to leave their work at the end of the year for adjustment, whereby the work is at a stand-still for several weeks at a time. This system is also productive of another evil similar, if report speaks true, to what was so ably described in Mr. Jeyaram Chetty's Report to the Madras Government in 1855 on the system of village accounts. He says "The Gumastahs are detained in the " " Huzur for more than two months, during which period they are required to draw " " up from their accounts various detailed statements and returns just at the fancy " " or whim of the Huzur Gumastahs under whom they are placed. I may also add " " here, that those who cannot afford to make the usual presents to the Huzur " " Gumastahs who examine their accounts are exposed to all sorts of inconvenience " " and annoyance: they are detained at the Cutcherry till a very late hour in the " " night &c." This was written nearly 40 years ago, and referred to another part of India, and it may not all be true of the present time or of Travancore, but it is certain that the accountants thoroughly dread the summons either to the Huzur or

The system of sending accounts to the Huzur for the annual adjustment causes much inconvenience and hardship.

to the head-office, and that they use every endeavour to evade attendance. The only reason why accountants might be required when the accounts of a sub-office are examined is to explain the entries, but if the accounts were clearly kept there would be no need for explanations; they would explain themselves.

1213. Thus, the system of accounts in the Forest Department requires alteration because under the present method.

(1) The expenditure in detail is not checked by competent officers, and too much power in paying money is left to the Aminadars and others.

(2) Arrears of un-adjusted balances are allowed to accumulate.

(3) The accounts are not ·sufficiently clear to allow of the attendance of accountants at the head-office and the Huzur being dispensed with.

Sooner or later the accounts must be kept in English, but, for the present, it would be a great gain to hold the Deputy or Assistant Conservators responsible for the expenditure in their respective ranges All payments should be passed through their hands, and the powers of the Aminadars and other subordinate officers to expend money should be curtailed. But this cannot be done unless more time is allowed for the checking of accounts before submission. If returns are called for with too breathless a haste, the object aimed at is defeated, and no proper check is kept on the expenditure.

1214. The correspondence carried on in the Department is chiefly in Malayalam. This to a certain extent is necessary, because the Amina- dars, and other officers in the outstations, are ignorant of English, but it is to be hoped that year by year the proportion of the English to the Malayalam letters will increase. The latter may be a very expressive language, but it is not the language of business, and its ambiguous phrases and long repetitions are calculated rather to confuse the hearer than to explain the meaning to him. Until lately, there were no tests for entering the Department, the consequence was that all those who could get no footing in other Departments endeavoured to enter the Forest Department. By a recent order, no candidate can obtain a pay of 10 rupees or upwards unless he has either matriculated or has passed one of the Revenue Test Examinations, though the promotion of the older officers is not interfered with. This is a step in the right direction, but unless we wish all our officers to learn their work at the expense of Government, we shall have to institute tests of our own, for which they must prepare before expecting to enter the Department. At present, almost all the officers of the Department are engaged in checking the sale of timber or in preparing the accounts, and in preventing smuggling, and no special knowledge is required of them, but, when forest work begins, then the executive officers will have to devote their attention to the growth of trees and how to treat them, to the study of forest entomology and the prevention of the ravages of insects, to surveying, engineering and other kindred subjects * some knowledge of which is absolutely necessary for the efficiency of the executive branch of the Department.

1215. The want of a code is very much felt in the Forest Department, and this for three reasons, first, because an officer entering the Department finds the greatest difficulty in knowing what is expected of him, what his duties are and how extensive his powers, and he is almost certain either to exceed his authority or to be constantly referring everything for orders. Many rulings and orders are scattered, as I am told, through a mass of correspondence, but they have never been collected, nor was it customary, when a certain ruling was issued, which affected a whole class of subordinates, to send a circular round to all of them, but only to inform that officer who had raised the question. The result may be imagined, and I have found officers of a dozen years service ignorant of some of the commonest rules of the Department. The second reason is that for want of definite orders even the older officers in the

---

* To those who are sceptical about the value of a study of Entomology I may mention that it has been calculated that M. Pasteur's researches into the diseases of silk worms have saved the French people a sum equal to the indemnity demanded from and paid by them after the Franco-Prussian war

Department have to refer to head quarters for instructions in settling matters of trivial importance which might be with perfect ease disposed of by them. Thus the correspondence is largely, and very unnecessarily, increased, and the settlement of trifling cases is delayed. The third reason is very similar, but refers to the Head of the Department. The powers of the Conservator are not laid down nor is it clearly known what works he can undertake without sanction, or those for which sanction is required. For want of proper instructions the expenditure on a given work may become largely increased, thus the repair of a building, if taken in hand promptly, might be effected for a small sum, but, if there is delay in obtaining sanction for expenditure the cost is much increased. The great advantages of a code are that it ensures uniformity and lays down definite instructions with clearness, thereby saving time and correspondence, and its want is felt more and more each year.

1216. Here again I must refer to the excessive centralization. In the Forest Department this is carried to extremes. In the old days when Forest work was in its infancy, and when the business was so light that every matter could be carefully enquired into by the Head of the Department and when, too, no definite decisions had been passed or precedents established on a variety of subjects, it was only proper that all matters should be referred to the Head of the Department for settlement, and his order was law which no one dared to dispute, but now things have very much altered. A number of precedents and decisions have become public property, and the people are aware of them. It is not therefore as difficult to arrive at a decision how to act, with so many rulings to guide one, as it formerly was. Judgment according to precedent is what is required rather than a careful weighing of all possible eventualities in order to form a precedent. At present the work has so largely increased that the whole time of the Conservator is occupied in deciding questions which might, as I have said, be equally well settled by others. Often, through pressure of work, the question cannot be taken up for weeks, much to the inconvenience of all concerned, and it may happen that the settlement of certain questions falls to an officer subordinate to the officer who has sent them up for reference. A Samprethy or Office Manager may practically hold the power of appeal over a Deputy or Assistant Conservator, though the latter, even supposing he was only his equal in rank, is in a far better position to judge the case fairly, being on the spot, and able to obtain information in addition to the evidence he takes down. Even the Head of the Department should use great caution in altering the decisions of his immediate subordinates, provided he can trust them, because they are much better situated to learn the truth than he is. An officer whose suggestions are not attended to, and whose orders are not upheld, loses the respect of his subordinates, and ceases to take an interest in his work. He is then quite useless and must be transferred. It is very pleasant for the lower subordinates to think that they have only one master, and can always get decisions against them by their immediate superiors reversed by him, but under such a system there can be no proper organisation at all. The tendency of the present day is to do away with centralization as much as possible, to select the immediate subordinates with caution, and to give them very considerable powers, definite instructions, and well defined responsibilities, to expect them, in fact, to carry out all the routine of the Department, leaving the Head leisure to consider and develop schemes for improvement, which he, and he only, can originate. In case of a Forest Department, to arrange for the collection of information on a variety of subjects, to watch progress elsewhere, and to adopt improvements therefrom, to organise schemes for conservancy or the extension of planting, to study the markets of foreign countries so as to regulate the prices of timber and other articles of forest produce, and generally, while thoroughly acquainting himself with all the details of his Department, to leave the routine to others. Unless this is done and if the Head is too much taken up with the work of the Department, he can find no leisure to think of improvements, and the Department can make no progress year after year.

Centralization in the Forest Department carried to extremes, thereby overburdening the Conservator with trivial work and leaving him no leisure to introduce measures of improvement.

(5)—*The Staff of the Forest Department.*

The Forest Staff

1217. The staff of the Department at the end of the year 1066 was as follows:—

193

Controlling Staff.

```
  1 Conservator 600/. and travelling allowance, averaging 120/.=720/. ⎫
  1 Asst. Conservator 150/. and        do.        do.      60/.=210/. ⎬1,410/.
  2    do.    do.  @ 100/. and        do.        do.      60/.=320/. ⎪
 17 Personal staff attached to above with batta    ...    ...  =160/. ⎭
```

Executive and Protective Staff.

```
  1 Head Aminadar @ 50/. and batta say 15/.          ...  =  65/. ⎫
  1 Cardamom do.  @ 35/.        do.        10/.       ...  =  45/. ⎪
 11 Range Aminadars @ 20/. to 30/. and batta         ...  =345/. ⎪
  1 Marahmat Overseer    ...    ...                   ...  =  30/. ⎬1,220/.
  1 Malayāttūr Reserve Inspector      ...    ...      ...  =  16/. ⎪
  1 Cardamom Vijārippukāran      ..    ...            ...  =  12/. ⎪
107 Clerks and Peons &c. attached to above ...       ...  =707/. ⎭
```

Stationary Staff.

Depôts.

```
  1 Puliyara Superintendent  100/. and batta 10/.       =110/. ⎫
  1 Perukuda    do.        50/.            10/.         =  60/. ⎪
  1 Kumili      do.        30/.            10/.         =  40/. ⎬825.
 13 depôt and sub depôt Superintendents @ 8/. to 15/.  =157/. ⎪
 74 Clerks, Peons and Watchers attached to  do.        =158/. ⎭
```

Offices.

```
  1 Sumprethy at Quilon    ...    ...    ...    ...     =30 /. ⎫
  3 Aminadars at Malayāttūr, Kōni, and Nagercoil                ⎪
                               @ 25/. to 30/.    ...   =80 /. ⎬475½.
 46 Clerks and Peons attached to   do. .  ...    ...   =365½/. ⎭
```

Watch Stations.

```
 36 Pillays @ 8/. and 10/.    ...    ..    ...    ...  =320/. ⎫
 72 Watchers @ 4½/.    ...    ...    ...    ...    ...  =334/. ⎬660/.
 ───
392                  Total per month Rs.     ...  4,590½.
```

Besides those employed on the teak plantations and on special works for which sanction has to be obtained each year.

1218. The duties of the Conservator and his Assistants are to travel about the country, to hold enquiries, to check the measurement of timber at the depôts, to superintend auctions, to supervise the capture of elephants and the collection of monopolies, to prevent encroachment on the Reserves, and, generally, to control the management of their respective ranges on the part of the Assistant Conservators, and of the whole country in the case of the Conservator.

Duties of the Staff.

The Head Aminadar has lately been employed in working down the timber left by those teak-contractors who have failed or have died. He has also supervision over the other Aminadars especially in the matter of elephant capturing. The Cardamom Aminadar is in charge of the cardamom gardens near Kōnniyūr. The Aminadars have to catch elephants and stamp timber, and to carry out the instructions of their superiors. Theirs is no easy work, and especially does the stamping of the timber give them trouble, for they are summoned about from one part of their districts to another at short notice to effect this stamping. The Marahmat Overseer is entrusted with the execution of works, and he has much travelling to do. The Malayāttūr Reserve Inspector has charge of the Reserve in that neighbourhood, while the Cardamom Vijārippukāran is stationed at Kaddakal and has to collect the cardamoms found to the south of the Punalūr river, under the orders of the Neduvangād Aminadar.

The Depôt Superintendents are stationary officers, and, as a rule, are only concerned with the sale of timber at their depôts, but they are occasionally called on to hold enquiries. The heads of the offices are concerned with official correspondence, and the preparation of accounts. Lastly, the watch-station pillars and watchers are entrusted with the duty of checking the timber that passes their stations.

1219. It will thus be seen that by far the greatest part of the staff of the Forest Department is stationary at the depôts, offices, and watch-stations, and that the only really executive officers are the Aminadars and the personal staff attached to them, and the Reserve Inspector. The Cardamom Aminadar and Vijā-

The Forest Department employed in selling timber rather than in Forest conservancy.

49

rippukáraň and the Marahmut Overseer have special work, and can therefore hardly be called executive officers in the proper sense of the word. The Aminadars, too, are not employed on works of improvement, but were appointed merely to prevent smuggling and to catch elephants. Thus, the Department is concerned more with the sale of timber and the collection of revenue, than with the conservancy of the forests, and, as I have shown, works of improvement have not received the attention they merited.

1220. One curious anomaly that exists, which is the outcome of the system of centralization, is that while holding very considerable powers in the way of spending money, holding auctions &c., the Aminadars are not empowered to supervise the watch-stations within their respective divisions, nor to examine their books, nor to give orders to the pillays, and, as the Assistant Conservators can only visit these stations at long intervals, the pillays are left very much to their own devices. Whether this arrangement was followed because it was thought that the Aminadars, who have such extensive powers to spend money, could not be trusted to supervise the watch-stations, I cannot say, but certainly this system is open to objection.

Watch-station officers not subordinated to the Aminadars

(6) *Legislation and offences against the Rules of the Department.*

1221. A list of all the important notices and proclamations relating to the Forest Department which have been published in the Gazette will be given in the Appendix, and I have endeavoured in the foregoing pages to explain the rules now in force in the Department. One peculiarity common to all these proclamations is that they do not specify any definite punishment for a breach of their provisions, and, as a consequence, the procedure has been peculiar, though in the majority of cases it has been successful. When an attempt to smuggle timber has been made and the timber has been seized, it has invariably been the custom to confiscate it and sell it by auction. Against the order of the Conservator to this effect there has always been an appeal to the Dewan, who has almost always confirmed the decision of the head of the Forest Department. Nowadays, the people are inclined to contest such decisions, and to appeal to the Civil Courts, and it is therefore very necessary that a Forest Act should be passed to legalise these rulings. Already some revenue has been lost for want of this Act, and it is certain that our position is becoming more and more difficult every day. Every case given against us makes the people more bold, and unless the Act, which has been under consideration for the last seven years, is soon passed, the revenue will suffer considerably. We have, indeed, a Forest Act No. IV of 1063 but that provides only for the formation and protection of Reserves.

As there are no penal sections in the Forest proclamations issued, almost all breaches of rules have to be treated departmentally.

1222. In cases of smuggling the timber can generally be confiscated, but occasionally we experience very great difficulty in putting our orders into effect. Thus for instance, a D. P. W. contractor lately built a bridge of smuggled teak, and handed it over to the D. P. W. before we had ascertained the fact. All we can do now is to make him pay for it, and as he pleads poverty we shall have very great difficulty in getting the money (700 and odd rupees) from him. As matters now stand, no one can be criminally punished for offences against Departmental rules, unless the act done can be considered to fall within the scope of the Indian Penal Code. Until quite recently it was held that persons found in possession of teak or blackwood, Government monopolies, could be convicted of theft, unless they could prove that they had obtained possession of them in a lawful manner, but, by a late High Court ruling, it was decided that Government monopolies did not differ in any way from ordinary property, and that the burden of proof must fall on the prosecution. The hardship of this decision is due to the fact that Government property is scattered all over the country in the shape of teak and black-wood trees standing on Sirkar land as well as on private holdings, and the only protection we have had against their destruction is that if any one cut them down they would have to prove where they got the timber or be convicted of theft, whereas now we have to prove that it was stolen, so a very wide door is opened to fraud and legislation is very urgently needed. We have also very much trouble in obtaining adequate punishment in cases of destruction of forest for cultivation.

This system is often convenient and speedy, but sometimes it becomes inoperative, and legislation is much required.

1223. Offences against Departmental rules may be classed under three heads.

Summary of offences.

(1) Thefts of teak, blackwood, cardamoms or other Government monopolies.

(2) Smuggling, or the evasion of charges for timber &c.

(3) The destruction of trees in clearing land for cultivation.

1224. Thefts of teak and black-wood are by no means uncommon. One method is for a person to buy teak at auction, and to sell it again, retaining the "rahadāri" or pass, under cover of which he cuts down and uses or sells other trees of the same species. Sometimes the auctions are of timber which has been found on the hills, and is considered to be too far from a depôt to warrant expenditure in delivering it there, but, more generally, they are held at the depôts for the sale of the rejected timber lying there, and, to get the passes, the people are willing to pay high prices for this inferior wood, far above what might be expected for such a class of timber. At other times teak is used without any attempt at obtaining a pass, in the hope that information will not reach the ears of the Forest Officers. Should an enquiry be held, efforts will be made to borrow passes from persons who are known to hold them, or an attempt will be made to bribe the officers sent to make the enquiry. Logs are often stolen in transit down the rivers, generally with the cunnivance of the raftsmen, and the proximity of Cochin territory to many parts of North Travancore makes these thefts much easier. Cardamoms are much stolen, in South Travancore by people coming up from Tinnevelly and returning there again without our knowing anything about their visit. On the Cardamom Hills the ryots are the chief culprits, and they manage to get the spice carried down to their villages in the Cumbum valley, generally with the help of the hillmen. In North Travancore the Muthuvāns, Kāders, and other hill tribes do the smuggling, and they exchange what they have taken to collect for knives, salt and tobacco in the villages of the Coimbatore district. A little ivory is said to be smuggled each year by the Palleyar Hillmen and the villagers of the Cumbum valley.

Thefts of Government monopolies very frequent.

1225. Our charges on timber are frequently evaded. Sometimes permits are taken out for a certain number of logs, and a much larger number is collected Rafts of logs to the number allowed on the permit are then sent down on dark nights, and they can thus be frequently run past the watch stations without any one being aware of it. If they are caught, a small present to the watcher puts matters right, or, if this will not do, the permit is produced, and the excuse is made that the logs were carried away by floods. If the rafts reach their destination without discovery the permits can be used again for more timber. Most timber-merchants are in the habit of taking out a succession of permits, a new one being paid for when the last is nearly exhausted, so that it is very difficult to catch them out, though the timber brought down under the permits has probably been felled a year or two before they were issued. Sometimes trees is felled, and boats are hollowed out and used in the interior without any payment of seigniorage. Sometimes permits are taken out for "mani-maruthu" or other woods of the third class, and "venteak", or some timber for which a higher seigniorage should be paid, is passed in its place. In a word, the ways in which our charges can be evaded are numberless, and should the timber be caught, there always remains a last chance of escape by bribing the officer holding the enquiry. Smuggling is the result either of insufficient supervision or of a prospect of very great gain, and in our case it is the want of supervision in the forest, and the good chance of escape after detection that have so encouraged it, for the amount saved is so trifling compared with the cost of working out the timber that the chances of confiscation must be very small, or the attempt would not be made. There is also a great waste of valuable timber, for the holders of permits fell away merrily, and leave in the forest to rot all logs that do not seem to exactly suit their wants, or which they find, on trial, to be too distant from road or river for their removal to be remunerative.

Our charges on timber often evaded in various ways.

1226. The clearing of land for hill-cultivation destroys much valuable timber. When the hillmen and others commence operations they usually select land covered with heavy forest, as being more productive, and although the timber growing in this description of forest is not so valuable (with a few exceptions) as what is found on grass land, yet its total amount is much greater, and if there happen to be ānjili, thambagam, and white or red cedar trees growing there the loss is very

Destruction of timber and deterioration of the land by hill-cultivation.

great. When the clearings are abandoned, forest similar to what previously existed does not again spring up, but the land becomes covered with a tangle of thorns, canes, and trees of ephemeral character and no value. The soil too is much deteri-- orated by exposure to sun and rain. When the clearing is burnt off the fire frequently spreads to the adjoining grass land, and, as no attempts to put it out are made, it probably chars the stems of the valuable teak and other trees growing there, and spreads over immense areas. In clearing new land, hill-cultivators are often afraid to fell living trees, but they either girdle and kill them, or they heap wood round their stems and the scorching that results destroys their vitality. Dur- ing the course of my travels I have frequently met with teak trees growing in land, formerly cleared for cultivation, in places where they certainly would not have been found (i. e. in heavy forest), had the land not been cleared. This therefore might be used as an argument in favour of hill-cultivation, inasmuch as it encourages the growth of a valuable tree in place of others of less value. If the hill-cultivators, in carrying out their operations, sowed the seeds of teak intentionally and in quantity, or if those seeds were accidentally conveyed by them to their new home, and if, when they returned again to cultivate those lands, they never touched any of the resulting trees, there would be something to be said in favour of that system, but, as a matter of fact, they cut down all young trees, teak included, and while leaving the older ones unfelled they lop them severely, so that, after perhaps a century of this clearing at intervals, one or two trees per acre at the outside, and these much gnarled and twisted, are all that are left of the original sowings.

1227. Judging by the light way in which those who break the present Act IV of 63 are punished, it will be a long time before the Magis- trates, who try cases of this kind, realise how important it is to pass deterrent sentences. The forests have always been looked upon as the property of the people, and the belief still holds ground that, in spite of legislation, the villagers should be allowed to walk into the jungles and help themselves to whatever they have a fancy. Not long ago some men were prosecuted for entering a Reserve and cutting canes and reeds. The Magistrate in acquitting them, entirely in defiance of the Act, said that he did not consider they were doing any harm, not so much for instance as would be done by a wild elephant !

Magistrates very reluct- ant to punish persons for forest offences.

(7)—Financial results.

1228. The revenue and expenditure of the Forest Department for the last 20 years will be given in the Appendix, and it will be there seen that the net revenue has increased in that period from a little under one lac to more than two and a half lacs per annum. The increase would have been greater, but for the fact that we do not now get any credit for the elephants we capture, as was formerly the case, and this too, though all the work falls on the Forest Department, and though one of the objects for which the Forest Aminadars were first appointed was to catch elephants. These results are eminently satisfactory from the point of view of revenue, and my predecessor must be congratulated on the steady way in which the returns have been increased. Whether these results have not been purchased at too great a cost, and whether it would not have been better to spend more money on works of improvement and conservancy is a different matter, about which I shall have more to say later on.

Financial results very satisfactory

1229. To sum up then, forest management in the past may be thus criticised.

Summary.

(1) The financial results have been excellent, but .

(2) scarcely any attempt has been made to improve the forests, or even to conserve them in the same state in which they formerly were.

(3) Smuggling and the wasteful destruction of valuable timber have been very common, because the supervision was inadequate, and, especially, because the forests were left entirely to merchants and others to do as they liked in them.

(4) Smuggling and theft have also been encouraged by the system of excessive centralization and by the very great powers given to low-paid and yet practically independent subordinates.

PART II.

(1) *Conservancy and improvement.*

1230. In that most interesting of books " The Naturalist on the Amazon " the author, Mr. Bates, makes the remark that no country can make
So long as a country is covered with forests it can make no progress in civilisation.
any progress in civilisation so long as it is covered with forests, and this statement needs no proof. Until the ground is cleared, and light and air are admitted, no crops can be raised, and life stagnates in the dense shade, while the people that live in the depths of the forest, as in " darkest Africa," the Andaman Islands and Brazil itself are debased and ignorant. Thus, the first step in improving a country is to clear away its forests, and thus, although a great part of the surface of the globe was at one time covered with forests, by far the greater portion of them has been cut down. In fact they have been, treated as one writer says " as an enemy to be extirpated, rather than a friend to be encouraged." But a reaction has now set in. It is generally admitted that forests have their uses and cannot be dispensed with, a fact which is being brought more prominently into notice year by year.

1231. I suppose it is hardly necessary at the present time to explain how useful forests are. The fact that the growing scarcity of timber
Few people have any idea of the immense value of timber, or of the large quantity annually consumed.
all over the world has drawn attention to the subject of forest conservancy is a sufficient proof that their utility is admitted. Yet, few people have any idea how valuable timber is or why. It has been estimated by Mulhall in his Dictionary of Statistics that the value of timber consumed yearly in Europe is £. 190,250,000 and in the United States £. 77,400,000, the quantity used in each country being 41 and 58 cubic feet per head respectively. The Madras Mail § has quoted figures to show that the money spent in furniture in England alone since 1840 was 958 millions sterling, or 24 shillings per head per annum. In 1880 the total forest produce of California realised 160 millions sterling, or double the value of the wheat raised and ten times the value of the gold and silver annually produced in that auriferous country. " The city of Paris burns the timber of 50,000 acres, yearly " " requiring an area of one million acres of forest to keep up the supply. (Mulhall.) " † Thus, the value of all the timber used throughout the world must reach an enormous figure, far beyond what might be supposed, and the importance of an abundant timber-supply cannot be denied. This is especially the case in a country poor as regards accumulated wealth, though rich in other ways, like Travancore. If by any chance the timber-supply of the State were to fail, it would be nothing less than a national calamity. Its cost would then be trebled at the very least, while there would be incalculable delay in obtaining what was required. Many people think that timber ought to be very cheap in Travancore because of the rapid growth of vegetation, but every green thing that springs up is not a teak tree, and though a teak plant will grow 10 feet high in a year, it takes 60–100 years for it to yield timber of useful dimensions. The large teak trees met with are centuries old and date back to a time when the history of this country is hidden in a mist of tradition, and were probably in existence before the reign of Aurungzebe.

1232. But forests are not merely valuable for their timber. They may be useful in other ways. In a lecture on Forestry delivered a year or
Various uses of forests.
two ago Dr. Schlich, Professor of Forestry at Cooper's Hill, indicated the utility of forests thus :—

(1) Forests supply timber, fuel, and other forest produce.

(2) They offer a convenient opportunity for the investment of capital, and for enterprise.

---

* A great part of the remarks in this chapter is taken from Broilliard's *Forest Organization*, Bagneris' *Sylviculture*, Fernandez' *Manual of Sylviculture* and the pages of the *Indian Forester*.

§ Oct. 26. 1888.

† Mulhall *Dict. of Statistics*.

(3) They produce a demand for labour in their management and working, as well as in a variety of industries which depend upon forests for their raw material.

(4) They reduce the temperature of the air and soil to a moderate extent, and render the climate more equable.

(5) They increase the relative humidity of the air, and reduce evaporation to a considerable extent.

(6) They tend to increase the rainfall.

(7) They help to regulate the water supply, insure a more sustained feeding of springs, tend to reduce violent sludge, and render the flow of water in rivers more continuous.

(8) They assist in preventing landslips, avalanches, the silting up of rivers and low lands, and arrest moving sands.

(9) They reduce the velocity of air currents, protect adjoining fields against cold and dry winds, and afford shelter to cattle, game, and useful birds.

(10) They assist in the production of oxygen and ozone.

(11) They may, under certain conditions, improve the healthiness of a country, and, under others, endanger it.

(12) They increase the artistic beauty of a country.

These various advantages may be grouped under three heads (a) Utility (b) Protection, and (c) Beauty.

1233. I have already drawn attention to the value of the forests for what they yield, but a point that is often overlooked is the employment given to thousands of persons both directly, in the felling, delivery, and sale of timber and other forest produce, and indirectly, when raw material from the forest is worked up into manufactured articles, like the "ceta" reed now being converted into paper at the Punalūr mills, and the lemon grass from which an oil is distilled. All those who are dependent for their daily food on the forests would be thrown out of employment if those forests were cut down, and the minor products were exhausted. The felling and delivery of timber also offers a good investment to capitalists, and in other countries extensive plantations have been opened as a speculation, and have proved very remunerative. The returns from this source are not so great as from agriculture, but trees can be grown on steep hill-sides, and in poor soils where it would be madness to attempt to grow crops. It is an admitted fact too that vegetation improves the soil, and trees have therefore been called the "pioneers of agriculture." The level plains of North West Canada * were at one time covered with pine forests, whose decayed remains now support the wheat fields of the Dominion, while the famous "Tchornozem" or black earth of Central Russia probably owes its fertility to the same cause. In Travancore the most ignorant hill-man knows that the land which has just given him a crop of paddy must be abandoned for some years till the scrub that grows up on it has attained a certain size, though he could not explain that this is because the plants send down their roots into the soil, and draw up nourishment which they deposit on the surface in the form of valuable leaf mould, containing the essential element of nitrogen, which grain crops require. Under all these circumstances it must be admitted that forests should be carefully preserved wherever this is possible, and that they should not be heedlessly destroyed.

Besides the actual value of the produce yielded forests contribute indirectly to the wealth of the country in many ways.

1234. The influence exerted by forests on the climate of a country, on the retention of moisture in the soil, and in the prevention of erosion and the silting up of rivers is enormous. In a dry and arid country trees cool the air and soil, increase the humidity of the air, prevent evaporation and increase the rainfall, and though in a moist climate like that of Travancore, their effects are not so marked in this direction owing to our heavy rainfall, the beneficial results are much greater in the prevention of land-slips, and in the regulation of the water supply, than in countries where the rainfall is light. It has been proved again and again beyond cavil that forests act like sponges in retaining the water that falls upon them, and in parting with it slowly. Thus, a regular

Beneficial effects of forests on the climate of a country, on the maintenance of the water supply, and on the conservancy of channels of navigation.

* Proceedings of the Royal Geographical Society. February 1892.

supply of cool water is always available in their neighbourhood, while, at a distance from them, the people can only obtain water with difficulty. Let those forests be cut down and all the springs will run dry. The force of the rain that falls on them is broken by the leaves, and instead of striking the ground with velocity, and washing away the soil, the drops fall softly from the trees and sink into the earth. Any one who is at all sceptical on this point need only travel about the hills of this State during the monsoon, and watch the condition of the roads which pass through forest, and of those which are bounded by cultivated or grass land. While he is in the forest he will find the roads but little affected by the heavy rain, but in the open he will see them cut up by the rush of water from the neighbouring fields, and, after a heavy shower, will find a deposit of mud and silt upon them often a foot in depth. A fall of 200 to 300 inches a year, of which as much as 18 inches has been known to fall in a single day, must in the nature of things produce an enormous effect on the surface of the country, and thousands of tons of soil must be washed down every year from the bare hill-sides into the rivers and backwaters of the low country, silting them up with gravel and sand. It is useless to go to great expense to reclaim shallow backwaters if no attempt is made to prevent the deposit of silt in the channels which are to convey the water away. If the steep slopes of the hills are cleared as at present, and the soil is allowed to be carried off to fill up these channels, great care will have to be taken to dredge them, or the channels will be year by year raised above the level of the adjoining fields, with the certainty that they will sooner or later burst their banks and cover the fields with a deposit of sand *. Lastly, trees give protection from wind, and raise the productiveness of fields.

1235. The experience of other countries teaches us that the land must not be entirely denuded of trees or the climate will be completely changed and the fertility of the soil ruined. A writer in the North American Review for Jan. 1879 says " Afghanistan," " Persia, Mesopotamia, Syria, Asia Minor, Greece, the South-" " ern islands of the Mediterranean, and the whole of North-" " ern Africa from Cairo to the western extremity of Morocco, " " countries which were once blessed with abundance and a glorious climate are now " " either absolute sand-wastes, or the abode of perennial droughts, hunger and " " wretchedness, and wherever statistical records have been preserved it is proved " " beyond the possibility of a doubt that their misfortunes commenced with the dis- " " appearance of their arboreal vegetation." Conversely, the beneficial effect of planting trees to arrest the advance of moving sands is well-known, and I need only mention the case of the Landes in Southern France. In the last century the sand from the sea shore urged east-ward by the force of the strong winds blowing across the Atlantic, was annually covering fields and overwhelming villages with shifting dunes, 100 to 200 ft. deep, but in the year 1789 the Engineer Bremontier commenced a system of erecting wattled fences and sowing belts of pine-trees and gorse across the path of the moving sands, and this system, carefully followed ever since, has been completely successful in saving a further area from ruin, while forests of valuable trees have been raised.§

The destruction of forests has brought desolation to many countries while afforestation has saved others from ruin.

1236. I think I need say no more to show that the preservation of forests is indispensable, on account of the value of their produce and of the protection they afford, as well as of their beneficial effect on the climate. If any one is at all sceptical on these points there is a whole library of books on the subject the perusal of which will soon remove his doubts. The question now to be asked is how is the timber and fuel supply of a country to be maintained and how can the forests of protection best be preserved. This may be done in two ways :—

Forests can be preserved in two ways.

(1) By planting to replace what is annually removed.

---

* Large areas of lowlying land in this State which are now submerged by brackish water were undoubtedly at one time occupied by fertile fields of paddy or by groves of trees, as may be seen by any observant traveller. Roots and stems of trees are frequently met with in these parts lying in the water, showing that at no very distant date these portions of the country were firm land. As examples I may mention the swamps about Karumádi and Kaypala. The channels which drained them no doubt were silted up and the water spread over the land on their banks. Such areas must be carefully distinguished from the lagoons and backwaters proper which adjoin the sea board, which were connected with the sea at one time and are gradually being silted up, as the Vembanád, Kayankulam and Paravúr lakes.

§ Indian Forester. XIV. pp. 1.

(2) By encouraging natural reproduction.

1237. Hitherto we have been depending entirely on planting for the future supply of timber in Travancore, and no effort has been made to encourage natural reproduction, although, owing to the large area of our forests and the remoteness of some of them, the hill slopes have been protected, and reproduction has taken place of its own accord in them. The extension of the plantations is an excellent step, but in our case teak alone has been planted, while no attempt has been made to replace the thousands of blackwood, ànjili, thambagam, vēnga, thēmbàvu and other valuable trees which are annually removed. The extension of the plantations moreover has not been regular, as it should have been if the object was to replace the annual removals, nor have the extensions been sufficiently large. As regards the other trees, we have been working on the assumption that the forests are full of them, and that when we clear away a few others will at once take their places, quite ignoring the fact that the useful trees only include about one-fifth of the total number, and that as we go on cutting down only these useful trees the interior species take their places, and the proportion of useful trees thus very rapidly diminishes. (I have noticed the remarkable instance of the disappearance of teak from the Idiyara valley in paras 936—938.) Again, when forests are worked without system and without the use of maps, it always happens that those portions most easy of access are denuded of good timber, while those at a distance are left untouched. If timber operations were distributed over the forests equally and regularly, the trees in the more distant portions would be felled as soon as they reached maturity, and their removal would allow of a second generation springing up, whereas at present, in the untouched forests, the older trees are left standing to a great age till in the course of time they decay and fall down. At the same time the trees in the more accessible parts are felled before they attain maturity, and such timber as comes to market is not so lasting. Thus, while there is loss through want of felling in one part there is loss through overfelling in another. In our case this loss has been further aggravated by the system of unsupervised felling on license which now exists. With our large area of forests it has been found quite impossible to control fellings which go on without restriction in every portion of the forest at the same time, and it would be quite impossible to control them (so long as the work is not concentrated) without the aid of a staff of enormous size. Upon the question of the best system to be followed in future both on general principles and in the particular case of Travancore a few words must be said.

The system of depending on planting to replace annual removals to be condemned.

1238. The system of depending entirely for the future timber-supply of the country on planting is one that has been much advocated, especially by those whose attention is turned to the subject for the first time, and in many respects it has much to recommend it. After visiting a plantation and seeing how cheaply the work can be done in comparison to the value of the timber raised on it, how much, in fact, nature does if we only assist her, it is very natural to wish to plant up with valuable trees a large portion of the waste land available, and to devote the rest of the waste land to agriculture, but though the cost of planting 100 acres with teak, for instance, is not heavy, the cost of planting with different trees even 100 square miles, which would not include all the steep land in Travancore unsuitable for agriculture, would be enormous, while the staff required to carry out the work would be unobtainable. The advantages of artificial reproduction are (1) Concentration of work. The same number of men can attend to the planting and management of a given number of trees if they are all located in the same spot better than if they are scattered about over a large area. (2) Reproduction is more certain for the same reason. (3) If we are at liberty to choose the site for the plantation where we like, we can select it so as to grow the most valuable trees, and in a spot whence they can be easily removed. (4) As the trees of each plantation will be of the same age the cost of felling and removing will be very light. (5) The returns, area for area, are much greater than in forest raised by natural regeneration where most of the species are of no value.

Advantages of artificial regeneration.

1239. The advantages of natural over artificial regeneration are as follows :— (1) A given area can be much more cheaply and easily re-forested by conservancy and protection than by planting. (2) The ravages of insects or of tree diseases are not likely to be nearly

Advantages of natural regeneration.

as disastrous in mixed forests as in plantations. This is a very important point, because it is now generally admitted that all trees or plants grown together in numbers are liable to disease. The wood of teak is attacked by a borer, and its leaves are eaten by a species of caterpillar; mahogany trees have their tops eaten by another caterpillar; the larch tree has its disease; the cocoanut suffers from a beetle which lives in its tender top; tea, coffee and cinchona are all liable to fungoid diseases or to the attacks of insects, which retard their growth if they do not permanently injure them. The spruce forests of Southern Germany * have lately been so devastated by the ravages of the "Nun" moth *Liparis monacha* that it has been found necessary to cut down the trees over thousands of acres that the timber might not be completely wasted, and the loss has been prodigious. It can readily be understood that when a forest is composed of mixed species, any disease or insect-pest that may appear does much less harm than when all the trees are of the same species and of the kind best suited to be the victim. In a mixed forest the chances are that only one or perhaps two species are attacked, and not only is the disease less likely to spread, but the damage done over equal areas will be much less. (3) The risk of failure is much less in the case of mixed than in pure forests. The latter have been planted, we will suppose, in a locality and under conditions which seem completely favourable, and for a time the growth may be excellent, but suddenly the trees will be found to be in an un-healthy state, perhaps after growing for 30 or 40 years, and they will have to be cut down, because the climate, or the soil, or the aspect, or something hardly thought of at the time of planting, was unsuitable. On the other hand, in the mixed forests raised by natural regeneration the trees are all of species acclimatised to that special locality, and descended from older ones growing there, and if any of them are found to languish, other species of equal value soon dominate them and take their places. (4) Though all plants and trees consume the same elements of the soil, some require a greater proportion of one and some of another element, and consequently a pure forest of a certain kind of tree will exhaust the soil of the element it chiefly requires, while a mixed forest will probably affect the proportions of the elements in the soil but little. Consequently, a greater amount of timber can be grown in mixed than in pure forests. †(5) As in plantations all the trees are of the same size there are no large trees to draw the others up, and consequently the trees in a plantation do not attain the same size as those in mixed forests, though there is greater uniformity among them. (6) Few trees are really gregarious in their wild state and even then only over small areas, it is therefore extremely probable that the growing of them alone in plantations may affect the character of the timber, though this is less likely to be the case with teak which is often found almost pure. (7) The system of natural regeneration does not require the uncovering of the soil, for the trees spring up in the forests, whereas the land has to be quite cleared before forming plantations, and on steep land the soil is not washed away under the former system. (8) If plantations are opened now we shall be able to get no timber of any size from them for 60 years, whereas if we give our attention to the smaller trees in the existing forests and encourage them we shall improve our timber supply in a very short time. (9) Finally, the great advantage that natural regeneration possesses over artificial is that under that system the steep slopes of the hills which must be retained under forest for climatic reasons can, not only be protected but made to yield a good return. The felling of trees in them, if judiciously carried out, is not only possible but is advisable, because a few trees taken away here and there as soon as they have arrived at maturity, make fewer inroads into the forest than would be caused by old trees crashing down after they had reached a great age.

1240. To sum up then, plantations are only to be recommended where land is expensive and difficult to obtain, where there is a good demand for small timber, where a special timber is in great request, and where the timber in question is naturally gregarious and is known to grow well alone. We must depend on natural regeneration for managing the forests that exist, not only because in the present state of our knowledge there is less risk in following this system, but because under no other system can we maintain our forests of pro-

We must in Travancore depend chiefly on natural regeneration for the future timber-supply of the State.

---

* Ind. Forester XVII. pp. 261.

† An alteration in the proportion of the elements of the soil may account for the gradual change in the species composing a forest to which I have referred in para 938.

tection. Even in France and Germany where land is scarcer and more valuable than in Travancore a greater amount of attention is paid to protection and natural re generation than to planting, and where planting is undertaken it is because this is the only way to afforest a barren moorland or sand-dune which no amount of protection would cover with trees. Here in Travancore we must not give up planting altogether, more especially as we have in the teak a tree which is known to be gregarious and also very valuable. We should endeavour to make a regular addition to the acreage of our plantations every year, and to prevent the risk of fungoid and other diseases spreading we should mix our teak with other trees, or break the plantations up with belts or groves of other trees which are of value and deserve attention, but we must chiefly depend on natural regeneration for the future timber supply of the country.

1241. The area of waste land at the disposal of Government occupies probably two-thirds of Travancore, and it is obviously unnecessary to set aside so large a proportion of the land for the supply of timber and fuel for the inhabitants of the rest of the country. What percentage of the area of a country should be reserved must depend on a number of circumstances. First among these is the character of the forests themselves. If they are rich and full of valuable timber, a smaller area will suffice than if the trees were of little value. Then, the requirements of the people must be considered. If they are backward in the scale of civilisation, all that they need may be enough fuel to cook their food, and sufficent wood to make their bows and arrows and spearhandles. On the other hand, they may require, as in America, a large quantity of timber per head per annum. Then again, we have to consider not only the requirements of the present day and of the present population, but have also to estimate the possible needs of a constantly increasing population, accumulating wealth, and advancing in civilisation. At the same time, we may hope that our forests will improve. under careful management, and, therefore, the same area a hundred years hence will be able to yield far more than it does now. In the determination of this question we may get much assistance from the experience of other countries. It is roughly estimated that from one-fourth to one-seventh of the area of a country should be reserved for the production of timber and fuel for the rest, and if we look at the figures given in para 895 of the area of different countries under forest, we shall find that this estimate corresponds with the deductions to be drawn from those figures. Of the countries there mentioned Russia, Scandinavia, Austria and Germany, with more than 24 per cent of the land covered with forests, export timber, while France and all others import it. France with an area of 17 per cent under forests spends 6 millions sterling a year on imported timber, and another 5 or 6 per cent would hardly make her independent of other countries. England annually imports timber to the value of nearly 20 millions sterling, and, if iron for building and coal for fuel were not so abundant, a very much larger quantity would be required. Taking the area of Travancore at 7,000 square miles, we may accept it as certain that an area of 1,500 square miles of forests should be sufficient to supply all the requirements of the State.

About ⅕ to ¼ the area of a country should be under forests. In Travancore an area of 1,500 square miles should be set aside for reservation.

1242. It will be no hardship to set aside so large an area for reservation, because there is so much more land in the State from which the people will be able to draw their supplies of fuel and small timber. In this respect, Travancore is unlike many other portions of India, where the area of forests of every kind is far less than 20 per cent of the whole and where it is necessary to insist on the most careful forest conservancy, more for the benefit of the future than of the present, and there the complaints of the people who do not, or will not, see this, are most bitter. As Sir M. E. Grant Duff pointed out " in regulating the proper proportion of trees, care must be taken neither to " sacrifice the present to the future, nor the future to the present. If we have too " many trees, we sacrifice the cultivator and grazier of the 19th Century to those " of the 20th. If we have too few, the cultivator and grazier of the 20th Century " will be, in many districts, simply non-existent. " The area I have indicated is, then, to be set aside for the production of timber for the people, any surplus not required being available for sale. As the trees in a forest grow year by year,

The object of Forest Conservancy to obtain the largest sustained annual yield while carefully preserving the value of the forest from depreciation.

there is an annual increase in their volume which varies with the species of tree and its rapidity of growth &c. This annual increment may be regarded as a sort of interest on the capital-value of the forest itself, which may be spent each year without depreciating the value of the forests. In one respect indeed, timber differs from money, in that, if it is not utilised, it is lost through decay, while money at interest goes on accumulating if it is not spent. Thus, we must steer between two courses, between excessive felling on the one hand which means trenching on capital, and, on the other, leaving a forest quite untouched which means loss of the annual increment. In other words, all reserved forests should be worked to yield the whole or nearly the whole quantity of timber by which they are calculated to increase annually, but not more than that. And here it may be well to enunciate the first axiom of Forestry. It is this— *No system of Forest Conservancy is deserving of the name which does not provide for the maintenance of the forest itself, and the increase and improvement in character of its annual yield.*

1243. Many people suppose that the main object of forest management is to
A large or increasing revenue no test whatever of proper forest-management. show a large profit, and that the proper test of efficient administration is an increasing revenue. But to believe this, we must also believe that the more we cut down the trees the faster will others, Hydra-like, spring up to take their places, which is manifestly absurd. It is true that when the demand for timber is very small, it is an advantage to remove and sell the mature trees to make room for others, but the number of trees removed must not exceed the annual increment, or the forest will deteriorate. *It is easy to show that the test of revenue is not only unreliable, but that the doctrine has done much harm in practice. Take the case of two countries, one with very fine forests, an excellent climate for vegetation, and fortunate in having a good sale for its timber, and another with exhausted, or naturally poor, forests, and no demand. Though the returns from the former might be much greater, yet the forests of the latter might be much better managed with regard to the future wants of the country. Even the doctrine that an increasing revenue must be aimed at will not bear scrutiny, because revenue depends on many circumstances. A decrease in the returns may be due to a fall in prices or a less demand, rather than to a smaller quantity of timber brought to market. The test of revenue alone is therefore illusory and must be abandoned, for to be of any use it would also be necessary to take stock of all the timber in the forests, to see that we have at the end of each year as valuable a stock as we had at its beginning. The real objects of forest conservancy are (1) the maintenance and improvement of the forests for future wants : (2) the regular and cheap supply of forest produce to the people in the present : (3) revenue. Objects (1) and (2) often clash with (3) because, if timber &c. is to be given to the people cheap, the returns cannot be so large as they might otherwise be. It is easy now to see how forest officers in Europe lay so much stress on maps and working plans, because they know that, unless these are properly drawn up and considered, the maintenance and improvement of the forests cannot be secured.

1244. Remembering then that the slopes of the hills must be preserved under
The reserves to include all the steep land as forests of protection, and, if this is not sufficient, the forest-land adjacent to the hills. forest, and that forest in such positions can be worked for timber, my task was to begin by selecting our reserves on the hill-slopes, and if I could not find a sufficient area in such localities to choose land at the foot of the hills and adjacent to the steep land. It is often thrown in our teeth that the forest officer, like another William of Normandy, would desire to convert the whole country into a deer-park, but nothing is farther from the truth. Knowing how much land will be wanted for the requirements of the country he works in, he sets aside that much and no more, and though he regrets to see valuable trees, which have taken generations to attain maturity, burnt and wasted, yet he knows too well that agriculture pays better on level land than the growing of timber, and that if he attempts to reserve too large an area, he will not be able to supervise it properly. It is also essential that the reserves should be well distributed, and

* In Europe, where all the trees of a forest are valuable, the annual increment runs from 32 to 80 cubic feet per acre or even more and is generally set down at ½ a tree per acre a year. Here, where most of the trees are of no value at present the quantity of timber we may remove without injuring the forests must be much less. See Campbell Walker's Forest Management pp. 3, 28.

should be so selected as to meet the wants of the people in the several large centres of population. Upwards of a dozen years ago Dewan Nanoo Pillay suggested to my predecessor to reserve a strip of forest, of 3 or 4 miles in width, from the north, of Travancore to Cape Comorin, and though it was impossible to carry out this suggestion literally, yet the idea was perfectly correct.

Existing reserves. 1245. At the present time the following are the chief reserves that have been proclaimed or are under settlement.

Beginning from the North we have .

(1) The *Malayāttūr Reserve* of 345 square miles, bounded on the north by the Travancore-Cochin boundary, on the east by the territory of the Pūnnyātta chief, on the south by the Periyār river, and on the west by the limits of cultivation. This reserve consists almost entirely of evergreen forest lying at an elevation between 200 and 1,000 feet. Formerly, teak of large size used to grow here on the drier hills and ridges covered with deciduous forest, but the best trees have now been almost all cut down. Within the last few years, large numbers of white cedar, ven-teak,* thambagam, and ānjili trees have been removed from these forests, but owing to obstructions in the river, which prevent the use of bamboos to assist in floating heavy logs, the heavier timbers have not been felled. The Muthuvāns have in parts cleared the finest land which is in consequence now covered with thorny scrub. This reserve supplies Cochin and Alleppey with timber, and it is well defined and protected by the broad Periyār river which runs along its southern and south-western side.

(2) *The Kōnni Reserve* of 300 square miles bounded on the north by the Pamba branch of the Valiya-ār, on the east by a high range of hills, on the south by the Acchankōvil river, and on the west by a line running from the teak plantation bungalow near Kōnni due north to Rājampāra, a few miles east of Peranāda. This reserve also consists almost entirely of evergreen forest, with deciduous forest on the ridges containing a good quantity of teak of small dimensions, the larger trees having been cut down, and it lies at an elevation between 200 and 1,000 feet. Many thousand acres of it have been cleared for cultivation and since abandoned, and many logs have been felled herefrom, but much valuable timber still remains. This reserve supplies logs to Cochin, Alleppey, Cottayam and Quilon.

(3) *The Kulatthuppura Reserve* of 121 square miles bounded on the north by the Quilon-Shencotta road, on the east by the Travancore-Tinnevelly boundary, on the south and west by the Kulatthupura river. Nearly all of this reserve too is covered with evergreen forest, but it includes several coffee estates which are private property, and numerous plantain gardens which must be excluded from the reserve. This reserve supplies Alleppey, Quilon and Trevandrum with timber. The elevation is the same as that of the other two. These three reserves aggregating 766 square miles comprise the greater part of the land set aside for forest conservancy, the other two reserves, mentioned in para 1148, lying in the low country, and not being of much value or large extent.

1246. As yet no reserve has been named in the South where forest conservancy is probably a matter of more pressing importance than **Other blocks of land to be reserved.** it is any where else. In July 1886 I recommended the reservation of a block of forest aggregating some 15 square miles and covered with valuable timber, but my predecessor considered that a larger area should have been selected there, and he was no doubt right, but the forest in that neighbourhood is so much intermixed with land sold for coffee cultivation that a further selection was not an easy matter, on account of various claims that would have been put in, and no land has therefore been set aside in that neighbourhood, as I had not time to go over the ground again. Again in North Travancore I recommended the reservation of the steep slopes of the hills between Kōthamaugalam and Arrakulam, and also of similar land lying between the Pamba and Arathu branches of the Valiya ār. These blocks of forest should all be reserved now, together with the extensive deciduous forests to the south of the Acchankōvil river and north and west of the land granted to Mr. Huxham

---

* Thambagam is a heavy timber, but the trees removed have been hollowed out for boats.

(see para 860.) Another block of forest land that will probably be required lies to the west of Madatthura, and south-east of Punalūr, in an angle of the Kulatthuppura river. The accompanying map will show approximately the 776 square miles now already set aside, and the other portions of forest that should also be reserved. If the total of 1,500 square miles which we require is not made up on the hill slopes, blocks of land in the low country, either covered with forest or suitable for growing it and useless for cultivation, must be selected, and we shall pay special attention to the preservation of the groves of teak which are to be found in many places in North Travancore. Thus, the suggestion of Dewan Nanoo Pillay will be nearly carried out, for we shall have a line of forest-reserves running from the north to the south of the State, supplying all the important towns and villages with timber and fuel, although the reserved area will not be of uniform breadth, nor will the forest be of equal value throughout, and although the territories of petty chiefs, or the occurrence of unsuitable land, will break its continuity.

1247. With a sufficient area set aside for the supply of timber for the State, the next thing to be considered is how to treat the reserved area, bearing in mind that in this area will be included the steep hill-slopes or forests of protection, which must not be felled in large blocks for fear of the effect this might have on the soil or climate. It is not sufficient to exclude the hillmen and those who live by shifting cultivation. By attending to this point we can prevent the useless destruction of forest, but we want something more than this if we would make the most of our forests, and that is that all portions of the forest should be equally worked, and that one part should not be exhausted of timber while another is left untouched. In order to ensure this, either all the valuable trees in a forest must be counted and only a certain number must be removed each year, and these at equal intervals all over the forest, or the forest must be divided into blocks each of which must be made to yield its produce in turn while the others are closed to all work. The former method is the most accurate, but it is inapplicable where the area is large, on account of the impossibility of counting every tree, and where the number of species of trees is great, the difficulty is vastly increased. In Travancore the method to be followed is that of apportioning out certain blocks to be cut over every year or group of years. In deciding on the details of the system to be followed we shall obtain much assistance by examining what has happened in other countries.

In managing the reserves it is not sufficient to exclude squatters; steps must be taken to ensure a sustained yield.

1248. Up to the middle of the 17th century the forests of France were worked in the same way that our forests now are, that is to say the most useful trees were felled and removed as they were wanted, regardless of the future timber-supply of the country; but, as the difficulty of obtaining good timber increased, it became a matter of the greatest importance to regulate the fellings. Accordingly, by the royal edict of 1669 it was ordered that all State forests should be worked à tire et aire, by which equal areas were to be clean-felled in consecutive order. Thus, suppose a forest contained 2,000 acres, and that it was calculated that the trees in it took 200 years to reach maturity, then 10 acres had to be clean-felled every year (8 trees an acre being left as standards), and afterwards, each individual block of 10 acres was not again touched for another 200 years. The chief advantage of this system was the facility and cheapness with which the timber could be worked out, but it had many disadvantages. When a piece of land is cleared of forest and then left to itself, the trees that grow up are very rarely of the same species as those that grew there at the time of felling. The species that spring up are chiefly those whose seeds are light and are blown about by the wind, or are succulent and attractive to birds and are therefore carried about by them. Those species which seed very freely have the advantage, as well as those which grow rapidly and can suppress the slower-growing kinds. Unless the ground happened to be well-stocked with small plants or seeds of the better kinds of trees when the forest was felled, there would be little chance of their being introduced afterwards, except from the adjoining forest, and as the blocks of forest were to be felled in conse-. cutive order the chances of seed of good trees being blown into each block diminished every year. At the end of the 200 years instead of having a fine forest in the block first-felled, the area would be covered with scrub and worthless trees, with trees of good kinds sparingly scattered through it, forming a very irregular forest. It was

The method of working forests "à tire et aire" was found unsuitable in France, and would not succeed in Travancore.

found too that the soil suffered from exposure when the system of clean felling was followed. All the evils incidental to such a method of working in Europe would be aggravated in this country where the heat of the sun is so much greater, the rainfall heavier, and the growth of ephemeral vegetation more rank. Such a system would therefore be quite unsuitable for Travancore.

1249. Finding that this system of clean fellings over equal areas in consecutive order was not a success, the French Forest officers have gradually elaborated another system which is calculated to give, under ordinary circumstances, the best results both in quantity and quality of timber. This is known as the Natural method.

The Natural method though it yields the best results cannot for the present be followed in Travancore.

Under this system a forest is first of all treated to a light cutting which lets in the sun-light sufficiently to encourage the small plants (which would languish in deep shade) to grow freely, but not so much as to unduly expose them. After an interval of about 20 years the forest is treated to a second cutting which lets in the light still more and allows the small trees to grow with vigour. Finally, after another interval of about the same time, the trees of the original forest still standing are all felled, and the land is thus left covered with trees of from a few months to 40 years old. The forest is then left to itself (except when it is necessary to make thinnings in order to give the more vigorous trees more room,) and at the end of the rotation of, say, 200 years, the trees in it are all from 160 to 200 years old, or practically of the same age. The same operations are again repeated, the complete sowing of the area being thus ensured, as well as the protection of the young trees, while the soil is never completely exposed. Although, as I have said, this method yields the best results it is not always applicable. It is necessary that the forest where it is to be tried should consist of trees more or less regular as regards size and age, and that it should be easily accessible, and should contain nothing but valuable trees and these of only a few species. This system further requires a skilled and numerous staff for its proper management. It is therefore not at present suitable for adoption in this country.

1250. The system which must be followed in Travancore is the Jardinage or Selection system which is in vogue in other parts of India, and is still employed in Europe in mountainous districts or where the forests are irregular or mixed. *"The Selection method is "simply the exploitations of primitive humanity generalised into a system." In early times people used to help themselves to such timber as they needed, and their wants being small, they seldom felled any trees, but depended on fallen wood for fuel and even for building their houses. The forest did not therefore in any way suffer, but remained always fully stocked, so long as the number of trees removed was restricted. But when a large demand for timber sprang up, thousands of trees were felled within a very small area, and thus it has happened, in Travancore as well as elsewhere, that in some localities certain trees have been exterminated, and even the whole of the forest has been cleared away, while the more distant forests have been left untouched. The Selection method aims at restricting the annual fellings to what the forests can yield without deterioration, and at distributing the fellings equally over the forests.

The Jardinage or Selection system the only system suitable for Travancore.

1251. It is necessary to explain the details of this system by an example. Let us suppose that we have to deal with a forest of 1,000 acres stocked with trees of all ages, each species of which requires § 100 years to come to maturity, and that each acre contains 50 mature trees. Then to keep the forest in its present condition, we must not fell more than half a tree an acre each year or 500 trees in all, equally distributed over the whole area. But this, if insisted on exactly, would occupy much time and be a very costly system of felling, besides requiring much supervision, and as, after attaining maturity, trees go on growing, for many years before beginning to decay, there is no harm in felling more trees in one locality (up to a certain limit,

The Selection system illustrated by an example.

---

* Broilliard's Forest organization pp. 170.

§ This is a mere assumption, because some trees are very slow-growing like the kambagam (*Hopea parviflora*) which probably takes 200 years to attain even a moderate size, while others like the cotton (*Bombax malabaricum*) and chini (*Tetrameles nudiflora*) grow three times as fast.

and provided they are pretty well distributed) and then leaving that locality intact for some time. Thus, instead of felling ¼ tree an acre over the thousand acres each year, we may fell 5 trees an acre every ten years. Dividing the thousand acres out then into 10 blocks of 100 acres each, we may each year fell 500 trees from one block, leaving the other 900 acres intact. This saves supervision, reduces the cost of exploiting the timber, and does not diminish the value of the forest. We may also either increase the number of blocks and reduce the number of trees to be removed on each occasion, or we may divide the area into fewer blocks and remove more trees at a time in each block. Circumstances must lead us to decide in each case, but in any case we require to know * the age at which the trees are most suitable for felling, and the average number of mature trees per acre.

1252. In drawing up our working-plans, then, the first thing to decide is how many trees per acre can be removed at a time without endan-
Method of calculating the yield of a forest described.
gering the yield of the forest. Suppose we accepted the French standard, and decided that no tree with a diameter of less t' an 2 feet at 4½ feet from the ground (or say 15 vannams quarter-girth at 2 koles from the ground) should be felled, we should have then to ascertain the average number of trees of this size to the acre throughout the reserve under consideration. In grass-forests the number would probably be from 20 to 30, in the ever-green forests from 60 to 150, and it would be necessary to eliminate from these figures all worthless species, for which there would be no demand. Having finally decided on the proper number upon which to base our calculations, and making due allowance for areas occupied by bare grass or rock, and for those portions of forest that had been already worked, we should then have to ascertain by counting the rings in the stems of those species that occurred in the reserve how long each species would take to attain the given size. In the case of Anjili we should probably find that 30 years was sufficient, but that kumbagam required 100 years before it was large enough to be felled. Again taking an average, we should eventually find that the reserve contained perhaps 10 valuable trees of the required dimensions to the acre and that 70 years was the average time needed for them to attain the proper size. The yield would then be ⅟th of a tree a year, and knowing the acreage of the reserve, we should then know how many trees could be annually removed from it without causing it to be unduly robbed of its timber. This estimated annual yield would have to be constantly revised, and a careful register would have to be kept of the actual number of trees removed each year from the reserve.§

1253. Besides keeping the number of trees to be felled below a certain figure, it would also be necessary to see that the fellings were, as far
The fellings to be equally distributed over the forests.
as possible, equally distributed over the reserve, otherwise the new system of working would be no better than the old for the nearer forests would, as at present, be denuded of their timber while the more remote were left untouched. It would be necessary then while confining the fellings to a portion of the reserve, say one-tenth, to distribute them equally over that area, and this would be one of the most difficult tasks to accomplish, especially if the felling were done on the permit system, as the permit-holders would naturally endeavour to take as many trees as possible from the same locality to save themselves work. To prevent this it would be advisable to cut the forest up into blocks of small area, from each of which a given number of trees might be removed.

---

* The most suitable age for felling depends on various circumstances. Timber of large dimensions is more valuable, because it is suitable for a greater variety of purposes, and if we require the most useful timber we can get, regardless of what it may cost us, we should not fell our trees till they had attained a great age. We should, for instance, leave all our teak to grow for 200 years or more, in fact until it showed signs of beginning to decay. But, if we have regard to the revenue to be obtained, then we must fell at a much earlier age, in fact as soon as our trees have reached the smallest size at which the timber will be generally useful, because trees grow slowly after reaching a certain age, that is to say maturity, and therefore five trees of moderate size may be grown in 400 years and they will contain much more timber than one large tree 400 years old, to say nothing of the gain by interest on the value of the younger trees. If, as is probable, we decide to fell our teak trees when they will give the best returns and at the same time be useful for ordinary purposes, than they will never be allowed to attain large dimensions, and it will become yearly more difficult to obtain the huge logs for which Travancore was once famous.

§ The minimum diameter mentioned above cannot always be adhered to without any deviation because the soil and climate varies with the locality, and in some places such a size as I have mentioned might never be attained. Thus, in South Travancore the kol-teak never grows to more than 1 foot in diameter, if so much. A very wide discretion must therefore be allowed to the officer who draws up the working plan, on this as on other points.

1254. The great danger to be avoided under this system is that of filling too much, and it will therefore be most necessary not to take too sanguine a view of the production of the forest, so as to be quite certain that no more than the actual increment of timber is removed each year. In the Madras Mail of August 24 1889 a writer, comparing the working of forests in India and France, sounded a note of warning on the extermination of the more valuable species of trees. Although the most careful arrangements may be made that no more trees should be felled than will be replaced by the annual increment of growth, it may happen, if the forest contains many worthless species, that as all the trees felled will be of the more valuable kinds, the forest will in time contain nothing but worthless trees, unless precautions are taken to ensure the reproduction of the valuable species by sowing their seeds or even planting them, and by removing the worthless kinds. In a word, if we interfere with nature's arrangements so far as to remove part of what she produces, we must assist her by replacing the equivalent of that part. Every effort too must be made to introduce to public notice the little known species in order to economise the more valuable.

Care to be taken not to overfell and not to stop short at felling only valuable species.

1255. It will be seen from the above that the difficulty of drawing up a correct working-plan will not be small. Our interior forests are so extensive and so little known, that many of our data must be obtained by guess-work, and time may prove that our inferences were not justifiable. Still we must make an attempt at forest organization on the lines I have indicated, and hope for the best. The plans must be drawn up by an officer of experience, and arrangements must be made to check their working and to correct them from time to time. If we find we have overestimated the yield and are over-cutting a forest we must at once alter our programme and cut less, and vice versâ.

Working-plans to be drawn up with care.

1256. It will probably be advisable, as in other countries, to keep a sort of reserve-fund in case of emergencies. This may be done by setting aside a fourth or other portion of the reserve in calculating the total area to be cut over. A sudden demand may at any time arise for timber for railways, ships or other special purposes, and unless such a reserve-fund were available we should have to remove more than the annual increment of timber, and in doing so we should permanently interfere with the yield of the forest. On the other hand, if no such special demand occurred, the reserved fund would not be lost, but would go to increase the value of the forest, and at the time of the revision of the estimate of annual yield, that yield could be proportionately increased.

Necessity to maintain a reserve-fund for cases of emergency.

1257. The advantages of the Selection method described above are that reproduction and the maintenance of a leaf-canopy are ensured, and thus it is the only system possible where the forests are irregular and the country mountainous. Its disadvantages are that the quality of the timber is inferior to that grown under the Natural method, much damage to the remaining trees is caused through felling trees growing in a forest, while the removal of the timber through standing forest does a certain amount of harm, and lastly, the quantity of the produce is less. In countries where the forests are not situated in mountainous districts, and where the maintenance of a leaf-canopy is not necessary to prevent erosion, but where they have to be worked on the Selection system on account of their irregular character, it is the object of the Forest Officers in charge to convert these irregular into regular forests by bringing the trees left standing to more or less of one age and size, with the view of ultimately working them on the Natural method, and thereby obtaining better returns and better timber. In Travancore we may desire by and by to effect these alterations in certain localities, but for the present we must concern ourselves with working all our forests on the Selection method, and with bringing them into good condition. This method indeed has as its chief object the improvement of the forest, regardless of revenue, and, properly speaking, no more should be removed from it than the decayed, dying and crowded trees which would certainly not represent its best material, but we could not so far sacrifice the present to the future as to abstain from removing a certain quantity of good timber when we have it in fair abundance.

Advantages and disadvantages of the Selection method.

1258. When each reserve has been selected and settled it will, of course, be necessary to demarcate with a very wide boundary all those parts where people are likely to trespass, that there may be no excuse on the plea of ignorance. Rivers make good boundaries if at the same time paths are cut along them, by which the guards may patrol. Ridges make the best boundaries of all, because the wide paths cut along them act also as fire-traces, and because the surrounding forest can generally be seen well from them. Where such natural boundaries cannot be found, the line must be run as straight as possible from point to point. It must be cut out to a width of 25 feet, and be demarcated by masonry pillars, heaps of stones, posts sunk in the ground, or boards nailed to trees as may be found convenient. These boundaries must always be kept clear and be patrolled by the guards who must be provided with uniforms.

The reserves to be carefully demarcated.

1259. I have explained that, as regards the working of the reserves, each of them would be cut up into blocks or " coupes " only one of which would be worked at a time, all the others being closed in order to concentrate the work and exercise a better supervision over it. The officer in charge would be one who was thoroughly trust-worthy and well-informed, for on him would devolve the carrying out of the working scheme. Whether the trees in the coupe were felled by Government agency or by holders of permits, his duty would be to see that the axe-men felled only such trees as he considered might be felled, and these ought, if possible, to have been all previously marked. During the working season he would remain entirely in the forest controlling the felling. When not otherwise engaged he would be employed in cutting paths through the forest, making slides, removing obstructions in the rivers, supervising improvement-thinnings, collecting and sowing the seeds of the more valuable species of trees, and preventing the felling of timber in other parts of the reserve. He and his staff would be comfortably stationed at a semi-permanent camp surrounded by an elephant-trench one of which would be constructed in each portion of the reserve. These camps should each contain two or three thatched mud huts, built sufficiently strong to admit of easy repair in case travellers through the forest wished to remain there for the night, at times when they were not in use. In the feverish season, when the timber-men leave the forests, the forest-officers would also move to the more open country where their services would be utilised in various ways. Especial attention would be paid to the exclusion of fire from the deciduous forests of the reserve, and I doubt not that when the hillmen come to see the utility of the preventive measures to check conflagrations they would, as in other parts of India, take the greatest interest in preventing the burning of the forests.

Description of the working of the reserves in detail.

1260. The question will naturally be asked what is to become of the hillmen who now live inside some of the reserves. Are they to be summarily ejected? Their numbers are about 8,000 and of these not more than 1,000, at the very outside, will be found living within the reserves, even when all the reserves have been selected. These people do comparatively little harm now, in fact, I may say, none, unless they fell virgin forest. For their own wants they require perhaps 300 acres a year for the 1,000 souls.* On the other hand they are very useful in building huts, cutting paths, collecting wax and cardamoms, assisting in the capture of elephants, tracking and acting as guides. They should therefore be allowed to remain where they are. I am told that on the Anamallies the Kåders have been allowed to remain in the forests on condition that they fell no new forest, but confine themselves to those parts which had been cleared by them before. This arrangement has been very successful, and these hillmen are contented and prove themselves of great service to the forest-officers. Some such restriction should be introduced here under heavy penalties of fine or even expulsion for its infringe-

Treatment of the hillmen residing in the reserves.

* I have taken Mr. Painter's estimate given at para 80 of the Forest Commission's Report in 1884, though I think this rather low. It would also be found necessary in practice to give the people more land than would suffice for a twelve years' rotation. The following is a rough estimate of requirements : a man will eat 1 edangali of rice or 2 edangalis of paddy a day, or 73 parras of paddy in the year. He, his wife and 8 children will consume 200 parras, and for cloths, salt &c. they will need another 80 parras=280 in all, the produce of 7 parras of land or 2 acres, if the year is favourable.

ment. The hillmen should also be prohibited from selling any of their grain to outsiders. In some parts of the hills adjacent to the low country it is customary for Mahomedan traders to supply the hillmen with knives, cloths, and grain for sowing, on condition that they pay them back in grain, with enormous interest, at the time of harvest. The hillmen thus are obliged to clear a very large area of land, and it is obvious that if this were permitted inside the reserves the forest would rapidly disappear. Some system of registering the names and ages of all the hillmen in the reserve would have to be adopted, in fact they should place themselves entirely under the Forest Department and agree to carry out the orders given them. Under these conditions there would be no objection to their remaining in the reserves.

1261. As all the reserves will be worked on carefully considered schemes for securing the largest yield from them, it will be seen that they *The reserves to be placed entirely in the hands of the Forest Department.* must be placed entirely in the hands of the Forest Department, and that no interference can be permitted from outside. Many intelligent persons who have not thoroughly studied the subject, consider the rules for the maintenance of reserves far too stringent, and would have them thrown open to the public for the removal of minor produce. But the very fact of persons entering a reserve endangers its working. Fuel-gatherers frequently carry in fire with them and valuable trees are burnt down: when they cannot find dead fuel handy these persons take to cutting down small trees and hiding them till they are dry enough to remove. The owners of cattle and sheep drive their animals into a forest, and not only is all the grass eaten down, but small plants are either nibbled or broken off, and the ground is rendered hard by the trampling of many feet. Men gathering reeds, or cutting grass, or collecting minor produce may do mischief in a variety of ways, by cutting down small trees for huts, chopping branches for the easier passage of their loads &c. In a word, although the produce they come to remove may be of little value, the damage they may do in collecting it is quite a sufficient reason for prohibiting their entering the reserves. Would Government for an instant allow persons to walk into their treasuries, and help themselves to a few cash when they felt so disposed, and yet a few cash here or a few cash there would not be so great a loss as the produce in the shape of fuel, grass &c. removed by the villagers. Fortunately, in Travancore there are so few cattle owned by the villagers on the outskirts of the forests, and there is so much land outside the reserves where fuel can be obtained, that it is really no hardship to any one to close the reserves.

1262. It is obvious from what I have said that a considerable amount of work in the way of surveying will have to be done in the reserves. *The reserves to be surveyed and mapped.* We have a map of Travancore on a scale of one mile to the inch which is fairly accurate, but we shall need maps of the reserves on a larger scale in order to show the streams, hills and valleys. Each reserve again will have to be cut up into blocks the area of which must be accurately ascertained, as our calculations will be based on area, and these blocks must again be cut up into compartments for more detailed working. Then again, if the blocks yield unequally, as is very probable, it may be necessary when the returns of a number of years have been compared, to alter the areas of some of the blocks so as to equalise their bearing capabilities. All roads and paths will also have to be entered in the maps, so that there will be constant work in this direction. Some of this may be done by the superior officers in their own districts, but the larger work of extensive surveys must be entrusted to a special officer who can devote his whole time to it.

1263. In the foregoing paras I have explained the best method of conserving and improving so much of the existing forests as is required *In addition to the conservancy of the reserves some planting recommended.* for the timber supply of the country, both present and future. Most of these reserves will run along the slope of the hills, but some will be in the low country. In addition to the maintenance of these reserves, it will be advisable to extend the plantations of teak already opened at Kōnniyūr, Malayattūr and Ariyankāvu and possibly to commence plantations of teak and other species of trees in other localities. The extension of the existing plantations, as I have shown in para 1155, has not been at all regular, and the cost has therefore been greater. It would have been in every way better if an equal area had been planted each year, and indeed it was through no fault of my predecessor that this was not done, as he constantly recommended extension. We are at present delivering 20,000 logs of teak a year which represents at 60 trees an acre the the produce of 333 acres of plantations, or at 10 trees to the acre (see para 963) 2,000 acres of natural forest.* Supposing that half of this quantity is supplied by the natural forest it would still be our duty to plant between 150 and 200 acres a year so long as we remove timber at this rate. It would also probably pay well to plant Anjili and red cedar, both of them valuable trees of rapid growth.

1264. The question of fuel is one that must sooner or later command the attention of Government. The price of fuel in Trevandrum and *Fuel-plantations and* other large towns has been steadily increasing, and it has now *reserves will be required* to be brought for sale in such places from a considerable dis- *are long.* tance. The reserve at Mukanni mala has been set aside for the production of firewood, but, beyond this, no attemps have been made to increase the supply. Before long the matter will become one of such pressing necessity as to require the commencement of plantations, or the closing of other reserves throughout the country. The population of Trevandrum was, at the time of the 1881 census 41,173, and allowing " 1 lb. per head a day or 365 lbs. a year in addition to the " | " wood grown in gardens and hedges near the town,"† we find that 6,713 tons a year are required for the fuel supply of the capital. According to Sir D. Brandis the yield of natural forest in Malabar may be set down at ¼ ton an acre a year, and for casuarina plantations at 2 tons per acre, on a 15 years rotation, so that Trevandrum requires for its fuel-supply the produce of 13,426 acres of natural forest or 3,356¼ acres of plantations.

1265. In addition to the supply of fuel for the towns, a very considerable quantity will be required for the railways which it is now un- *Railway fuel and sleep-* der contemplation to construct. It is estimated that the two *er supply.* sections from Quilon to Shencottah, and from Trevandrum to Nagercoil will each be about 50 miles long or 100 miles in all. According to Sir D. Brandis (para 110) the estimated consumption of wood for fuel is 100 tons a mile. We should therefore require to keep for these purpose 5,000 acres of plantations or 20,000 acres of natural forest to maintain the supply. In Ceylon and Madras coal, which is generally calculated to possess three times the heating power of wood, is often used, but in Travancore coal is expensive, so that we should be entirely dependent on our forests or on plantations for the supply of fuel for the railways. These figures show that the question will need attention before very long, and that we shall probably have to commence fuel plantations to assist the natural production of the forests. Similarly, a large supply of timber will be annually required for sleepers for the railwa,. The 100 miles of line would require 1,76,000 sleepers, and as the life of each sleeper would probably not be more than 6 to 8 years,§ from 22,000 to 30,000 would be required each year.

1266. The money spent on works of improvement, whether plantations, buildings, roads or other objects is not at present nearly *More money to be spent* sufficient, and considering what benefits we derive from our *on works of improvement.* forests in the shape of free firewood, cheap timber and large revenue, we should in justice to our successors, make more return to them. The forests have been handed down to us as a valuable property to be enjoyed by us and passed on to our successors in good condition, and if we merely make all we can out of them and spend as little as possible beyond the cost of working out the timber we require, our successors will have every right to complain of our selfishness. In European countries it is generally agreed that 50 per cent of the profits of the | forests should be returned again to them each year in the shape of improvements. Let us see what is being done in Travancore in this direction. In the year 1066 ‡ the revenue from the Forest Department was 5,59,012 Rs. and the expenditure

---

* Some of this teak is cut from private compounds and from kol-teak forests where the trees grow very close together, but the above figures are correct for trees cut from the mixed forests of the interior.
† Brandis Suggestions for Forest Administration in the Madras Presidency, para 313, and 130.
§ Vincent's Report on the Ceylon Forests para 453.
Administration Report for 1066.

2,89,666 Rs. showing a profit of 2,69,346 Rs. To this should be added the sum of 5,066 Rs. spent on plantations, and 4,985 Rs. on works of improvement, chiefly buildings. Thus the real profit was 2,79,397 Rs. and the amount that should have been spent on improvements is 1,39,698 Rs. Instead of that the actual amount spent, as shown above, was only 10,051 Rs. or 3½ per cent of the profits.†

1267. I have been speaking so far only of the area it is proposed to reserve, and of the plantations which we shall probably find it advisable to create. What then of the very extensive area outside the reserves and plantations, which belongs to Government and contains valuable timber? As the land which I have suggested should be reserved will be quite sufficient for the wants of the people no money should be expended on the area excluded. Not only would we not object to the clearing of this land, but we would in every way approve of it, as this would tend to make the country more healthy and more accessible. At the same time it would be a pity to waste the valuable trees scattered through the unreserved area. Under the present system, all trees of small size except the royalties may be felled for firewood or for other purposes, provided they are not sawn up or used as timber. In this way vast numbers of small anjili, thambagam and other valuable trees are cut down, while at the same time large trees of the worthless kinds are preserved, by the rule prohibiting their being cut down after they have attained a quarter girth of 10 vannams. What we should wish to do is to preserve the better kinds from their infancy, and to care less about the inferior kinds. We should therefore prohibit the felling or cutting of a few of the very best kinds, even as plants, in just the same way as we prohibit the cutting of teak and black-wood.* The poorer kinds might be cut for firewood or agricultural implements whatever their size might be, but if used for buildings or sawn a seigniorage should be levied as at present. The result of this would be to get rid of all the inferior kinds and also to bring them into general notice and use, whilst preserving all the good kinds, even outside the reserves. In order that this plan should be successful two conditions are necessary, (1) that the number of reserved trees should be small, and (2) that the timbers of these trees and their leaves &c. should be easily recognizable. For reservation § we should choose the following in addition to the 4 royal timbers:—

> Anjili *Artocarpus hirsuta.*
> Kambagam *Hopea parviflora.*
> Vēnga *Pterocarpus marsupium.*
> Thēmbavu *Terminalia tomentosa,*
> Thēvathāram *Cedrela toona.*
> Irūl *Xylia dolatriformis.*

Of the other trees mentioned in Part I. I have omitted white cedar and jack which are essentially jungle-trees (except when planted), found only in the evergreen forests, and therefore they would not be likely to suffer much from lopping. Ven-teak and the other trees mentioned are too widely distributed, and not of sufficient value to warrant their being added to the list, though if they were found to be disappearing, one or more of them might be included.

1268. Another work of improvement may here be indicated. In para 961 I have shown that teak in the low-country is riddled by a species of borer which renders the wood almost valueless, and that the borer attacks trees which have been lopped during the growing season. Each one of these bored trees is a nursery for the propagation of future generations of borers, and the sooner they are all felled the better, as their presence is a danger to the sound trees, in that the moths that produce the borers might take to laying their eggs in the bark of the latter trees. The timber indeed will be of little value, but it will cover the cost of felling, even if it is only sold for firewood.

---

† In the same way, in 1067, the total net revenue was 3,22,940 Rs., and the expenditure on works of improvement 11,634 Rs. or 3½ per cent.

* It will not, I think, be denied that the groves of pure teak found in North Travancore owe their existence to the rule prohibiting the felling of this valuable tree, while no such restriction was passed on others, the royal timbers excepted.

§ We should not interfere with the lopping and felling of these trees in private compounds, and in this respect the reserved trees would differ from the royalties.

(2) Felling and delivery.

1269.. If the suggestions I have made in the foregoing paras are carried out, a certain area will be set apart for the production of timber and fuel both in reserves and plantations, and the rest of the country will be left for occupation by the people. In the last chapter I have described the system of felling of (1) the royal timbers, and (2) all other species, now in force in the country, whether in reserved forests or elsewhere. In future the method to be followed must differ according as the locality in which the trees are growing lies within or without the reserves. First as regards the royal timbers. These are now felled almost exclusively by contractors, wherever they may be found But in future the royal timbers growing within the reserves should not be felled on contract, or at all events not in the same way as at present, because we cannot allow contractors to enter the forests at all points and fell just such trees as suits them. The same rules which will regulate the felling of ordinary trees must be applied also to the removal of the royalties. The work must be concentrated, the felling of teak and the other royal trees must only be allowed over a portion of the reserves at one time, and the number of trees to ne removed must be restricted. It will be better therefore to place the felling of these royal trees under Government supervision and carry it out by Government agency. With this view, I have inserted a clause in the new agreements made with contractors, binding them to cease felling in any given localities when called upon to do so. Work done by Government is admittedly more expensive than work done on contract, but it is much better carried out, and the waste is less. Its whole success depends on the thoroughness of the supervision.

Royal trees growing in the reserves to be felled by Government agency.

1270. Outside the reserves the royal trees should be felled, as at present, on contract,* provision being made that immature trees of small size shall not be removed from Government waste land. In compounds and private holdings the case is different. The people cannot grow garden-produce and at the same time leave the teak untouched, because the trees would overshadow and dwarf the plants growing beneath them. The system of lopping the teak in the wet weather was introduced with the view of preserving the trees, while removing the causes that hindered cultivation, and, like most compromises, both objects which it was designed to assist have been only partially successful. The teak trees deprived of their leaves cease to grow and fall a prey to the borer, while the space occupied by the stems of the trees reduces the area of cultivation. As, then, we ought to have sufficient land for the production of timber, set apart in our reserves and plantations, I see no object in forcing the owners of compounds to retain and preserve the teak and blackwood trees growing in them, to the detriment of their produce. Let the trees be cut down and removed, wherever land is taken up for permanent cultivation, however small these trees may be, and whether mature or not. But in order to induce the people to preserve the royal trees growing on their land, if they can do so without damaging them and without interfering with their cultivation, it is proposed to allow the owners a certain share of the selling price of the trees, when they are removed from their properties, provided the land has been occupied by them for 10 years. Otherwise, if there was no such restriction, people might take up land on which teak was growing, and in a short time ask for compensation for trees that had been there for scores of years before their occupation.

Royal trees outside the reserves to be felled on contract. How to deal with teak and blackwood in compounds.

1271. The removal of teak and blackwood (as well as of all valuable trees) from the area outside the reserves, would very much reduce the work of the Forest Department, in as much as it would tend to concentrate its operations, and to reduce the opportunities for smuggling. A great part of the time of all the superior staff is taken up in enquiring into complaints as to whether A did or did not fell Government timber without a permit and without paying for it, and convert it to his own use, as stated in such and such a petition. If we got rid of all the useful trees scattered so thinly over the country that it is impossible to preserve them from

The removal of teak and blackwood from the area outside the reserves will much facilitate our work.

* By reducing the number of contractors, retaining only the best of them, dispensing with advances, and strengthening the supervision, the evils mentioned in the last chapter will be avoided.

54

theft, we should be able to dispense with the staff now engaged in trying to protect them, and to devote all our attention to our legitimate work of forest protection and conservancy.

1272. The concentration of work is an object so desirable, that it would be well to carry out some scheme for bringing to market all the *Proposed treatment of teak and blackwood trees standing outside the reserves.* teak and blackwood timber growing in the low country outside the reserves A staff might be appointed, to begin from one end of the State, for instance from the north, and west of the road from Cottayam to Angâmala and on to Trichûr, and to go steadily through the country felling all those royal trees which were of a saleable size, wherever they were found to be too scattered to be properly looked after. In some places small groves of teak (kâvû) might be met with, of too limited an area to be worth forming into reserves, but which it would be a pity to cut down. In such cases the trees could be counted and measured, and handed over to the charge of a neighbouring land-owner for safe custody, in return for a small monthly stipend. In private compounds the teak &c. would be felled, unless the owner wished the trees to be preserved, in which case he would be made to sign a paper promising to preserve the trees specified. Lists would be sent in from time to time showing the number and position of the groves preserved and the number of trees in each, also of the trees growing in private compounds, and left standing at the request of the owners. A list would also be submitted of all the trees felled and lying about the country, and these would be sold by auction from time to time where they lay, or be sent down to the nearest depôt. Thus, all the royal trees of a saleable size scattered about the country would be placed under protection, and consequently there would be no chance of smuggling, while a very considerable revenue would be obtained by the sale of the trees felled. This enumeration would not be nearly so difficult a task as might be supposed. The number of teak and blackwood trees in the cultivated portion of the country is really very small, and is diminishing every year, and, in a somewhat similar case, the D. P. W have carried out an enumeration of all the trees on the road sides without a hitch.

1273. As regards other trees beside the royalties in the reserves, we must be guided by circumstances as to whether they should be felled *Other trees but royalties growing in the reserves to be felled by Government agency or on permit.* by Government agency or on permit. Undoubtedly, the former is the best system, and it is the one which must be adopted, wherever this is possible, but for the present it may be advisable to allow permit-holders to fell the commoner kinds of timber under careful supervision. We may perhaps mark and fell the trees and allow merchants to come and buy them and remove them. In order that felling by Government agency might be thoroughly successful, it would be necessary that an unlimited demand for every kind of timber should exist, so that we should be able to fell any species of tree in any locality, with the certainty of being able to dispose of it. Now, in our case there is an unlimited demand for about half a dozen of the best timbers, but for all the others an uncertain demand or none at all exists. It will probably be better, at all events for the present, to fell teak, blackwood and those timbers for which there is an unlimited demand ourselves, and to allow permit-holders to cut other species, as there is a demand for them or not, but only in the localities pointed out to them.

1274. Outside the reserves, timber would be cut on permit as at present, the royalties only being cut on contract. In some localities we *Outside the reserves, the trees to be cut on permit, except where we might decide to fell certain trees by Government agency alone* might find it advisable to refuse permits for some or all of the "reserved woods" mentioned in para 1267, but, as a rule, our aim should be to place as few restrictions as possible on the felling of timber outside the reserves, so that there might be less anxiety on the part of the people to fell from the reserved area.

1275. When the reserves have all been selected, it may probably be well to stop all felling in them until some working-schemes have been *The reserves to be closed for a time when all have been selected.* devised, and until the work of demarcating them has been completed. This will of course produce a great rush for timber outside the reserved area, and we may expect that in a very short time the unreserved forests will be denuded of timber, which is a result not at all undesirable. But, until all the reserves have been selected and demarcated, at all events all the reserves in one place, we cannot stop felling within them, or the forests, not yet chosen but suitable for reservation, would be quite cleared out before they had been closed.

1276. The maintenance of the present depôts, the opening of others, or the closing of all must depend on circumstances, and no decision about them can be given now. Those depôts at which sawn timber is sold, as explained in para 1175, have given very good returns, but for the last eighteen months the receipts have fallen off, while the revenue from permits has largely increased. It is not easy to say what is the cause of this. It may be that the people are finding out that there are many other timbers which they can get at a cheap rate on permit, which are as good or nearly as good as thambagam, vênga and thêmbâvu, for which the Government charge a heavy " mêl-lâvam " at the depôts. If this is the reason, the fall of revenue from the depôts is a good sign, as showing that the people are becoming familiarized with the inferior timbers. On the other hand, the fall of revenue may be due to a really lessened demand in Tinnevelly (the destination of most of our timber) owing to famine or some other cause, or it may be the result of the opening of some cheaper market where timber similar to what we sell can be purchased at rates lower than ours. Lastly, it may be due to a falling off in the character of the timber recently supplied. Some or all of these causes may have operated to produce the fall in revenue from the depôts, and we cannot say if the receipts will increase again or still further diminish. Similarly, we cannot lay down any definite rule as to whether depôts should be altogether closed or not. Even if we cease to supply the public from them, we shall probably find it advisable to maintain them for the supply of timber for Government works. If there is any change in locality, the change will be in the direction of the interior, as there are many reasons why it is desirable to locate the depôts near the forests which feed them.

The depôt system to be maintained, and probably also the depôts now existing.

1277. Permits are at present granted only from the Quilon office, though the timber to be felled may be growing in the extreme horth of Travancore. This gives those people who wish to take out permits very great trouble, if they happen to be living at a distance from Quilon. The time has now arrived when powers to grant permits and receive seigniorage should be delegated to Deputy Conservators and in some instances to Assistant Conservators, not only that the people may be saved inconvenience, but that the work of the Conservator may be lightened. Such concessions, and the redress of grievances similar to those mentioned in para 1181, give us no extra trouble, while materially assisting merchants in their trade, and I have found that when the facilities for procuring and removing timber are increased, the people have less inducements for smuggling, (which is often due to the difficulty of getting their requirements honestly, as much as to the possibility of escaping detection) and the number of permits taken out increases.

Permits to be granted by Deputy Conservators, as well as by the Conservator.

1278. I have already explained that all timber growing on private land should be free of seigniorage, save only the royal timbers. The principle that the land may be granted to the people for cultivation, while the trees growing on it are retained by Government is repugnant to European ideas, though it has some support in the local system of taxing the fruit-bearing trees in compounds and not the land itself, a system which would generally be regarded as an inducement to slovenly cultivation. Indeed, the retention of a claim to all the teak and blackwood growing on private properties can only be defended on the grounds that the value of the property that would be surrendered is so great, and that this has always been the custom. It is now proposed to give owners of land a share of the value of the royal trees grown by them on it, and also, in cases of land taken up in future for cultivation, provision is being made that if the trees growing thereon are not removed by Government within a given time they shall become the property of the occupant. It follows, as a matter of course, that all trees growing on land already in the possession of the people will in a short time become their property, if not removed by Government, for they have only to

All timber growing on private property will cease to belong to Government under the proposed rules, whether all claims to it are given up or not.

give notice to have them removed, and if not removed they will become indisputably theirs, and can be transported without charge by any route or to any place. It is only a matter of time therefore, when the charges on timber grown on private land if removed by water cease to be leviable, whether the charges are definitely abolished or not. It would therefore be better to make the concession gracefully, and to abolish those anomalous charges at once.

1279. It will be seen from what I have said that all improvements effected, and all charges naturally occurring will tend in one direction,

Eventually all land not included in the reserves and plantations will cease to give us revenue, and we shall pay no more attention to them.

viz: towards making the reserves and plantations the only source of the timber and fuel supply of the country. The reserved areas, carefully guarded from encroachment, and worked so as to give a sustained yield, which should gradually increasing in amount from one period to another, will be a valuable property, sufficient for all the wants of the State and the people. But the area not set aside will become rapidly denuded of its timber, for, though we shall endeavour to improve the character of the timber on it by passing rules for the preservation of the best species, the facilities for felling will be so much greater here than in the reserves, that, by preference, permits will always be taken out for all the trees growing on it, and they will be cut down and removed. The people too will take up land in the unreserved area, and on their holdings the timber not removed by Government will become their property, and we shall get no revenue therefrom. From all unreserved waste land it will be to the interest of Government to cut down and sell all the royal trees growing on it which are too scattered to warrant their inclusion in a reserve, and though we propose at present to give the growers of teak and blackwood only a share of its selling price when removed from their properties, it will probably be found adisable after some time to give up all claims to these trees, and to abolish the monopoly. We shall indeed be at liberty to extend the reserves, and to take in land on which Government trees are found to be growing in abundance, or where they are likely to thrive, but, as what is not included in the reserves will cease to yield any revenue, it will not be necessary to give any attention to it, and we shall confine ourselves to the reserved areas. As a natural consequence, all or most of the watch stations (alluded to in para 1180) will be abolished, and work will be less in the low country and more in the forests.

(3) Selling.

1280. It is not proposed to make any alteration in the manner of selling the royalties now in force, which I have described in para 1185.

The present system of selling the royalties by contract recommended.

Messrs Wallibhoyi and Co. will probably take a new contract from us, and if they are reluctant to do so, there are others who are anxious to purchase all our teak and blackwood. But our forests cannot long continue to supply such large quantities of these timbers, especially the former, and, after a few years, the supply will be very much diminished. It is probable then that ro more will be brought to market than can be used in the country, and very little will be exported. The demand in the country is rapidly increasing, and this, not only because teak is the best wood for all general purposes, but because its price has remained stationary while the prices of the other useful woods has rapidly risen on account of the greater difficulty of procuring them.

1281. The rates for teak and blackwood must from time to time be raised as the demand for them increases, until they approximate to the

The selling price of the royalties will gradually rise.

figures at which foreign timber can be imported, which must of course be the limit which can be obtained. The present selling rate of teak being from 10 to 17 Rs. per candy, equal to about 10 annas to 1 rupee per cubic foot, we may expect to be able to eventually realise double this sum, for teak could not be imported under 2 Rs. a cubic foot. The increase in charge must be gradual so long as we have a large number of logs for sale, the Bombay firm not being in a position to pay a high figure for timber which has to be exported to a distance, though all the teak they sell locally fetches a very handsome price, and far above what we receive from them.

1282. The same remarks apply to other woods. The present methods of sale will probably be maintained, but the prices will gradually rise. If we did not export any timber at all, but consumed the whole of the supply in the country, we might charge about double our present rates, but as the present supply exceeds the local demand, our rates are kept down to a figure at which merchants will • purchase our timber for export. As the supply falls off we must increase our rates, and, neglecting the foreign markets, make local purchasers pay a much higher price than they do at present. It may seem hard that a comparatively high rate should be asked for timber in a country where it grows so easily, but, unless I am greatly mistaken, it will soon be admitted by all that our forests have been very much over-cut, and that the number of candies which can be felled annually, without detriment to the forests, is much smaller than is now being felled, and the price must be raised in order to pay for the cost of protection. Again, the Government share of the selling price is much less than in other countries, where it equals the cost of working the timber out of the forests, our charges being only about one-quarter the cost of exploitation in the case of the more valuable woods, the royalties excepted.

The price of other timbers will also rise. Reason.

1283. The export duty on timber should be abolished. At present it varies from 6½ annas to 1¼ rupees per candy, and in the case of inferior timbers this is very heavy. What the object of a duty on Government produce can be I do not know. It already pays our dues, before it is removed from the forests, and then it is subjected to this second charge. If its object is to keep the price of timber in the country low by protection, this object is not attained in the case of the more valuable species, which are all exported from Travancore in large quantities, while it is effectual in checking the export of inferior and little known kinds which we wish the public abroad, as well as at home, to make a trial of. If its object is to check smuggling, this object · is not always secured, as the customs officers are generally too anxious to secure revenue to ask whether the timber has been properly obtained or not. In any case the remeasuring, and the opportunities of delaying and bullying the merchants that this custom gives, certainly discourage, rather than assist, trade.

All export duties on timber to be abolished.

1284. It has been suggested that our system of measurement should be altered, and that cubic feet should be used instead of candies and fractions of a candy. But the Travancore measure is a very convenient one, and as timber will be more and more required for home-consumption, it seems less necessary to make any alteration.

The existing system of measurement to be maintained, at least for the present.

1285. If the depôt system is maintained, care must be taken not to increase the number of depôts too much, or their chief object will be defeated. Numerous depôts are no doubt convenient for the people, but, if too numerous, they compete with each other, and they require much supervision. The question of auction sales or daily sales at depôts has been much debated. Col. Campbell Walker told me he had tried the daily sale system and had found it a failure. Mr. Vincent on the other hand condemns auction sales and highly recommends daily sales.* It appears to me that the choice depends on circumstances. If the officer in charge of the depôt is thoroughly impartial and honest, daily sales may be held, but there is always a temptation to favour one purchaser before another, and, even if the measurements are carefully taken and entered in the stock-book on arrival, there are many ways of showing favouritism, as may be imagined. If the Superintendent of the depôt is dishonest, the loss to Government may be enormous. In any case therefore, depôts at which daily sales are held, require very close supervision and the strictest adherence to rules. Auction sales, as Mr. Vincent points out, unless the state of the market is carefully watched, may often result in a loss. Timber, we will say, is brought down from the forests and sold in large lots for what it will fetch. Only the wealthy merchants can afford to hold large stocks, consequently there are few bidders and they combine to keep the price down, arranging beforehand what quantity each will take. Probably the best way for the present,

Relative advantages of auction-sales and daily-sales at fixed rates discussed.

---

* Ceylon Forest Report, paras 293—295.

is to continue the daily sales under careful supervision, and to hold frequent auctions of all the refuse timber, put up in small lots to attract the smaller purchasers.

1286. But the daily sale of timber at fixed rates to every comer is a business that does not really belong to the Forest Department. In

The daily sale of timber at depôts not the business of the Forest Department but of the timber-merchant. India this work is often undertaken by the Forest Department, though it is part of the trade of the timber-merchant, because there are not enough timber-merchants possessed of capital sufficient to buy the timber at fair rates from Government for retail sale to the consumer. The proper duties of the Forest Officer are to protect and grow the timber, and to fell and deliver as much as the forest in his charge can produce without depreciation, and the more he occupies himself with the details of sale, the less can he attend to the management of his forests. Indeed, in most parts of Europe, and in some places in North India, the timber is sold standing, and the purchaser comes in and fells it for removal, not as the permit-holder does with us, helping himself to such trees as he likes and rejecting others, but cutting only those trees that had been previously marked and measured. This we cannot expect to do with our present staff, but we may hope to do away with the daily sale system in time, and to replace it by a system of auctions, at comparatively short intervals, of the timber felled by the Department and delivered at the depôts, or, better still, of standing timber.

1287. The auction sale system cannot be generally adopted until we have raised up a class of wealthy timber merchants or middlemen

Middlemen to be by all means encouraged. to retail the timber, and until the reserves have all been selected so that there can be no smuggling from the forests. Much has been said against the middleman for the low prices at which he purchases his timber, and the high rates he charges for it. As regards the first complaint, if timber is sold at outright auction and the number of purchasers is small, the price realised will naturally be low. The proper step to take to prevent this, is not to force the sale, but to encourage more merchants to come forward as purchasers by facilitating matters for them, by holding auctions regularly, by selling the timber in small lots at first, and by consulting their wishes wherever this is possible. It is obvious that if the auctions are held at irregular and long intervals, if the depôts are suddenly changed, opened in one year and closed the next, or if the system is altered at short notice, only the large merchants who have much capital can afford to buy, and the purchase of timber, instead of being a regular business, becomes a speculation, in which only low rates of purchase and high selling rates will pay; again, a merchant often finds no sale for timber which he may have purchased at a high figure or in large quantities, and losses in one direction must be made up by extra profits in another. One of the great wants of the country is seasoned timber. The majority of buildings are constructed of green wood, and, as a consequence, doors and windows shrink or expand with every change of weather. If the timber were seasoned this would not be the case, but until the public realise this, and are prepared to pay extra for the seasoning, the middlemen will not take the trouble to stack their timber.* We must therefore encourage the merchants to establish a regular trade while we gradually withdraw from the retail sale of timber.

(4) *Miscellaneous.*

1288. The method of capturing wild elephants has been fully described in the last chapter, and I have also shown there that this

For the capturing of wild elephants more trained animals required. system of catching them in pits is probably better suited to the country than any other, owing to the small size of the herds and the broken character of the forests frequented by them. The mortality among the captured animals is due to delay in removing them from the pits, and this can only be prevented by keeping a sufficient number of trained elephants in readiness to escort them to the training houses, as soon as the

---

* The length of time allowed for seasoning timber in European countries is surprising, and for those trades in which the timber used must be absolutely free from liability to expansion or contraction, it must be kept for not less than 5 years in a cool place. See Indian Forester Vol. XIII. pp. 283.

wild animals fall into the pits. I have already addressed Government on the
necessity of increasing our staff of trained elephants. ·

1289. Considerable improvement may be made in the method of collecting
cardamoms. At present this. spice is collected and delivered to
Government, and the growers or collectors receive one fixed
rate viz: two fifths of the selling price, all over the country,
with the exception of some parts of North Travancore, where
the right to collect is put up to auction, and the person who engages to deliver the
greatest quantity is accepted, and any quantity that he can gather in excess of the
stipulated amount is his own. Now, for cardamoms grown in gardens it is quite
right that a fixed rate should be given, but in the case of wild cardamoms which
are scattered unevenly through the forests, an adherence to one rate in all cases probably causes loss. The auction system is the right one, but it should be tried in a
different way. No one should be allowed to take any of the spice for his own use,
so long as it is a monoply. It should all be delivered to Government. The person
who offers to deliver the greatest quantity at the cheapest rate should get the contract. To explain, some of the planters collect and deliver cardamoms at the ⅘ths
rate, but as the price given is now very low, they content themselves with gathering the spice on their estates or close round them, where it can be picked cheaply.
If a higher price were given, they would send farther to collect and more would be
gathered and delivered to Government, and the extra quantity, which is now stolen
by thieves, would more than compensate for the increased price paid. There would
be another advantage in this, that the contractors would know exactly what rate
they would get for what they delivered, whereas at present, if prices fall, they may
be considerable losers, unless they leave a very large margin for prices to fluctuate.

The right to collect wild cardamoms should be auctioned each year in the different forest areas.

1290. I have shown in para 1202 that the rate given is nominally ⅘ths of
the selling price, but in reality it is between ¼ and ½. Again, it
is a great hardship that the money is not paid to the cardamom
ryots immediately on the delivery of their spice. It is true
that they receive advances, but the balance is not paid them
for months, and all this entails the keeping of further accounts. It would not be
difficult to fix on a price to be paid to the ryots, proportioned to the price of the
spice in the London market at the end of December each year, the payments being
made about February. The price of the spice would not vary much between
December and May, when the auction is usually held.*

The payment for cardamoms from gardens should be made immediately on delivery.

1291. In the new Forest Act, wax is not declared a monopoly, but will be
treated as ordinary forest produce. People who have bee hives in
their own gardens, will be allowed to use the wax without being charged with appropriating Government property, but in
reserves, as well as in other lands belonging to Government,
the right to collect will only be given to a few contractors. This seems a better
arrangement than the present one. Dammer and lac have also been excluded from
the list of monopolies.

Wax, dammer and lac no longer classed as monopolies.

1292. I have shown in para 1205 that fuel is practically free all through
Travancore, except in the town of Quilon.† That an exception
should be made to the disadvantage of this one place seems
very hard, more especially as the charges on fuel are aimed
at the two large mills situated there, which do immense good
by finding occupation for upwards of 1000 persons, and which therefore deserve
support rather than discouragement. The question of levying a tax on firewood is
really a matter affecting the revenue rather than the forests, so long as there is
such a large extent of waste land, owned by the Government, covered with scrub. If
the Government are so well off that they can afford to present the people with fuel free
of charge, thereby resigning a lac or so of revenue each year, no one can accuse them
of want of liberality, but if the forests do not yield such large returns as they would
otherwise do, this should not be a matter of surprise.

Fuel should be taxed all over the country, and not only in certain places.

---

*.For further particulars see Appendix I.
† Since writing the above, the duty on firewood at Quilon has been abolished.

1293. I fancy that most people agree in thinking that all fuel grown on
Government land should be subjected to a charge, but that

Fuel-reserves to be formed, and all fuel taken from them to be sold.

when grown on private land it should be free. The difficulty is to distinguish between the two. Under the present system no attempt is made to charge for the one, and to pass the other free. Fuel in logs rafted down the rivers is liable to tax wherever it comes from, and fuel brought to the town of Quilon is similarly penalised, wherever it may have been grown. Government land and private land in the low country are so much intermixed that no distinction can be made. The only way to carry out this principle is to charge for fuel on the spot where it is grown, in other words to form fuel-reserves in the neighbourhood of all the large towns, to protect them, and to charge for all fuel taken out of them, but not for what may be taken from unprotected Government lands. At first, no doubt, these fuel-reserves would hardly pay for their up-keep, as the people would gather fuel from the unreserved waste land, but this would soon be exhausted, and they would then find it cheaper to purchase their firewood from these reserves, than to go 6 or 8 miles further to get it from waste land in the interior. Fuel-reserves must be chosen ere long or the price of fuel will rise rapidly.

1294. In paras 1208 and 1213 I have spoken of the cumbersome system of accounts in use in the country, the want of proper checking of

System of accounts wants revision.

the expenditure, and the opportunity given for advances to be lost sight of, and to remain unrecovered. I need not discuss the subject further. Every one admits the faults of the system, and sooner or later it must be revised and altered. A curious custom exists in the Forest Department which I think I have not mentioned before, of allowing our accountants to make out the annual statement of receipts and expenditure, but this statement is not embodied in the Administration Report. The account-branch in the Huzur put it on one side, and from the detailed accounts submitted to them, a different statement is prepared, in which some items are struck out and others inserted. If our method is incorrect, it would be much simpler for us to be shown where the account was wrong, that we might be able to prepare a correct statement another year, and so not lose all the fruits of the labour of preparation.

(5) *Staff.*

1295. I have, in para 1217, given a list of the staff of the Forest Department at the end of the year 1066. Before it can be well adapted for

intervals 21.

The controlling staff is under-manned and must be strengthened.

the work that it will have to do, its " personnel " must be considerably changed. It will be seen from that list and from what I have said elsewhere, that the controlling staff is very much under-manned, and, as a result, the officers in the rest of the Department have far too much power, and, as they are poorly paid and exposed to numberless temptations, it is not very surprising if they consult their own interests rather than those of Government. Even, therefore, if it was not proposed to make any alteration in the method of working, the controlling staff would have to be increased. But, as I have shown, there will be a large amount of work of a technical nature to be done, such as choosing reserves, cutting roads, building huts, clearing boundaries, supervising timber-felling, arranging working-plans, surveying and planting. This can only be done under the supervision of well educated and intelligent officers. The controlling staff will have to be as follows :—

Controlling Staff.

1 Conservator on 700/. to 1,000/. and batta averaging 150/.	1,000/.
2 Deputy Conservators and 1 Superintendent of Surveys on 350/. to 550/. and batta 100/.	1,650/.
3 Assistant Conservators on 150/. to 250/. and batta 60/.	780/.
26 peons attached to above @ 9/.	234/.
	3,664.

1296. For administrative purposes the country should be divided into three parts, as recommended by the Forest Commission of 1884. The northernmost should include all North Travancore as far south as a line drawn from Alleppey to Cottayam, and then along the road from Cottayam to the Gudalûr frontier, but including also that part of the Peermerd plateau which lies to the south of the road. The central range should include all the country south of the above line as far as the Vâmanapuram river from its mouth up to Pâllode, and east of Pâllode the range of hills dividing the valleys of the Pâllode and Kulatthurpura rivers. The southern range would embrace the rest of Travancore. The Conservator and Deputy Conservators should be located at Trevandrum, Quilon and Cottayam, one in each Range and the Assistants at Malayâttûr, Kônniyûr, and either Nagercoil or Puliyara.

Division of the country into ranges, and the location of the superior officers.

1297. Below the Assistant Conservators should be Rangers, of whom 5 or 6 would be required. They should be arranged in four grades drawing from 50 to 100 Rs., with a horse allowance and batta amounting to 30 Rs. a month. Their duties would be to superintend the working of the reserves and teak plantations, and they should be English speaking officers with some technical knowledge. At present, our great want is a class of officers who know something of forest work, and who can carry out instructions without having a superior officer at their elbows to teach them. Our Aminadars can write reports and pass timber, but when told to lay down nurseries, to cut roads or boundaries, or to draw up a sketch, they know nothing of these matters, while, as for calculating the cubical contents of standing trees, this is quite beyond them. Until we get some reliable officers of this class it will be most difficult to carry out any works of improvement. Each of these Rangers would have 1 writer and about 8 forest guards under him. Below the Rangers should come Foresters (Aminadars) probably about 8 in number drawing from 15/. to 35/. with consolidated batta of 15/. They would be employed more in the low country than in the forests, in supervising land at the disposal of Government but not reserved, in holding enquiries, and in carrying out special work. Each of them would require a clerk and 2 or 3 peons as they have now. The Cardamom Aminadar would be one of these, but he would have a larger staff than the others. The establishment under this head would therefore stand as follows :—

The strength of the Executive and Protective Staff.

Executive and Protective Staff.

6 Rangers @ 100/. including batta	600/.
8 Foresters @ 40/. do. do.	320/.
14 Clerks @ 10/. do. do.	140/.
100 Peons and forest guards @ 9/.	900/.
	1,960/.

1298. The Stationary Staff would be reduced rather than increased. On the depôts an expenditure of 800/. would probably be sufficient, to be afterwards reduced when there was less selling of timber in sawn materials, and when daily sales had been discontinued. The pay of the subordinates attached to the offices would be about 800/. Most of the watch-stations would eventually be abolished, as the timber in the low country disappeared, and the forests were protected by forest guards. But, for the present, the watch-stations would have to be retained, otherwise there would be great robbery of timber growing outside the reserves. Placing the expenditure under this head at 400/. instead of 660/. we get 2,000/. which sum would afterwards be capable of reduction. The whole expenditure would then stand somewhat as follows :—

The Stationary Staff.

Controlling Staff	3,664/.
Executive and Protective Staff ...	1,960/.
Stationary Staff	2,000/.
	7,624/.

1299. These figures are only approximate, and are intended rather to indicate what direction a reorganization should take, than the exact number and character of the appointments recommended. Instead of having a large number of badly paid officers with no supervision over them, it is intended that the controlling staff should be strengthened, while the Rangers would be much better paid than the Aminadars are now. There would then be very much closer supervision over the felling operations,* for even if the felling is done by merchants and not by Government Agency it will be centralised, and will be closely superintended both by the Rangers, and by the Deputy or Assistant Conservators above them again. This will make the maintenance of so many watch-stations unnecessary. No considerable increase in the cost of office establishments is anticipated, because this is the easiest work of all, and these officers should therefore not be so highly paid as those who have to undergo hardships in the forests. It may be however, that when we come to change the Department from an almost purely Vernacular Department to one in which English is chiefly used, we may have to pay more for the services of better educated men·.

In reorganizing the Dept. the great object is to improve the supervision.

1300. In one respect I feel that I have probably underestimated the expenditure that will be necessary, and that is in the batta allowed, especially to the subordinate officers. The difficulties of travelling in the forests are vastly greater.than are experienced by those whose tours are confined to the low country, and the cost of transport is double as great. As it is of the first importance that our officers should be encouraged to travel, the most effectual way to do this is to give them liberal rates of batta and mileage.

Travelling allowance should be liberal for the Forest Department.

1301. A uniform should be worn by Executive officers especially all forest guards and peons. Clerks and others employed in offices need not wear uniform, but, if deputed for enquiries, or put on special work, they should be supplied with some badge or symbol of office to show that they have authority to act. A case occurred lately in which a depôt watch-pillay's authority was successfully defied, because he had nothing to show that he was a Government officer. I have also heard that people, pretending to be Government servants, sometimes arrest carts on frivolous pretences, and demand a present from the owners for releasing them. This could not happen if Government officers were supplied with badges of office.

A uniform to be worn by the Executive officers.

1302. As the work expected of the controlling Staff and the Rangers and other better paid officers of the Executive Staff will be more technical than formerly they will be obliged to possess some knowledge of the following subjects, and to pass examinations in them as may be afterwards decided.

The controlling and executive Staff should be required to pass certain tests.

1. Surveying.

2. The names, uses, and habit of the growth of the principal trees in the country, and be able to identify them.

3. The general principles of Forest organisation.

4. Sylviculture.

5. The measurement of trees and timber, the making of roads &c.

6. Law, including the Forest Act, portions of the Penal Code &c., also the Forest Code, Rules under the Act &c.

1303. Forest work is, as a rule, most distasteful to natives, and to the Nairs especially, even at the present date consider that anything is better than a forest life. They people the jungle with imaginary animals, tigers with horses' heads, and demons of malignant character, with "Gorgons and Hydras and Chimœras dire"

Forest work very distasteful to natives, and efforts should be made to remedy this.

---

* The magnitude of the felling operations may be inferred from the fact that we pay more than 2 lacs of rupees to contractors for felling timber, and sell permits to merchants to fell 40,000 candies a year. All this timber is felled now practically without supervision. It is merely checked before removal, but the contractors and merchants now fell how and where they like, within certain limits.

who, they imagine, stand ready to slay any one who trespasses within their domains, and they enter the forest with fear and trembling. It must be admitted that in times past no facilities were given for camping, while the forest paths were, and are, of the roughest, and many have died from malaria brought on by exposure, or through accidents. As we pay more attention to work in the forests, we may hope to make better arrangments for the comfort of the subordinates, by cutting roads, forming camping grounds, and studying to complete all our work in the forests during the time that they are free from malaria. There should then be less distaste among the natives of a jungle-life, which those few who have tried it much have got to like.

(6) Legislation.

1304. As a new Forest Act closely following those in force in British India is about to become law, I need not make many remarks on the subjects here. The Act is in some respects weaker than the Madras Act. Our Forest officers will not have all the powers conferred on Forest officers in British territory, nor enjoy the same immunity from prosecution. Nevertheless the Act will do much good in the hands of those Magistrates who take an intelligent view of the subject. It is to be hoped that they will use their full powers to put down hill-cultivation and smuggling.

A new Forest Act is likely to be passed soon.

(7) Financial results.

1305. As explained in para 1226 the financial results of the past management of the forests have been most satisfactory from the point of view of Revenue, but unfortunately this has been the only object aimed at. No attention has been paid to conservancy, and no attempt has been made to replace the annual removal of immense quantities of timber, if we except the planting of about 600 acres of teak in 24 years, an area which would hardly supply two years' deliveries at the present rate of felling,* were the trees all fit for the axe. Side by side with this absence of any attempt at reproduction, there have been admittedly heavy fellings of timber, other than teak, far in excess of the annual growths. Thus, the large revenue secured has been obtained by sacrificing part of the capital of the forests, whose existence we owe to the absence of much demand for other timbers besides teak until within the last 20 years, and to the care formerly shewn in felling teak of large size only.§

Hitherto there has been a very heavy drain on the forests, and scarcely any attempt has been made to replace removals.

1306. We have now arrived at a time when the number of trees annually felled must be reduced, while at the same time the expenditure on conservancy and works of improvement must be increased. Instead of realising more than 3 lacs a year we may find that the revenue only exceeds the expenditure by a small amount. Doubtless we shall be aided by the rise in price of timber, which must occur, by a greater demand for inferior kinds which are abundant and at present valueless, and by a general charge for fuel all over the country, but these will hardly compensate for the loss of revenue from teak and other valuable woods. When the reserves have all been selected, and when a commencement has been made at working them in rotation, it is probable that they will yield but a small revenue for the first few years, while the cost of Settlement, Demarcation, and Survey will just at that time be very heavy. It may be therefore that the revenue will hardly cover the expenditure for a short time. After that, as prices rise, as we find new uses for our inferior timber, and as we discover the value

The time has arrived to restrict the fellings, and to replace what is annually removed and the sooner this is done the better.

---

* In 1887 the number of teak trees felled was about 24,000, and taking a full stocking of the plantations at 60 trees per acre, this would represent the produce of 400 acres.

§ The net revenue from the Travancore forests in 1891-2 was 3,11,306 Rs. though practically nothing was charged for fuel or minor forest produce, and the area was about 2,000 square miles. In the year 1890-1 the Madras Forests covering an area of 17,823 square miles gave a net revenue of 5½ lacs, and the Board of Revenue commenting on this said " It would be a mistake to consider the leaving of so large a surplus as this a matter " " for unalloyed satisfaction : in these early days of forest exploitation, the forests need all the money that can " " be spent on them, and a surplus of nearly one-third of the total receipts is obviously much greater than it " " should be. For this starving of the forests, the present weakness of the staff must be held responsible, and " " with the full staff now sanctioned, therefore, this should be remedied, and a plentiful investment of capital " " in buildings, roads, demarcation, survey, &c. should be made." Ind. Forester, XVIII, 193.

of natural products now neglected, the revenue will again rise, and will steadily increase from year to year, and we shall have the satisfaction of knowing that we are no longer drawing on the capital of our forests, and that their value is annually increasing. But if we continue, as at present, to cut down our timber without regard to future wants, the returns for a few years may be very good, but it will then be impossible, when an attempt at conservancy is made, to work the forests at a profit, while their value will have been permanently reduced. There can be but one opinion as to the expediency of restricting fellings to certain areas in rotation, limiting the outturn, and working our forests on a system which has proved most successful elsewhere. But it will require some patience to face the opposition of merchants who find their hitherto unrestricted fellings carefully supervised, to bear the complaints of the D. P. W. and Marahmut Department when they cannot get timber to their liking, and to offer explanations to Government when the revenue falls off. Nevertheless the forests must be worked on a scientific system sooner or later, and the sooner this is done, the better for the prosperity and happiness of this favoured land.

APPENDIX L

THE CARDAMOM MONOPOLY.

The cardamoms of commerce consist of the capsules and seeds of a Zingiberaceous plant (Elettaria cardamomum) found wild in the evergreen forests of Travancore, Cochin and Malabar at elevations between 500 and 4,000 ft. above sea-level.

2. The average annual production of this State for the last 19 years has been not quite 203 candies (of 600 lbs.) realizing an average of 1,020 rupees per candy or 2,07,116 rupees a year.

3. They have therefore yielded a very considerable revenue, and whereas in British India cardamoms are collected by the Forest Department, it was thought worth while in Travancore, as explained in para 1136, to form a separate Department in the year 1044 M. E. for the collection of the spice and the prevention of smuggling.

4. By far the greater part of the cardamoms produced is obtained from gardens which have been in existence from a very remote period, and which are still among the most productive. It may therefore be concluded that this kind of cultivation does not exhaust the soil, provided that the plants are not allowed to overbear, nor the shade above them to get too thin.

5. From statistics obtained from the Cardamom Office in 1064 M. E. I ascertained, that beside 12 gardens in the neighbourhood of Thodupura, owned by Government, and extending over 8430 parras* (=1200 acres), there were 2,479 gardens aggregating 3,07,525 parras (=about 44,000§ acres) held by 1,197 persons, and situated on the Cardamom Hills and near Thodupura. Of the area in private hands, 14,370 parras (say 2,000 acres) lie near Thodupura, 3,800 parras (about 500 acres) at Mlāpāra on the Peermerd plateau, and 3,105 parras (about 450 acres) at Peermerd, the produce of all these lands being classed as "Kanni" cardamoms (see para 1201 of this Report). The other 2,86,250 parras (say 41,000 acres) are on the Cardamom Hills proper, which produce the "Magara elam." These last named gardens lie along the edge of the plateau overhanging the Kambam valley in Madura, and are almost all owned by ryots living in the valley below, that is to say in British territory.

6. Owing to the large revenue accruing to Government, and to the fact that the gardens are considered to be in a certain sense private property, (Government retaining however the ownership of the land itself and the right to oust the occupant if the crop is not delivered to its officers) cardamoms have been treated as a monopoly up to the present time.

7. It has often been proposed to abolish the monopoly on the grounds that the age for monopolies has passed by, and that this cultivation should be left to private enterprise. While conceding the point that monopolies would be quite out of place in European countries, I am of opinion that in a State like Travancore where there is very little accumulated wealth in the hands of the people, it is the duty of Government to obtain revenue by assisting the people to raise produce, rather than to stand

---

* A parra of cardamom garden = 64 perakkams, each 10 ft. square, or 6,400 square ft. in all, or rather more than ⅓ of an acre.

§ This estimate is probably too high. I much doubt if there are more than 20,000 acres on the Cardamom Hills at the very outside, and 4,000 acres, all told, yielding "kanni-ēlam."

aside and allow them to grow crops unassisted, upon which or upon the land that produced the crops a tax has afterwards to be imposed. That the people are in want of assistance may be inferred from the collapse of pepper cultivation after the abolition of the monopoly in 1036, M. E. and the recent failure to get them to take up land for tea-cultivation, even though some encouragement was offered.

8. In my Memorandum on the Cardamom monopoly forwarded to Government with my letter No. 557 of 17th November 1888 when I was in charge of the Cardamom Department, I showed that if the monopoly were abolished, no other system would bring in so large a revenue, while the difficulty of realising the revenue, owing to the greater number of the ryots living in British territory, would be immense. If an export-duty were levied, five-sixths of the spice would be smuggled into Madura by some of the numerous paths leading out of Travancore. If it were decided to collect a land-tax, it would be necessary, first of all, to survey the gardens at great expense, while the ryots would never agree to a tax of 5 Rs. an acre (the lowest rate that could be charged), and lastly, the whole revenue from wild cardamoms would be lost, save in the case of reserved forests. Supposing that an assessment were charged on the produce of each garden, taking an average of several years, the only advantage this method would have over a land-tax would be that the cost of survey would be saved, while it would be an encouragement to slovenly cultivation. Finally, Government might take over all the gardens and give the holders compensation, but it would not be easy to come to terms with the ryots, while the difficulty of collection would be considerable when the Government no longer had their assistance.

9. Under the monopoly system the ryots undertake all the trouble of collecting the spice, and the Government is only charged with the work of protection. I consider therefore that the monopoly should be maintained as long as possible, some modifications being made in the present system, so as to encourage the ryots to keep their gardens in better order that the crops may be more equal.

10. Through the kindness of Mr. Maltby, Superintendent of the Cardamom Hills I have obtained from the Cardamom office the following returns of the crops collected by his Department* for 38 years from 1030 to 1067 M. E. the figures are for thulams of about 20 English lbs. each :—

Year.	Kanni.	Magaram.	Total.	Year.	Kanni.	Magaram	Total.	Year.	Kanni.	Magaram.	Total.
1030	214	1,244	1,458	1043	289	4,516	4,805	1056	923	4,289	5,212
1031	873	15,564	16,137	1044	163	4,229	4,397	1057	889	3,669	4,488
1032	327	8,130	8,457	1045	213	1,560	2,173	1058	556	974	1,530
1033	301	4,253	4,554	1046	377	6,131	6,508	1059	1,417	6,764	8,181
1034	261	3,940	4,201	1047	211	8,914	9,125	1060	1,899	11,274	13,173
1035	210	1,734	1,944	1048	182	5,063	5,245	1061	920	3,043	3,963
1036	447	4,519	4,966	1049	383	5,310	5,701	1062	1,145	1,644	3,780
1037	120	1,548	1,668	1050	264	3,867	4,131	1063	1,717	4,151	5,864
1038	284	1,556	1,840	1051	654	6,975	7,639	1064	1,262	3,344	4,606
1039	354	3,353	3,707	1052	336	856	1,192	1065	732	1,283	2,017
1040	209	1,407	1,616	1053	493	3,274	3,772	1066	1,844	6,698	8,544
1041	89	766	855	1054	1,069	2,822	3,891	1067	1,749	9,306	11,055
1042	158	759	917	1055	737	6,241	6,978				

11. The first thing that strikes one on looking at these figures is the extraordinary fluctuation in the quantity gathered from year to year. Thus, in 1030 M.E. 1834-5 only 1,458 thulams were obtained, but in the following year the figures rose to 16,137, the highest on record, while in 1041 M. E. 1865-6 they were only 855. Another thing to be noted is that the fluctuations are much greater in the figures for the "Magaram" than in those for the "Kanni" cardamoms. Further, there is no periodicity in the large and small crops. Most plants yielding fruit give alternate good and bad crops, but, with cardamoms, a good crop may or may not be followed by a poor one, while the third crop instead of being better is often much worse.

* And by the Cardamom branch of the Forest Department from 1030 to 1041.

12. Now, it is well known that the cardamom crop depends very much on the seasons, showery weather in March, April and May being very favourable, and dry weather being prejudicial to the crop, but even granting that the weather about Thodupura (in the "Kanni" district) is more propitious, and that the rains fall more regularly there than on the Cardamom Hills, this cannot, I think, account for all the fluctuations in the returns of "Magara-élam." The reason I believe to be this.

13. On the Cardamom Hills the cost of cooly labour has very much increased during the last 20 years, owing to various causes, and it is now so heavy that it does not pay the ryots to pick their crops unless the season is very favourable and the prices fairly high. Weeding also is much more neglected than formerly. The ryots visit their gardens a few months before crop commences, and if it promises to be a good one they proceed with the weeding, and when the time comes, they carefully collect the spice. If however the crop is likely to be poor, no weeding is done, and only the crop from the best parts is gathered, the rest being lost. Thus Government loses considerably, and we cannot be surprised at this behaviour of the ryots, for at the rates now given them they would otherwise lose heavily.

14. On the gardens in the "Kanni" district the case is different : a large area is cultivated by Government and these gardens are kept clean, and the produce is gathered each year, whatever the selling price of the spice may be. The cost of labour moreover has not appreciably increased, consequently the private gardens are weeded, and the whole crop is regularly picked.

15. The lesson to be learnt from this is that the ryots on the Cardamom Hills proper should receive better terms than at present, to enable them to pick their crops regularly and deliver them to Government, without losing money over the transaction.

16. Up to a dozen years ago the whole cardamom crop of the world was produced in Travancore, Cochin and Malabar, and in a bad year the prices would be good, and in a good year bad, thus the annual returns tended to an average. But the high prices ruling from 1880 to 1885 stimulated the cultivation of cardamoms in Ceylon, and before long large supplies were shipped from that island,[*] the result being a rapid fall in prices, the demand not having kept pace with the increased supply. In $\frac{1083\ \text{m. e.}}{1887\text{-}8}$ the prices were the lowest on record : the following year, owing, I believe, to a smaller crop in Ceylon, the prices were much better, but since then they have declined again, though the crops in Travancore have been small until last year. It is obvious therefore that the future prospects of cardamoms depend entirely on production in other countries. If this is large, the prices must remain low, but if it is small, prices will again rise.

17. I understand that present prices do not pay the Ceylon growers and there is not likely to be any extension of cultivation for some time. Opinions are divided as to what would happen if the price again rose : some say that Ceylon would again increase her output, while others consider that the area in that island suitable for growing cardamoms is limited, and that the highest possible yield has been already reached.

18. But, if not grown more extensively in Ceylon, it is quite possible that cardamoms may be found to thrive well in Burmah or other countries, so that we must hope that the prices will not again rise so high as to stimulate extension of cultivation, but that they may keep steady at a moderate figure. We must also endeavour to secure a more regular yield.

19. To ensure this we must pay the ryots better, and, as stated in my memorandum, I recommend

    (1) That the monopoly be maintained.

    (2) That the ryots' share be raised from two-fifths to one-half, or a sliding scale be introduced.

---

* Exports from Ceylon rose from 143 cwts. in 1880-1 to 2,684 cwts. in 1887-8. Coorg and Mysore also contribute more than formerly to the world's supply.

(3) That mēlvāram and other unnecessary deductions be discontinued.

(4) That delivery should be taken on the hills.

(5) That the price to be given should be notified to the ryots before the commencement of each crop, and that the spice should be paid for on delivery.

(6) That, if this system results in a loss, the monopoly be abolished, and an assessment be charged. That the cardamom forests be reserved, and that the ryots be ordered to demarcate their gardens to prevent encroachment: also, that, if the assessment is not paid for two years in succession, the ryot should lose all claim to his garden.

20. The following are the figures given me by the Commercial Agent at Alleppey of the total number of candies sold, the average price realised, and the total amount obtained. These differ slightly from the figures given in the Administration Report, which also include "mēlvāram" or landtax, and other small charges.

Statement showing the sale of Travancore cardamoms.

Year	Cardamoms in candies of 600 E. lbs.	Average price per candy in Rs.	Total amount realized.	Remarks
1049	192	1,048	2,00,815	It may be noted that the 2nd column multiplied by the 3rd does not bring out exactly the 4th column. This is owing to the system of placing in the 2nd column a round figure and omitting fractions. If the number of pounds exceeds half a candy it is put down as one, but if under half a candy it is omitted.
1050	144	1,034	1,49,296	
1051	275	838	2,30,268	
1052	47	1,600	74,692	
1053	133	1,710	2,28,520	
1054	140	2,353	3,28,176	
1055	248	1,966	4,87,596	The average price is obtained by dividing the average of the 4th column by the average of the 2nd
1056	188	1,889	3,44,920	
1057	158	1,427	2,25,855	
1058	62	1,825	1,13,397	
1059	303	1,015	3,08,601	
1060	484	769	3,72,278	
1061	148	682	1,01,101	
1062	88	863	75,892	
1063	256	492	1,26,059	
1064	176	776	1,36,018	
1065	84	530	49,787	
1066	326	534	1,74,847	
1067	400	519	2,07,681	
Average.	203	1,020	2,07,116	

APPENDIX II.

THE HILL TRIBES OF TRAVANCORE.

The Hillmen of Travancore number between 8,000 and 10,000 persons, who live scattered through the forests of the State from the extreme south to the confines of Cochin. Preferring those parts of the country which are least inhabited and therefore abound in game, they retire before the approach of civilisation, and are to be found most numerous where the absence of competition gives them greater freedom and more room to carry out their cultivation. At the same time they like to be within reach of bazaars where they can procure salt, cloths and knives, and other necessaries in exchange for forest produce.

2. These peoples are divided into 12 or 14 tribes who live apart, and whose members do not intermarry with those of other tribes. Each tribe or clan has a certain tract of country which is considered to belong to it, and even each village of a tribe has its land allotted to it, and no one would dare to encroach on the land assigned to another clan or another village without permission. Thus the Urālies and Mannāns look on the Periyār as the line of demarcation between them, the former living to the west and the latter to the east of that river, save where in one or two places each clan has allowed the other to clear land on its side of the river for temporary occupation.

3. Though split up into so many tribes these Hillmen probably sprang from 2 or at most 3 sources. The Kānies, Malayadayars, Kochivālans, Hill Pandārans, Urālies, Vishavāns, Ullādans and Kādars are dark-skinned, and many of them have short noses and thick lips and possess African features. It is probable, therefore, that they are descended from the original inhabitants of the country, and that their ancestors took refuge in the forests to escape being reduced to slavery like their Pulleyar congeners. The Arayans are, as far as my observation goes, fairer-skinned and more intelligent, and they are possibly descended from a superior race, a supposition which is borne out by the fact that the Hill Pandārans owe a sort of allegiance to them. All these tribes speak Malayalam.

4. The Muthuvāns, Mannāns and Palliyar, on the other hand, speak a language much more like Tamil than Malayalam, and they have admittedly immigrated from the Tamil country to the Travancore hills at a comparatively recent date. Moreover they intermarry with the Tamils of the plains even at the present time.

5. As regards appearance, the Muthuvāns, who claim the superiority over all the other tribes are probably the tallest, and have the best features of all, with aquiline noses, beards and moustachios. The Mannāns and Palliyar have, as a rule, little hair on their faces, but they are pleasant looking, bright and quick. I have not come much in contact with the Arayans, but the other tribes are quieter and slower, just as the Malayalee is more leisurely in his movements than the Tamil.

6. The men of these Hill-tribes are often very sturdily built and muscular, from the abundance of food they obtain and the healthy lives they lead, though often living in unhealthy localities. This is especially the case with the Kādars and Kānies. Their senses are, from constant use, keenly developed, and they can hear sounds and see objects which other people would not notice. For the same reason they can bear fatigue, and endure hunger and thirst, more readily than natives from the low country.

7. The numbers of the various Hill-tribes are probably decreasing. This we know to be the case in some cases. For instance there were not long ago 5 or 6 camps of Mannāns to the south of Kumili, whereas now there are only two; the Muthuvāns of Nēriamangalam as well as the Mannāns of the Cardamom Hills all say that they were more numerous formerly, while some of the smaller tribes like the Vishavāns and Hill Pandārans are certainly dying out, and as they will not inter-

marry with other tribes, it is only a question of time when their clan will ha
to exist. On the other hand, the Palliyar of the Cardamom Hills are increasing.

8. Smallpox carries off very many, and it is rare to see an old man. W
this scourge does appear they generally leave the sick to take care of themsel
with a little food handy, and themselves move away to some other place, hop
thereby to escape the contagion. Cholera also sometimes appears, and fever car
off a certain number. As they have no medicines to counteract the effects of th
diseases nature just takes her course, and the mortality is great.

9. With the exception of the Kâdars who can scarcely be said to be Hill-mer
Travancore, as they live more often in British territory or Cochin, and the l
Pandârans, all the tribes clear land and raise crops of paddy or râgi, as the case m
be. In addition to these grains, they grow plantains, tapioca, pumpkins of vari
kinds, yams and chillies, so that, as a general rule, they are very well off. Unl
some accident happens, the supply of grain, which is carefully stored in granaries
in huts in trees, lasts till the next crop,.but if the quantity is insufficient, the Hill
go out and dig for yams and other roots, or collect the fruit of various trees, thou
it cannot be said that any of them are very pleasant to eat.

10. All the Hillmen eat fish, which they catch by nets, by lines or by poisoni
The Urâlies who live near the Periyâr are especially clever fishermen. The flesh
most animals is acceptable to them, and some of them have a particular liking
that of the large black monkey (Semnopithecus Johnii), but with the except
of some Kânies and Arayans, none of them will eat, or even touch, the flesh of l
bison. The Hillmen often possess guns, but some of the Kânies still use the b
and arrow.

11. Besides the grain they grow, most of the Hill tribes raise a little bha
and tobacco, the leaves of which they dry and smoke, but as they do not know h
to prepare them they are always very glad to get any properly cured tobacco t
may be offered them. Many of them are sadly addicted to the use of opium, wh
they have got into the habit of taking to alleviate pain, or to mitigate the bad effo
of fever.

12. The Hillmen are as a rule truthful, and much more reliable than men fr
the low country. From living such a free and easy life they are independent, a
do not like to be driven. They have therefore to a certain extent to be humour
They are said to be very moral, and most of them make a point of sending away th
women folk on the arrival of any stranger at a village. The Mannâns are less p
ticular. They are good tempered and easily pleased, and they may often be hea
shouting with laughter at the anecdotes or remarks of some one of their number.

13. For a person travelling from place to place, and not staying long at a
one spot it is not easy to collect much information about the religion of the Hillme
They seem to worship some beneficent Deity, who is supposed to inhabit one of t
neighbouring hills, and whose favour they are desirous of retaining, but they p
more attention to mitigating the anger of malignant demons who are supposed to
ever on the watch to do them harm. Thus, as I was travelling once on the Card
mom Hills, I wanted to gather the fruit of the common Hill Cycas (C. circinal
when a Mannân stopped me, and said that one of their number had once eat
some of the fruit there and got very ill after it, so that a demon must live in th
place. On another occasion I wished to enter a patch of forest on the crest of t
Kaliâni-pâra ridge, but I could get no one to follow me. I was told as a reason, th
once on a time a Mannân had entered the same bit of forest and had never be
seen again, and that this forest was haunted by a very powerful demon to who
they made yearly sacrifices. Many pieces of forest are often seen on the hills le
untouched when the surrounding land has been cleared, and this is because they a
supposed to be each inhabited by some spirit. There seem also to be some traces
manes-worship. Mr. Munro tells me that the Mannâns who are said to be descen
ed from men of various trades from the Tamil country, on certain days do pooja
the tools of their ancestors.

14. All the tribes are broken up into small "kudies" or villages varying
number from. 5 to 20 or 30 families, and each village is led by a headman, wh

usually the oldest member. But among the Muthuvāns, and Mannāns, and perhaps some others, the headship is hereditary, and here perhaps the office carries with it more power, the headman among the Kānies being nothing more than the leader. Each village is independent of the others, save perhaps among the tribes mentioned, who often owe some sort of allegiance to one particular headman, thus the Muthuvāns of all the Nēriamangalam hills look on Bāka Muthuvān as their chief, and the Mannāns of the Cardamom Hills are nominally under the head of the Varakil Mannāns.

15. As the cultivation carried on by the Hillmen is not permanent, they are obliged to change their homes at short intervals. Where the soil is fertile, they remain perhaps 2 or 3 years in one place, sowing grain one year, and then obtaining a crop of tapioca or plantains the next year from the same land. Most of the tribes move every year, clearing a bit of forest in January, reaping the grain in September and clearing another piece of forest the following year. As the changes are so frequent it is not worth while for them to build permanent houses, their homes are therefore constructed merely of reeds very neatly put up, and generally clean, which last just one year. Some of the tribes prefer to have their huts of small size scattered about their clearings, each family occupying a separate building, but others, like the Mannāns, build one or two large houses with many rooms in some central place. The Kānies vary their cultivation and thus remain several years in one spot, and they have in many places planted jack, arecanut and other fruitbearing trees, the produce of which they sell in the bazaars, but their huts are all temporary.

16. The Arayans alone build more permanent homes, always selecting some steep hillslope away from elephant tracks. Here they terrace the hill-side, build houses with mud walls, and plant useful trees around them, but they shift their cultivation from spot to spot in the neighbourhood.

17. As regards clothing, most of the tribes living on the lower slopes of the hills wear little else but the loin-cloth, but the Mannāns and still more the Muthuvāns, whose homes are in a colder climate, wear heavier upper cloths as well, and are glad to get coats or blankets.

18. All the Hillmen are expert trackers, and from their knowledge of the country, the facility with which they use their knives in clearing paths, and their endurance they are invaluable as guides to any one travelling in the forests. They are also much in request for running up huts, which they do in a surprisingly short time. The Forest Department employ them also to collect ivory, dammer, and cardamoms, and are entirely dependent on them for getting wax and honey from the lofty trees and precipices on which the bees swarm.

19. Though these Hill tribes have many characteristics in common, they really differ from each other very considerably, thus the Urālies are better cultivators than the others and know much more about trees, while the Palliyar are better sportsmen. The most curious tribe is that of the Hill Pandārans. They live on the fruit of the Cycas, on fish, and fruit, and the pith of the sago palm, and on any roots they can dig up. They do not clear land nor sow paddy, and they live generally in caves. They are exceedingly timid and no inducement can make them come out and show themselves. They always fly at the approach of other human beings.

20. The following is an estimate of their numbers beginning from South Travancore.

Kānies from the south up to the Chenthrōni valley

On the Palli and Parali rivers	11	villages	=300	persons.
On the Kōtba	„	22	„	=400	„
On the Ney	„	22	„	=400	„
On the Karumana	„	14	„	=300	„
On the Vāmanapuram	„	8	„	=300	„
On the Kalleda	„	3	„	=300	„

2,000

Palliyar.
On the Kalleda and Acchankōvil rivers 2 villages ... 50 pers

Malayadayar.
On the Kakkād (Rāni) river near Nāṇāttapāra 30

Hill Pandārans.
On the Bamba river (Rāni) 3 or 4 gangs 100

Kochivālans.
On the Valiyār (Rāni) 2 villages 50

Ullādans.
On the Pālāyi river 100

Arayans.
Scattered along the foot of the hills in numerous
   camps from the Valiyār to Thodupura ... ... 4,000 ,,

Vishuvāns.
At the foot of the Hills on the Periyār 3 gangs ... 50 ,,

Urālies.
On the Hills to the west of the Periyār 23 villages ... 700 ,,

Palliyar.
Of the Cardamom Hills near Vandamettu 3 villages... 200 ,,

Mannāns.
On the hills east of the Periyār up to the foot of the
   High Range.
   1. Sundra Pandy Mannāns   2 villages   70.
   2. Varakīl Mannāns (including
      .   Udamanshola)   12   ,,   . 600.
   3. Lower Periyār Mannāns   2   ,,   100.
   4. Kōvar Mannāns   1   ,,   ·   50.
                                                   ———
                                                   820   ,,

Muthuvāns.
   1. Bāka Muthuvāns to the west of High Range 200.
   2. Sānthapāra   do.   do. south   do.   100.
   3. Thēvikolam   do.   on the High Range   100.
   4. Anjināda   do.   do. east of   do.   400.
                                                   ———
                                                   800   ,,
                                                   ———————
                             Total.........8,900   ,,

LIST OF TRAVANCORE TREES.§

Names of trees.	English and Vernacular names.	Habitat.	Remarks.
N. O. DILLENIACEÆ.			
1. Dillenia pentagyna (Roxb)	E. Naythåkku. M. Punna, kodapunna. K. Petti punna.	Deciduous forest 0–3000 ft. abundant every where.	Wood reddish, heavy and strong: used elsewhere, but not in Travancore.
N. O. MAGNOLIACEÆ.			
2. Micholia champaca (Linn)...	E. and M. Champak. T. Chambagam.	Open forest about 1000 ft. rare as a wild tree.	Wood very durable, not used in Travancore. Bark and root medicinal.
N. O. ANONACEÆ.			
3. Unona pannosa (Dals)		Evergreen forest 1000–4000 ft. very common.	Wood tough.
4. Polyalthia longifolia (Benth & H.)	T. Assothi. M. Cherana, aruna.	Planted, doubtfully indigenous: wild in cultivated places.	Wood very elastic and tough, bends very easily.
5. P. coffeoides. (Benth & H)	K. Villa.	Evergreen forests 1000–3000 ft. abundant	Medium sized tree.
6. P. fragrans (Benth & H.) ...	M. Nedu nar. K. Chals, udambatti, kodanyi.	Evergreen forests 0–2000 ft. abundant.	Stems elastic and straight, much used for masts.
7. P. cerasoides (Benth and H.)	... ...	Do. do.	Small tree.
8. Popowia Beddomeana (H. ?. and T.)	... ...	Evergreen forests 3000–3000 ft.	Do. do.
9. Goniothalamus Thwaitesii (H. f. and T.)	... ...	Evergreen forests 3000–4000 ft.	Do. do.
10. G. Wightii (H. f. and T.) ...	... ...	Evergreen forests 3000–5000 ft. common.	Do. do:
11. Mitrephora Heynæana (Thwaites).	... ...	Lower elevations.	Do. do.
12. Xylopia parvifolia (Hook f)	... ...	Evergreen forests 0—2000 ft.	Medium sized tree.
13. Miliusa Wightiana (H. f. and T.)	... ...	Evergreen forests 3000—5000 ft.	Do. do.
14. M. velutina (H. f. and T.)	M. Kåna kayitha	Central Travancore deciduous forests 0—2000 ft.	Wood very elastic, used for shafts.

§ Taken from Hooker's Flora of British India, with some additions of new trees and others hitherto known to occur in Ceylon and other countries only. This list includes trees of 30 ft. high and upwards.

* E = English,   T = Tamil,   M = Malayalam,   K = Hillman's

Names of trees.	English and Vernacular names.	Habitat.	Remarks.
15. Saccopetalum tomentosum (H. f. & T.)	...	Evergreen forest ...	Small tree.
16. Alphonsea zeylanica (H. f. & T.)	...	Evergreen forest 2000—3000 ft.	Do.
17. Orophea uniflora (H. f. & T.)	...	Evergreen forest 0—4000 ft.	Do.
18. O. Thomsoni (Bedd)	...	Evergreen forest 1000—4000 ft.	Do.
19. O. erythrocarpa (Bedd)	...	Evergreen forest 3000 ft.	Wood tough.
20. Bocagea Dalzelii (H. f. & T.)	T. Nedu nalta / K. Kanakayitha.	Evergreen forest 0—3000 ft.	Wood tough and elastic, used for shafts; the leaves contain tannin.
N. O. BERBERIDEÆ.			
21. Berberis nepalensis (Spreng.)	E. Nepaul barberry / K. Maranths.	Evergreen forest 5000—6000 ft. N. Travancore.	Wood bright yellow, used for inlaying and as a dye in N. India. Hillmen consider the bark a remedy for snake bite.
N. O. CAPPARIDEÆ.			
22. Crataeva religiosa (Forst.)	T. Kalā māvalangoi / M. Nir māthalam	Abundant on river banks from 0—5000 ft.	Bark, leaf and root are used medicinally: timber used in other countries.
N. O. VIOLARIEÆ.			
23. Alsodeia zeylanica (Arnt)	...	Evergreen forests 0—1000 ft.	Small tree.
N. O. BIXINEÆ.			
24. Cochlospermum Gossypium (DC.)	T. Tanaku ... / M. Appakodakka.	Deciduous forests 0—1000 ft. common.	Cotton from the capsules is useful for stuffing. The gum is official.
25. Scolopia crenata (Clos)	K. Charala ...	Evergreen forests 2000—4000 ft.	Wood white and strong.
26. S. Gærtneri (Thwaites)	...	Evergreen forests 0—1500 ft.	Small tree.
27. Flacourtia cataphracta (Roxb.)	T. Vayangkarai / M. Thalira. / K. Chalanga.	Evergreen forests 0—4000 ft. abundant.	Wood reddish very hard. Fruit edible.
28. Hydnocarpus Wightiana (Blume)	T. and M. Maravetti / K. Kodi, ntrvetti.	Evergreen forests 0—1000 ft. very abundant. Planted.	A lamp oil is extracted from the seeds and largely used.
29. H. alpina (Wight)	...	Evergreen forests 0—8000 ft. rare.	Wood is useful and splits readily.
30. Asteriastigma macrocarpa (Bedd)	T. Vellei nāngu	Evergreen forests 0—3000 ft. common.	Wood white and soft.
N. O. PITTOSPOREÆ.			
31. Pittosporum tetraspermum (W. & A.)	K. Kacchapatta	Evergreen forests 5000 ft.	Small tree.
32. P. nilghirense (W. & A.)	...	Do.	Do.

N. O. POLYGALÆ.

33.	Xanthophyllum flavescens (Roxb.)	T. Muttei ... M. Madaku.	Evergreen forests 0—4000 ft., very common...	Small tree.

N. O. GUTTIFERÆ.

34.	Garcinia cambogia (Desrouss.)	E. Gamboge T. Kodakkappuli. M. Kodapuli, pinaru. K. Chigiri, kodagan.	Evergreen forests 0—3000 ft., abundant. Much planted in the low country.	Wood elastic, lemon coloured and useful. The fruit is eaten largely.
35.	G. echinocarpa (Thwaites.)	T. Madul ... K. Pura.	Evergreen orests of S. Travancore only, 2000—4000 ft. locally common.	Medium sized tree.
36.	G. Morella (Desrouss.)	T. Makki ... M. Valogam.	Evergreen forests of S. Travancore, 2000 ft. not common.	Yields an excellent pigment.
37.	G. Wightii (T. Ander.)	M. Puli maranga K. Koli vala.	Banks of rivers in N. Travancore 0—500 ft.	Pigment probably good.
38.	G. travancorica (Bedd.)		Evergreen forests 3000—5000 ft., S. Travancore, local.	Pigment said to be good.
39.	G. xanthochymus (Hook f.)	K. Ana vaya	Evergreen forests 2000—4000 ft. N. Travancore, local.	Small tree.
40.	Calophyllum tomentosum (Wight.)	T. Malampunna M. Punnapey. K. Viri.	Evergreen forests 0—5000 ft. common.	Stem very straight, wood red with a long fibre. Much used for masts.
41.	C. Wightianum (Wall.)	T. Sirapunna M. Parupunna. cherupunna.	Banks of rivers 0—500 ft.	An oil is extracted from the fruit.
42.	† Mesua ferrea (Linn.)	T. and M. Nangu K. Peri.	Evergreen forests from 0—6000 ft., abundant.	Wood exceedingly hard and heavy, excellent for building.
43.	Poeciloneuron indicum (Bedd.)	T. Puthangkolli M. Vayla.	Evergreen forests from 0—3000 ft., abundant locally.	Wood very hard and heavy, excellent for building.
44.	P. pauciflorum (Bedd)	T. Puthangkolli	Evergreen forests from 2000—4000 ft., rare, S. Travancore.	Wood hard and useful.

N. O. TERNSTRŒMIACEÆ.

45.	Ternstrœmia Japonica (Thunb)		Evergreen forests about 4000 ft.	Small tree.

† Col. Beddome distinguishes 3 species of Mesua found in S. India, M. coromandelina, M. speciosa and M. ferrea, but Sir J. Hooker considers them all to be varieties of the same plant. Accepting this classification, the varieties are nevertheless very strongly marked, the "kara nanga" or broad leaved variety with small flowers and fruit has the strongest timber, the M. ferrea of Beddome. The "nanga," the M. coromandelina of Beddome, has larger flowers and fruit, but small and narrow leaves; while the least strong is Beddome's M. speciosa with long leaves and large showy flowers. "air nanga."

Names of trees.	English and Vernacular names.	Habitat.	Remarks.
46. Eurya japonica (Thunb) ..	K. Attalhuvarei	Evergreen forest, 3000—5000 ft.	Small tree.
47. Gordonia obtusa (Wall)	K. Alangi, tla	Evergreen forest 3000—7000 ft. common.	Wood white and easily worked, useful for buildings.
N. O. DIPTEROCARPEÆ			
48. Dipterocarpus turbinatus (Gærtn.)	T. Ennei M. Kalpayin. K. Varangu : velayani.	Evergreen forests 0—3000 ft. common very large tree.	Wood brownish, soft but useful for building. An oil exudes from the trunk which is used medicinally.
49. D. sp. nov	K. Ear anjili ...	Evergreen forests 0—1000 ft. local.	A very lofty tree used for boats.
50. Vatica Roxburghiana (Blume)	K. Vellei payin ...	Evergreen forests 0—1000 ft.	Wood said to be useful.
51. Hopea parviflora (Bedd) .	T. Kongu ... M. Thambagam, kambagam. K. Pongu.	Evergreen forests 0—3000 ft.	Wood hard heavy and durable, much used.
52. H. Wightiana (Wall)	K. Ilapongu	Evergreen forests 0—2000 ft. plains.	Wood strong hard and heavy.
53. H. racophloea (Dyer)	T. Karung kongu K. Neduvali kongu.	Evergreen forests 1000—3000 ft local.	Wood yellow, hard, heavy and very durable.
54. Vateria indica (Linn)	K. White dammer ... T. Vella kunthirikam. M. Payin, thelli.	Evergreen forests 0—3000 ft. common, much planted.	Wood white, soft and useless. A gum exudes which is useful as a varnish.
55. Balanocarpus eroaa (Bedd)	...	Evergreen forests 1000—3000 ft.	Large tree, wood good.
56. B. utilis (Bedd) ...	...	Do. do.	Do. do.
N. O. MALVACEÆ			
57. Hibiscus tiliaceus (Linn) ...	M. Nir parutti ...	Low country on the sides of canals and rivers : planted.	A small tree. The bark yields a fibre.
58. Bombax malabaricum (DC.)	E. Cotton tree T. & M. Ilavu.	Very abundant everywhere 0—4000 ft.	An immense tree, wood soft and light, used for boats and tea boxes.
59. Cullenia excelsa (Wight)...	T. Vadegin § K. Ear ayani.	Evergreen forests 2000—4000 ft common.	A very large tree, wood white and useless.
N. O. STERCULIACEÆ.			
60. Sterculia fœtida (Linn) ...	T. Pinari ...	Deciduous forest 0—2000 ft.	Wood light and useful, seeds edible.
61. S. urens (Roxb-) ...	...	Deciduous forest 0—2000 ft.	Wood white and soft, a gum exudes from the stem, seeds edible.

§ Called kār anjili by the Peermerd planters.

No. & Species	Vernacular names	Forest & locality	Remarks
62. S. villosa (Roxb)	T. Muruthan, M. Vakka.	Deciduous forest 0—2000 ft., common.	Yields a splendid fibre used for elephant ropes. Wood soft and useless.
63. S. guttata (Roxb)	T. Thondi, M. Kâvalam.	Deciduous forest 0—2000 ft	Wood useless. Fibre useful.
64. S. nobilis (R. B.)	Do.	In the low country.	Wood soft. Small tree.
65. S. alata (Roxb.)	T. Anei thondi, kithondi, M. Porla.	Evergreen forests about 500 ft.	A very lofty tree.
66. Heritiera Papilio (Bedd)...	... ...	Evergreen forests about 2000—4000 ft.	A large tree.
67. Pterospermum rubiginosum (Heyne).	T. Chinna polavu, M. Mala vîram, K. Ponangha.	Evergreen forests 0—3000 ft.	A very handsome tree. Wood hard and useful.
68. P. Heyneanum (Wall)	T. Polavu, M. Malavîram, thopali, palaka unam, K. Nay unam.	Evergreen forests 0—1000 ft. Much planted.	Wood white and soft.
69. P. glabrescens (W. and A.)	... ...	Evergreen forests 2000 ft.	Flowers white and handsome.
70. Leptonychia moacurroides (Bedd)	... ...	Evergreen forests 2000—3000 ft.	A small tree.
N. O. Tiliaceæ.			
71. Berrya Ammonilla (Roxb)	... ...	Evergreen forests 2000 ft. Deciduous forest 0—3000 ft.	Wood brown and very elastic. Wood excellent for axe and tool handles.
72. Grewia tiliæfolia (Vahl) ...	T. Unu, M. Chadicha, K. Thadicha, unal.		
73. Elæocarpus serratus (Linn)	T. Ulang kârai, utizzocham, M. Valiya kâra.	Evergreen forests 0—2000 ft.	Fruit edible.
74. E. oblongus (Gærtn)	K. Nâvâdi, pulanthi	...	Wood white and tough.
75. E. tuberculatus (Roxb)		Do.	Medium sized tree.
76. E. Monocera (Cav)		Evergreen forest 4000 ft.	Do.
77. E. ferrugineus (Wight)		Evergreen forests 5000 ft.	Do.
78. E. Munronii (Mast)		Evergreen forests 3000 ft.	Do.
N. O. Rutaceæ.			
79. Evodia Roxburghiana (Benth)	...	Evergreen forests 1000—3000 ft.	Small tree. Leaves scented.
80. Zanthoxylum Rhetsa (DC.)	M. Mullilam	Evergreen forests 0—1000 ft.	Do. do.

Names of trees.	English and Vernacular names.	Habitat.	Remarks.
81. Toddalia bilocularis (W. & A.) ...	... ...	Evergreen forests, 3000 ft.	Small tree. Leaves scented.
82. Acronychia laurifolia (Blume) ...	M. Mätta näri ...	Evergreen forests 0—1000 ft.	Do. do.
83. Murraya exotica (Linn)	... ...	Evergreen forests.	Do. do.
84. M. Kœnigii (Spreng)	T. Kariveppilei ...	Evergreen forests 2000—4000 ft.	Fruit yields a gum. do.
85. Clausena indica (Oliv) ...	K. Gorakkatta, katti, katta pannachi, katta koti.	Evergreen forests 2000—4000 ft. N. Travancore.	Fruit edible. Leaves eaten in curries, sweet scented.
86. Atalantia racemosa (W. & A.)	T. Katta narngam ...	Evergreen forests 2000—5000 ft.	Wood hard. Leaves scented.
87. Citrus medica (Linn) ...	E. Wild orange ... ... T. Katta narngam.	Do.	Wood white and hard do.
88. Ægle marmelos (Correa) ...	B. Bael T. Vilvam. M. Kavalum	In dry places, much planted.	Fruit used medicinally. do.
N. O. Simarubeæ.			
89. Ailantus excelsa (Roxb) ...	T. Peru	Doubtfully indigenous, much planted.	An ornamental tree, wood light not durable.
90. A. malabarica (DC.)	T. Mattipalei, pongiliyam ... M. Padalihavetti. K. Thetsu	Evergreen forests 0—1000 ft. N. Travancore.	A very large tree. Resin fragrant when burnt, and used medicinally.
91. Samadera indica (Gœrtn) ...	M. Karingotta	Evergreen forests 0—1000 ft. much planted.	A small tree, a medicinal oil is made from the seeds.
N. O. Ochnaceæ.			
92. Ochna Wightiana (Wall) ...	... ...	Evergreen forests 0—1000 ft.	Small tree.
93. Gomphia angustifolia (Vahl)	... ...	Do.	Do.
N. O. Burseraceæ.			
94. Protium caudatum (W. & A.) ...	T. Kilerei ... ... E. Black dammer	Deciduous forests 2000—3000 ft. Evergreen forests, 0—5000 ft. abundant.	Do. A lofty tree with a straight white stem producing the dammer of commerce.
95. Canarium strictum (Roxb)	T. Karungkunthrikam. M. Thelli. K. Kungiliyam, viraga.		
96. Filicium decipiens (Thwaites)	K. Väl maricha, nir väli, ntroli	Evergreen forests with light rainfall 1000—3000 ft.	Lofty and very handsome tree.

37

N. O. MELIACEÆ.

97. Melia Azadirachta (Linn)	E. Margosa, neem T. & M. Vēmbu	Doubtfully indigenous, much planted in the low country.	Small tree, wood hard and excellent, a medicinal oil is made from the seeds.
98. M. Azedarach (Linn)	E. Persian lilac T. & M. Malei vēmbu.	Abundant in deciduous forest, and on old cultivation.	Small tree. Very fast growing. Wood excellent. An oil is made from the seeds.
99. Dysoxylum malabaricum (Bedd)	E. White cedar T. and M. Vellei agil.	Evergreen forests 0—3000 ft. abundant and well distributed.	A very lofty tree, wood pale, yellow and sweet scented, used for oil casks and largely felled.
100. D. sp. nov.	M. Kār agil	Evergreen forests 0—3000 ft. rare and local. Central Travancore.	A large tree.
101. D. Beddomei (Hiern)	T. Adanthei	Evergreen forests 3000 ft. Peermard hills.	A very large tree.
102. Aglaia Roxburghiana (Miq)	T. Chokkala	Evergreen forests 0—4000 ft. Ariencharu forests.	Medium sized tree, wood red and strong, used for bandy wheels. Fruit edible.
103. A. minutiflora (Bedd)	K. Nīr mtkni	Evergreen forests 3000 ft.	A small tree with sweet scented wood.
104. Lansium anamalayanum (Bedd)	T. Santhana viri M. Vandakanni. K. Thēvathali.	Evergreen forests 2000—4000 ft.	Medium sized tree, wood sweet scented.
105. Amoora Rohituka (W. & A.)		Evergreen forests 2000—3000 ft.	A small tree, an oil is made from the seeds in Bengal.
106. Walsura piscidia (Roxb)	T. Valsura		Small tree. Bark used for poisoning fish.
107. Heynea trijuga (Roxb)	M. Korakidi	Evergreen forests 2000—6000 ft.	Medium sized tree, wood scented.
108. Beddomea indica (Hook f)		Evergreen forests 1000—4000 ft.	Small tree.
109. D. simplicifolia (Bedd)		Evergreen forests 1000—4000 ft.	Small tree.
110. Soymida febrifuga (Adr Juss)	T. Shem	Evergreen forests. Rare.	Medium sized tree. Wood red and durable. Bark used as a febrifuge.
111. Chickrassia tabularis (Adr Juss)	E. Chittagong wood K. Malei rēppa.	Evergreen forests, 0—3000 ft. Rare.	Large tree. Wood scented, prettily veined and useful for furniture.
112. Cedrela toona (Roxb)	E. Red cedar T. Santhana vēmbu. M. Maihagirivēmbu, thēvrahārum. K. Vodi vēmbu.	Evergreen forests 0—4000 ft. Common on Peermard.	An immense tree. Wood scented, red and easily worked. Much valued. Bark used as a febrifuge.

N. O. CHAILLETIACEÆ.

113. Chailletia gelonioides (Hook f)		Evergreen forests 0—4000 ft.	Small tree.

N. O. OLACINEÆ.

114. Anacolosa densiflora (Bedd)	T. Katta vekkali M. Kalmatukkam. K. Kāna madakku, kānayam.	Evergreen forests 0—3000 ft.	Large tree with a very straight stem.
115. Gomphandra axillaris (Wall)		Evergreen forests 0—4000 ft. common	Small tree.
116. Apodytes Benthamiana (Wight)		Evergreen forests.	Medium sized tree.
117. A. Beddomei (Mast)		do.	Do.

Names of trees.	English and Vernacular names.	Habitat.	Remarks.
118. Mappia fœtida (Miers)	T. Arali	Evergreen forests 0—4000 ft.	Large tree. Flowers very maladorous.
119. M. ovata (Miers)		Do.	Small tree.
120. M. oblonga (Miers)	K. Chorla, pilipicohu...	Do.	Do.
N. O. ILICINEÆ.			
121. Ilex denticulata (Wall)		Evergreen forests 6000—8000 ft.	Large tree.
122. I. Wightiana (Wall)		Do. 2000—4000 ft.	
N. O. CELASTRINEÆ.			
123. Euonymus crenulatus (Wall)		Evergreen forests 2000—4000 ft.	Small tree. Wood white, hard and close grained
124. Glyptopetalum zeylanicum (Thwaites)		Do.	Small tree.
125. Microtropis densiflora (Wight)		Do.	Do.
126. Lophopetalum Wightianum (Arn)	T. Vengalkallei M. Vengkotta, karuka. K. Vengkadavan.	Evergreen forests and on river banks from 0—3000 ft	Very lofty tree. Wood light, white and useful, durable if smoked.
127. Kurrimia paniculata (Wall)		Evergreen forests 3000—6000 ft. abundant on Peermerd.	Very lofty tree.
128. Elæodendron glaucum (Pers)	T. Karuvali	Evergreen forests 1000—3000 ft.	Very large tree with hard red wood.
N. O. RHAMNEÆ.			
129. Zizyphus Jujuba (Lamk)	E. Jujube T. Ellandi. M. Elantha.	Deciduous forest in South Travancore Planted.	Small thorny tree, wood reddish and strong, fruit edible.
130. Z. glabrata (Heyne)	T. Karuhová	Do.	Small tree, unarmed.
131. Z. xylopyrus (Willd)	T. Kottei M. Kotta.	Deciduous forest throughout Travancore at low elevations.	Small thorny tree. Wood yellow, hard and durable.
132. Z. rugosa (Lamk)	M. Thodali	Do.	Small thorny tree.
133. Colubrina asiatica (Brong)		Do.	Small tree, unarmed.
N. O. AMPELIDEÆ.			
134. Leea sambucina (Willd)	T. Nyokku M. Nyeru, manipernndi.	Evergreen forest 0—3000 ft. common.	Straggling weak tree.

251

N. O. SAPINDACEÆ.

135.	Hemigyrosa deficiens (Bedd)	...	...Evergreen forests 2000—3000 ft.	Small tree.
136.	Allophyllus Cobbe (Blume)		...Low country.	Do.
137.	Schleichera trijuga (Willd)	E. Ceylon oak / V. Puvan. / M. Puvam.	Deciduous forests 0—2000 ft.	A large handsome tree. Wood hard, used for oil mills. An oil is extracted from the seeds.
138.	Sapindus trifoliatus (Linn)	E. Soap nut / T. Meringa kottei. / K. China: shothali : oitha vanji.	..Deciduous forests 0—2000 ft. much planted.	Small handsome tree: the nuts are used for washing.
139.	Nephelium Longana (Camb)	E. Longan. ... / T. Kalia pavan. / K. Rhem povan : molei.	. Evergreen forests, 0 - 4000 ft. very abundant.	Large tree: wood red, hard and durable, fruit eaten in China where the tree is cultivated.
140.	N. stipulaceum (Bedd) ...	K. Kanam mayili ...	...Evergreen forests 0—3000 ft. local.	Medium sized tree, wood hard and serviceable.
141.	Harpullia cupanoides (Roxb)	K. Chitila madaku ..	...Evergreen forests 2000—4000 ft.	Small tree.
142.	Dodonæa viscosa (Linn)	M. Virali ...	Open forest from 0—6000 ft.	Small tree. Wood hard and close grained, yellow.
143.	Turpinia pomifera (DC.) ...	K. Unatharavie / K. Pambavetti · anatha.	...Evergreen forests 3000—4000 ft. common.	Medium sized tree.

N. O. SABIACEÆ.

144.	Meliosma Wightii (Planch)		Evergreen forests 4000—8000 ft.	Small tree.
145.	M. simplicifolia (Roxb) ...	K. Kusavi : kalavi	Evergreen forests 2000—4000 ft.	Do:
146.	M. Arnottiana (Wight) ...	Do. do.	Do.	Do.

N. O. ANACARDIACEÆ.

147.	Mangifera indica (Linn)..	E. Mango ... / T. & M. Mavu.	...Evergreen forests 0—4000 ft. common, much planted.	Very large tree. Fruit edible. Wood white but not durable, much used for rough planking.
148.	Gluta travancorica (Bedd)	E. Red wood ... / T. Shenkottani : shenchanthanam.	Evergreen forests of South Travancore 0—4000 ft. common.	Very large tree with a bright red wood suitable for furniture but not strong.
149.	Buchanania latifolia (Roxb)	T. Morala ... / M. Mnaappéra. / K. Móra kéngi : móra.	Deciduous forests 0—3000 ft. common.	Small tree: wood tough and useful, seeds eaten, bark used in tanning.
150.	D. angustifolia (Roxb) ...	...	Do.	Small tree.
151.	D. lanceolata (Wight) ...	...	Do.	Do.
152.	Semecarpus indica (W. & A.)	T. Uchi	Evergreen forests, 0—2500 ft.	Medium sized tree of rapid growth. Heart
153.	Odina Wodier (Roxb) ...	M. Kalasan.	Deciduous forests 0—1000 ft., common. Planted.	wood red,... grained and good. The gum is used ...cinally.

Names of trees.	English and Vernacular names.	Habitat.	Remarks.
154. Semecarpus Anacardium (Linn)	E. Marking nut ... T. Shengkotta : thembárvi. K. Sámbiri	...Deciduous forest 0—2000 ft., common.	Small tree : fruit contains a black resin used for marking linen. Wood soft.
155. S. travancorica (Bedd) . .	T. Káita shengkotta ... K. Théa chára.	...Evergreen forests 1000—4000 ft.	Very large tree.
156. S. auriculata (Bedd) ...	T. Velleicharie ...	...Evergreen forests 0—2000 ft.	Medium sized tree.
157. Holigarna ferruginea (Marchand) ...	T. Chárvi · karun chárvi M. Chára.	...Evergreen forests, and beside water, 0—3000 ft.	Large tree, wood white, juice black and acrid, raising blisters.
158. H. Grahamii (Hook f)	... ...	...Evergreen forests, 2000—3000 ft.	Large tree with black, acrid juice.
159. Nothopegia Colebrookiana (Bland) ...	... ...	do.	Small tree with milky, acrid juice.
160. N. travancorica (Bedd) ...	... ...	do.	Small tree.
161. Spondias mangifera (Willd) ...	E. Hog plum ... T. and M. Ambalam. K. Ambayam : mámpuli	...Open forest 0—3500 ft., common, planted.	Medium sized tree, wood soft and useless, fruit edible, often pickled.
	N. O. LEGUMINOSÆ.		
162. Mundulea suberosa (Benth) ...	T. Pil áveram	...Open forest on the eastern slopes of the hills 0—2000 ft.	Small tree.
163. Erythrina stricta (Roxb)	E. Coral ... T. and M. Murukku.	...Deciduous forest 0—3000 ft.	Small tree, wood white and very soft but useful.
164. Butea frondosa (Roxb) ...	E. Bastard teak ... T. Porasam. M. Palásin samatha ; samatha. K. Pupalasu : mukkam póyam.	...Deciduous forest 0—3000 ft., North Travancore	Medium sized tree, juice and flowers used for dying : lac is obtained from this tree : wood brown and useful.
165. Dalbergia latifolia (Roxb)	E. Blackwood : rosewood T. Thothagathi. M. Eetti.	...Deciduous forest, 0—4000 ft., well distributed.	Large handsome tree, wood purplish black, hard, heavy and durable.
166. D. paniculata (Roxb) ...	E. Bastard blackwood ... T. Eravu. M. Velitha vitti : oita theli.	...Deciduous forest 0—3000 ft. common.	Large handsome tree ; wood white and useless.
167. Pterocarpus marsupium (Roxb)	T. Vengei ... M. Venga.	...Deciduous forest 0—3500 ft., common.	Large tree, wood brown, hard, heavy and durable. Highly valued. Produces gum kino.
168. Pongamia glabra (Vent) ...	T. Pungu ... M. Pongu.	...Evergreen forest and river banks, 0—3000 ft. Planted.	Small tree. Wood light and useful. A medicinal oil is obtained from the seeds.

No.	Botanical name	Vernacular names	Distribution	Remarks
169.	Ormosia travancorica (Bedd)	T. and M. Mala manjādi K. Kuni.	Evergreen forests, 0—3000 ft.	Large handsome tree.
170.	Cassia Fistula (Linn)	E. Pudding pipe T. and M. Konnei.	Decidnous forests, 0—2000 ft.	Small tree, wood dark brown, strong and durable
171.	C. siamea (Lam)	T. Manjakonnei	do. Planted extensively.	Small tree of fast growth, wood brown and elastic, makes excellent fuel.
172.	Cynometra ramiflora (Linn)	M. Irpa	Evergreen forests 0—3000 ft. planted.	Medium sized tree. Wood light brown, hard and durable.
173.	C. travancorica (Bedd)	...	Evergreen forests 2000—4000 ft.	Large handsome tree.
174.	Dialium ovoideum (Thwaites)	T. Malam pali	Evergreen forests 0—2000 ft.	Immense tree with strong, hard wood.
175.	Hardwickia pinnata (Roxb)	T. Maderpan sampivītni ... M. Shurbi: kolla : K. Urum.	Evergreen forests 0—3000 ft.	Immense tree. Wood pale and used for planks and boats. Stem yields a red medicinal resin.
176.	Tamarindus indica (Linn)	E. Tamarind T. and M. Puli.	Dry forests 0—2000 ft.	Immense tree. Fruit edible. Wood red, hard, heavy and durable, makes excellent fuel.
177.	Humboldtia unijuga (Bedd)	... ...	Evergreen forests 2000—4000 ft. Local.	Small tree.
178.	H. Vahliana (Wight)	M. Korutthi	Evergreen forests and river banks 0—2000 ft., common.	Medium sized tree.
179.	M. alata (Lawson)	K. Kunthāni	Evergreen forests 0—3000 ft. Local.	Small tree.
180.	Bauhinia malabarica (Roxb)	M. Arām puli K. Kokka vēli : ponnan : mīn puli.	Deciduous forests 0—2000 ft.	Medium sized tree, wood hard.
181.	Xylia dolabriformis (Benth)	M. Irūl : irummula : kada K. I'ungāli.	Open forest 0—2000 ft. Not found in South Travancore, common in North.	Large tree. Wood dark, red, hard, heavy and durable : much valued.
182.	Adenanther pavonina (Linn)	T. Anai kundamani M. Manjādi.	Open forest, doubtfully indigenous.	Medium sized tree. Wood red, hard and useful.
183.	Acrocarpos fraxinifolius (Wight)	T. Malei konnei K. Kurangādi : kurunjan.	Evergreen forests 2000—4000 ft. Feerwood and High range.	Immense tree, wood pink, light and useful.
184.	Acacia farnesiana (Willd)	T. Vedda valla.	Dry forests on eastern slopes 0—5000 ft. planted	Small thorny tree, wood tough and useful : a gem is produced.
185	A. planifrons (W. and A.)	E. Umbrella thorn. T. Salei.	Do. 0—1500 ft.	Small thorny tree, wood tough and hard, excellent fuel.
186.	A. arabica (Willd)	M. Babul. T. and M. Karuvēlam	Do. do.	Small thorny tree, yielding excellent fuel and a valuable gum.
187.	A. leucophloea (Willd)	T. Vel vēlam.	Do. do.	Small thorny tree yielding excellent fuel.
188.	A. Catechu (Willd)	T. Chhalei.	Do. do.	Small thorny tree, catechu is obtained from it.
189.	A. Sundra (DC.)	T. Karangāli	Do. do.	Do.
190.	A. ferruginea (DC.)	T. Vel vēlam	Do. do.	Small thorny tree, wood reddish-brown good for building

Name of trees.	English and Vernacular names.	Habitat.	Remarks.
191. A. Latronum (Willd) ...	E. Robber thorn / T. Odei : naal.	Dry forests on the eastern slopes 0—1500 ft.	Small thorny tree yielding excellent fuel.
192. Albizzia Lebbek (Benth) ...	T. Vágei.	Open forests 0—1000 ft. doubtfully indigenous. Planted in dry districts only.	Medium sized tree, wood reddish, brown, hard and durable. Yields a gum, not thorny.
193. A. odoratissima (Benth) ...	T. Karu vágei / M. Karu vágei. / K. Chittila vága.	Open forests 0—3000 ft. common.	Medium sized tree, wood dark and very good, much used.
194. A. procera (Benth) ...	T. Nalla vágei / M. Kotte vága : karunthagara.	Open forests 0—3000 ft. abundant, especially in North Travancore.	Medium sized tree, wood dark coloured and good. Fast growing.
195. A. stipulata (Boiv) ...	T. Pili vágei / M. Motta vága.	Open forests 0—5000 ft.	Medium sized tree, wood reddish, brown, good but not much used, very fast growing.
196. A. amara (Boivin) ...		Open forests 0—3000 ft. drier districts.	Medium sized tree, wood dark and durable.
197. A Wightii (Grah) ...	T. Chêla vágei	Open forests 0—3000 ft. wetter districts.	Medium sized tree, very fast growing, wood soft and useless.
198. Pithecolobium bigeminum (Benth) ...	T. Kal pákku / K. Panni vága : aithaparantha.	Open forests 0—3000 ft common.	Small tree, wood useless.
199. P. subcoriaceum (Thwaites) ...		Open forests, 5000—8000 ft.	Small tree.
200. Inga cynometroides (Bedd) ...		Evergreen forests, 3000—4000 ft.	Medium sized tree.
N. O. ROSACEÆ.			
201. Parinarium travancoricum (Bedd) ...		Evergreen forests 2000 ft. Central Travancore. Rare.	Small graceful tree.
202. Pygeum Wightianum (Blume) ...	T. Pálungkuchi / M. Nay kambagam : shettheri. / K. Mátta kongu : rottiyan.	Evergreen forests 2000—4000 ft. common.	Large tree. Smells strongly of prussic acid.
N. O. RHIZOPHOREÆ.			
203. Rhizophora mucronata (Lam) ...	E. Mangrove	Tidal backwaters, common.	Small tree.
204. R. conjugata (Linn) ...	E. Mangrove / M. Chiriya kanda.	do.	do.
205. Bruguiera eriopetala (W. & A.) ...		do.	do.
206. Carallia integerrima (DC.) ...	M. Varanga : vallayam	Evergreen forests 0—4000 ft., common and planted.	Large tree. Wood reddish but brittle, splits easily. Used for rough work.

H

207.	Weihea ceylanica (Baill) ...	...	Evergreen forests 0—2000 ft.	Small tree.
206.	Blepharistemma corymbosum (Wall) ...	...	do.	do.

N. O. COMBRETACEÆ.

209.	Terminalia belerica (Roxb) ...	T. and M. Thāni K. Adamaruthu.	Decidnous forests, 0—2000 ft., very common.	Immense tree with a straight trunk. Wood soft, white and not durable. Sometimes used for boats. Fruit sold as gallnuts.
210.	T. Chebula (Retz)	T. & M. Kadukka	Deciduous forests 0—4000 ft. Local. Common on Cardamom Hills.	Medium sized tree yielding the myrabolans of commerce. Wood brown, hard and durable.
211.	T. tomentosa (Bedd) ..	T. Karimaruthu M. Thēmbāvu.	Decidnous forests, 0—2000 ft., common.	Immense tree. Wood dark brown, hard, heavy and useful. Much used for house building.
212.	T. paniculata (Roth) ...	T. Ven maruthu M. Pu maruthu.	do. very abundant.	Immense tree. Wood hard and durable. Used.
213.	Anogeissus latifolia (Wall) ...	T. Vekkali ... M. Mala kanjiram	Decidnous forests of drier parts 0—4000 ft., South Travancore and Cardamom Hills.	Large tree. Wood hard and strong. Used for bandies.

N. O. MYRTACEÆ.

214.	Rhodomyrtus tomentosa (Wight) ...	E. Hill gooseberry K. Korātia.	Open forests, 5000—8000 ft.	Small tree. Fruit edible.
215.	Eugenia Munronii (Wight) ...	T. Ilambili ...	Open forests 1000—4000 ft.	Small tree. Wood hard.
216.	E. Beddomei (Duthie)...	...	Evergreen forests 5000 ft.	Large tree.
217.	E. hemispherica (Wight) ...	...	Evergreen forests 2000—4000 ft.	Do.
218.	E, lucta (Ham) ...	...	Evergreen forests 0—4000 ft.	Medium sized tree. Flowers handsome, crimson and lemon.
219.	E. Arnottiana (Wight) ...	T. Naval ... M. Naga. K. Ayri.	Evergreen forests 0—3000 ft.	Large tree. Wood hard. Fruit edible.
220.	E. Wightiana (Wight) ...	...	Do. do.	Small tree.
221.	E. ceylanica (Wight) ...	M. Nytra ...	Do. do.	Large tree.
222.	E. Gardneri (Thwaites) ...	...	Do. do.	Small tree.
223.	E. caryophyllæa (Wight) ...	...	Low country.	Do. Fruit edible.
224.	E. Nessiana (Wight) ...	...	Evergreen forests 0—4000 ft.	Do.
225.	E. Jambolana (Lam) ...	T. Naval ...	Do. do. much planted.	Large tree.
226.	E. Haynesiana (Wall) ...	...	Evergreen forests 0—4000 ft.	Small tree.
227.	E. Jossinia (Duthie) ...	...	Do. do.	Do.
228.	E. floccosa (Bedd) ...	...	Do. do. South Travancore	Large.

Names of trees.	English and Vernacular names.	Habitat.	Remarks.
229. E. calcadensis (Bedd) ...	...	Evergreen forests 0—4000 ft. South Travancore.	Small tree.
230. E. Rottleriana (W. & A.)	...	Evergreen forests.	Do.
231. E. Moóniana (Wight) ...	...	Do.	Do.
232. E. microphylla (Bedd) ...	...	Evergreen forests 4000—5000 ft. South Travancore.	Do.
233. Barringtonia recemosa (Blume)	T. Samudra... / M. Samstravádi.	On backwaters, abundant.	Small tree: wood strong and serviceable. Flowers handsome.
234. B. acutangula (Gærtn) ...	E. Indian oak / M. Atta péra, cherya samstravádi.	On river banks in the low country, common.	Small tree. Wood red, tough and strong. Much used elsewhere.
235. Careya arborea (Roxb) ...	T. Ayima ... / M. Pêra. / K. Poyu.	Deciduous forest 0—3000 ft., very common.	Small tree. Wood tough and durable, but not used, except for rough house building &c.
N. O. MELASTOMACEÆ.			
236. Memecylon amplexicaule (Roxb) ...	...	Evergreen forests, 2000—4000 ft.	Small tree. Wood white and very hard.
237. M. deccanense (C. B. Clarke)	...	do.	da.
238. M. Heyneanum (Benth) ...	T. and M. Kanaloi ... / K. Kkuyävu, pävan thetti.	do.	do.
239. M. edule (Roxb) ...	...	do.	do.
N. O. LYTHRACEÆ.			
240. Pemphis acidula(Forst) ...	...	On river banks and backwaters.	Small tree.
241. Largerstrœmia lanceolata (Wall) ...	T. Venthékku: vemvila / M. Senjil. / K. Venda: vengalam.	Open forests 0—3500 ft., abundant.	A large tree with a straight stem. Wood much used and exported, light brown.
242. L. Flos-Reginæ (Reis) ...	T. Pumarutha ... / M. Mani marutha; nir marutha. / K. Shem marutha.	Open forest and on river banks, 0—2500 ft., common, planted.	Large and handsome tree. Wood reddish brown, superior to vanteak.
N. O. SAMYDACEÆ.			
243. Casearia esculenta (Roxb) ...	...	Evergreen forests, 2000—4000 ft.	Small tree.
244. C. wynadensis (Bedd) ...	...	do.	do.
245. Homalium zeylanicum (Benth) ...	...	do.	Medium sized tree.

246.	H. travancoricum (Bedd)	... ...	...Evergreen forests, 2000—4000 ft.	Medium sized tree.
	N. O. DATISCACEÆ.			
247.	Tetrameles nudiflora (R. Br.)	...{ T. and M. Chīni ... \ M. Valla chīni, valla pana.	...Evergreen and open forests, 0—2500 feet, common.	Immense tree of rapid growth. Wood white and light. Much used for boats and tea boxes.
	N. O. ARALIACEÆ.			
248.	Aralia malabarica (Bedd)	... ...	...Evergreen forests, 0—3000 ft.	Small thorny tree.
249.	Polyscias acuminata (Seem)	... ...	...Evergreen forests, 4000—6000 ft.	Small tree.
250.	Heptapleurum racemosum (Bedd)	... ...	... do.	Large tree.
251.	H. venulosum (Seem)	... ...	...Evergreen forests, 0—6000 ft.	Small tree.
	N. O. CORNACEÆ.			
252.	Alangium Lamarckii (Thwaites)	...{ T. Alanji. .. \ M. Arinji.	...Open forests, 0—3000 ft.	Small tree. Wood dark brown and useful.
253.	Mastixia arborea (C. B. Clarke)	... ...	...Evergreen forests, 2000—7000 ft., common.	Large tree.
	N. O. CAPRIFOLIACEÆ.			
254.	Viburnum punctatum (Ham)	... ...	...Evergreen forests, 2000—4000 ft.	Small tree.
	N. O. RUBIACEÆ.			
255.	Sarcocephalus cordatus (Miq)	... K. Nellīni ..	...River banks and swampy places in the low country.	do.
256.	Anthocephalus Cadamba (Miq)	... M. Atta vanji ...	...River banks, common, planted.	Medium sized tree. Wood yellow, light and useful.
257.	Adina cordifolia (Hook)	... T. and M. Manjakadamba	...Evergreen and open forests, 0—3000 feet, common.	Immense tree: Wood yellow and much used.
258.	Stephegyne parvifolia (Korth)	..{ T. Chinnakadambu \ M. Sira kadambu. \ K. Kambli.	...Deciduous and open forests, 0—2000 feet.	Medium sized tree. Wood chestnut-coloured and useful.
259.	S. tubulosa (Hook)	... ...	... do.	Small tree.
260.	Nauclea missionis (Wall)	... ...	...On the banks of streams.	Large tree good.
261.	Hymenodictyon excelsum (Wall)	...{ T. Peranjili : naykadambu \ M. Vellakadamba.	...Deciduous forests, 0—3000 ft.	
262.	H. obovatum (Wall)	... ...	...Deciduous forests 0—2000 ft.	
263.	Wendlandia Notoniana (Wall)	... K. Pava : thōvara ...	... do.	

Names of trees.	English and Vernacular names.	Habitat.	Remarks.
264. W. angustifolia (Wight) ...	T. Kurs : påvetti ...	Deciduous forests 0—2000 ft.	Small tree.
265. Webera corymbosa (Willd) ...		Evergreen forests 0—3000 ft.	Do.
266. Byroophyllum tetrandrum (Hook) ...		Evergreen forests 3000 to 5000 ft.	Do.
267. Randia uliginosa (DC.) ...	T. and M. Kåru.	Deciduous forests 0—2000 ft.	Small thorny tree.
268. R. dumetorum (Lamk) ...	... ...	Do. do.	Do.
269. R. densiflora (Benth) ...	... ...	Do. ...	Do.
270. Canthium didymum (Roxb) ...	T. Irambaraithan	Evergreen forests 0—2000 ft.	Small tree. Wood hard.
271. C. neilgherrense (Wight)	{ T. Nenytti ... { K. Nalla manthana : kavel	Evergreen forests 2000—5000 ft.	Small tree. Wood very hard.
272. C. travancoricum (Bedd) ...		Do. do.	Do. do.
273. Ixora Notoniana (Wall) ...	{ T. Kalilambli. { K. Irumbaripi	Evergreen forests 0—4000 ft.	Small tree with very hard wood.
274. I. parviflora (Vahl) ...	T. Shulunda	Do. do.	Do.
275. Pavetta indica (Linn) ...	... ...	Do. do.	Do.
276. Morinda citrifolia (Linn) ...	T. Nūna ...	River banks in the low country.	Small tree.
277. M. tinctoria (Roxb) ...	T. Manjanātti	Low country, common everywhere.	Do. roots yield a valuable red dye, wood yellow and good.
N. O. COMPOSITÆ.			
278. Vernonia arborea (Ham) ...	{ T. Shatthi ... { M. Kadavari { K. Karanthei : kirana.	Edges of evergreen forest 2000—5000 ft.	Small, fast growing tree, with handsome foliage and flowers.
279. V. travancorica (Hook) ...	K. Thūupū ...	Do. do.	Do. do.
N. O. VACCINIACEÆ.			
280. Vaccinium Leschenaultii (Wight) ...	K. Kalava ...	Open forest, 5000—8000 ft.	Small tree. Fruit edible.
281. V. Bosrdilloni (Lawson) ...	K. Kalava ...	Evergreen forests, 2000—4000 ft. on river banks.	Small tree. Fruit edible.
N. O. ERICACEÆ.			
282. Rhododendron arboreum (Sm) ...	T. Billi	Open forests from 5000 ft. upwards.	Small tree with handsome flowers. Wood useful.
N. O. MYRSINEÆ.			
283. Mæsa indica (Wall) ...	K. Kirithi ...	Open forests from 1000—6000 ft. common.	Small tree.

No.	Species	Vernacular	Habitat	Remarks
284.	Myrsine capitellata (Wall)	...	...	Evergreen forests 6000—8000 ... Small tree. Wood good: fruit edible.
285.	Ardisia paniculata (Roxb)	...	...	Evergreen forests from 1000—4000 Small tree. feet.
286.	A. pauciflora (Heyne)	...	...	do.	do.
287.	A. rhomboidea (Wight)...	...	...	do.	do.
288.	Ægiceras majus (Gaertn)	...	...	Rivers and tidal backwaters.	Small tree with milky juice.

N. O. SAPOTACEÆ.

No.	Species	Vernacular	Habitat	Remarks
289.	Chrysophyllum Roxburghii (G. Don).	T. Katti iloppei / M. Pala.	Evergreen forests 2000—4000 ft. common.	Large tree with milky juice and red wood useful for shingles.
290.	Sideroxylon tomentosum (Roxb)	T. Palei	Evergreen forests 2000—4000 ft.	Small thorny tree, with milky juice and red wood. Fruit edible.
291.	Isonandra lanceolata (Wight)	...	do.	Small tree.
292.	Dichopsis elliptica (Benth)	T. Katti iloppei / M. Pala. / K. Kei pala.	do.	Very large tree with milky juice, and red, useful wood, good for building and shingles.
293.	Bassia longifolia (Linn)	T. Natti illuppei	Doubtfully indigenous, planted extensively in drier districts.	Large tree with milky juice and hard, durable wood. Flowers are dried and eaten. Seeds yield oil.
294.	B. malabarica (Bedd)	T. Atti illuppei	On river banks in the low country.	Small tree, juice milky.
295.	Mimusops Elengi (Linn)	T. Mahila: magadam... / M. Elenji.	Doubtfully indigenous, planted.	Large tree; fruit eaten; seeds yield oil.
296.	M. Roxburghiana (Wight)	T. Etna palei	Evergreen forests 2000—5000 ft.	Large tree: juice milky. Wood red and strong.

N. O. EBENACEÆ.

No.	Species	Vernacular	Habitat	Remarks
297.	Diospyros Embryopteris (Pers)	M. Panichi :.:	River banks and evergreen forests in the low country, common.	Medium sized tree with fairly strong wood. Fruit contains tannin and a gum used for fishing lines &c.
298.	D. Toposia (Ham)	T. Karun thuvarei	Evergreen forests 2000—4000 ft. common.	Medium sized tree with white, soft wood.
299.	D. foliolosa (Wall)	T. Vellei thuvarei.	Do. do.	Do.
300.	D. ovalifolia (Wight)	...	Do. do.	Do.
301.	D. Ebenum (Kœnig)	E. Ebony / T. Karenthell. / M. Karu: mushtimbi.	Evergreen forests 0—2000 ft. sparingly distributed.	Medium sized tree: heart wood black, hard and heavy : most valuable.
302.	D. microphylla (Bedd)	T. Chinna thuvarei / M. Ehicheriocha. / K. Chetrakali.	Evergreen forests 0—3000 ft. common	tree of very ornamental appearance. Wood said to be good.
303.	D. insignis (Thwaites)	T. Potta thuvarei	Do.	... to be good.

Names of trees.	English and Vernacular names.	Habitat.	Remarks.
304. D. Candolleana (Wight)...	... ...	Evergreen forests, 0—3000 ft., common.	Large tree.
305. D. paniculata (Dals) ...	... ...	Do. do. [mon.	Do.

N. O. Styracea.

306. Symplocos spicata (Roxb)	T. Kambli ratti	Evergreen forests 0—4000 ft., common	Small tree
307. S. oligandra (Bedd) ...	... ...	Evergreen forests 3000—5000 ft.	Do.
308. S. Gardneriana (Wight)...	... ...	Do.	Do.
309. S. pendula (Wight) ...	... ...	Evergreen forests 4000—6000 ft.	Do.
310. S. semilis (Clarke) ...	... ...	Do.	Do.

N. O. Oleaceæ.

311. Linociera malabarica (Wall)	K. Kal idalei	Evergreen forests 2000—4000 ft.	Small tree
312. L. Wightii (Clarke) ...	... ...	Evergreen forests 1000—3000 ft.	Do.
313. L. leprocarpa (Thwaites) ...	... ...	Do.	Do.
314. Olea dioica (Roxb) ...	{ T. Payar ... ... / M. Edana. ... / K. Paravu idalei: man idalei.	Open and deciduous forest, 0—2000 feet, common.	Large tree. Wood said to be good.
315. Ligustrum Roxburghii (Clarke)	... ...	Evergreen forests, 2000—4000 ft.	Small tree.
316. L. Perrottetii (A. DC.) ...	K. Pungu	do.	do.
317. L. Decaisnei (Clarke) ...	... ...	do.	do.

N. O. Apocynaceæ.

318. Hunteria corymbosa (Roxb)	... ...	Evergreen forests near Courtallum, 0—3000 ft.	Small tree. Wood close grained and fine. Juice milky.
319. Cerbera Odollam (Gærtn) ...	M. Othalam ...	On rivers and backwaters in the low country.	Small tree: juice milky: kernel of the fruit poisonous.
320. Alstonia scholaris (Brown) ...	{ T. Mukampalei ... / M. Ehla pala: kodapala	Deciduous and open forests, 0—3000 feet, common.	Immense tree of rapid growth. Juice milky: wood white, bitter light, used for packing cases.
321. Holarrhena antidysenterica (Wall) ...	{ T. Kodagapalei ... / K. Paunipalei	Deciduous forest, 0—2000 ft.	Small tree. Juice milky, fruit medicinal.
322. Tabernæmontana dichotoma (Roxb)...	{ T. Kandalei palei / M. Kunam pala.	Open forest, 0—3000 ft., common.	Small tree. Juice milky.

323.	Wrightia tinctoria (Brown)	T. Nila pālei ... M. Eoocha. K. Irum pāla : thonda pāla.	Deciduous and open forests 0—3000 feet.	Small tree. Juice milky. Wood white, hard and fine.
324.	W. tomentosa (Roem) ...		do.	Small tree. Juice milky.

N. O. LOGANIACEÆ.

325.	Fagræa obovata (Wall) ...		Evergreen forest 2000—5000 ft.	Scandent tree. Flowers showy white.
326.	Strychnos Nux-vomica (Linn)	E. Strychnine M. Kanjiram. T. Yetti.	Open and deciduous forest, 0—3000 feet, common.	Medium sized tree, wood bitter, strong, and durable, seeds largely exported.
327.	S. potatorum (Linn) ...	E. Clearing nut T. Thettankottai.	do.	Medium sized tree. Wood durable. Seeds used for clearing water.

N. O. BORAGINEÆ.

328.	Cordia Myxa (Linn)	R. Sebistan' plum T. Vidi. M. Virasham. K. Karudi.	Open and deciduous forests 0—4000 ft. Planted.	Small tree. Fruit mucilaginous.
329.	C. obliqua (Willd) ...		Do.	Do. wood al.
330.	C. monoica (Roxb) ...		Do.	
331.	C. Perottetii (Wight) .		Do.	
332.	C. octandra (A. DC.) ...			
333.	Ehretia lævis (Roxb) .	M. Chavandi	Evergreen forests 0—2000 ft.	
334.	E. Wightiana (Wall) ...		Do.	

N. O. SOLANACEÆ.

335.	Solanum verbascifolium (Linn)	T. Anei chandei ...	Evergreen forests 2000—40.	Small thorny tree.

N. O. BIGNONIACEÆ.

336.	Oroxylum indicum (Vent)	T. Arānthei : arandei M. Palaga payani.	Evergreen forests 0—3000	all tree. Wood soft.
337.	Dolichandrone Rheedii (Seem)		Open country and river banks, Plains.	Small t
338.	Stereospermum chelonoides (DC)	T. Pumbāthri M. Kāring kara. K. Karanyāvu.	Open and deciduous forests, 0—3000 ft.	Large tree. Wood yellow, useful and durable, but very hard.

Names of trees.	English and Vernacular names.	Habitat.	Remarks.
339. S. suaveolens (DC) ...		Deciduous forests, 0—3000 ft., common.	Large tree. Wood useful.
340. S. xylocarpum (Wight) ...	T. Malai uthi : sira kora	Do. do. [mon.	Do. tree of rapid growth. Wood very ornamental and valuable.
	M. Puthiri : vedang konnan : adang korna.		
341. Pajanelia Rheedii (DC) ...	M. Payāni	Plains.	
N. O. VERBENACEÆ.		‸ ft. com-	Small t
342. Callicarpa lanata (Linn). .	T. Vattalei palla ...	Edges of evergreen forest and open places 3000—4000 ft.	. tree.
	M. Thām perivellam : uma thăkka.		
	K. Pura.		
343. Tectona grandis (Linn) ...	K. Teak	Deciduous forest 0—3000 ft.	Large tree yielding splendid timber.
344. Premna tomentosa (Willd)	T. and M. Thākku.	Deciduous forest 0—500 ft.	Medium sized tree. Wood hard and close grained.
	... ...		
345. P. thyrsoidea (Wight)	... ...	Do.	Small tree.
346. Gmelina arborea (Linn) ..	T. Gumadi ...	Deciduous forest, 0—3000 ft.	Medium sized tree. Wood pale and useful for many purposes. Bark and root used medicinally.
	M. Kumbil.		
	K. Kumala.		
347. Vitex trifolia (Linn)	T. Nir noochi	Open forest 0—1000 ft.	Small tree.
348. V. Negundo (Linn) ...	M. Karu noochi.	Do.	Small tree. Wood dark brown and durable.
	T. Vellei noochi		
349. V. altissima (Linn)	T. Mayila ...	Open forest from 0—3000 ft. common.	Large tree. Wood brown, hard, durable, very useful.
	M. Mayilella.		
350. V. pubescens (Vahl) ...	Do.	Open forest 0—1000 ft.	Small tree. Wood brown and durable.
351. V. leucoxylon (Linn) ...	Do.	Do. do.	Small tree. Wood whitish and compact.
		Banks of streams rare.	
352. Clerodendron infortunatum (Gartn)...	T. Perugilei ...	Open forest 0—4000 ft. common.	Small tree. Timber worthless.
	M. Valia perivalam.		
N. O. MYRISTICEÆ.			
353. Myristica laurifolia (Hook f.)	K. Wild nutmeg ...	Evergreen forests 2000—5000 ft. common.	Large tree. Wood yellowish, soft and decays easily. Juice red.
	T. Malampadavu : jathikkay		
	K. Paithapana.		
354. M. malabarica (Lamk) ...	M. Pathiri.	Evergreen forests 0—1000 ft.	Medium sized tree. The mace of the fruit is collected under the name of ponnanpayin, juice red.
355. M. magnifica (Bedd) ...	T. Chēra pānu	Evergreen forests of Central and South Travancore, local, 0—1000 ft.	Large handsome tree, juice red.
	M. Chēra payin.		

356. M. nilagirica (Wall)	M. Chenalla .... K. Pānu : karayan.: undipāna.	Evergreen forests, 0—3000 ft., v~	~~~o red.
N. O. LAURINEÆ.			
357. Cryptocarya Wightiana (Thwaites) ...		Evergreen forests, 2000—6000 ft.	~er said to be good.
358. Apollonias Arnottii (Nees)	K. Chenthanam	do.	Lofty tree.
359. Beilschmiedia fagifolia (Nees) ...		Evergreen forests, 0—2000 ft.	Medium sized tree.
360. B. Wightii (Benth) ...		Evergreen forests, 2000—4000 ft.	do.
361. Cinnamomum zeylanicum (Breyn)	E. Wild cinnamon T. Karavā, M. Eringolam : elavangam : vayana. K. Lavanga.	Evergreen forests, 0—4000 ft., abundant.	Large tree, leaves scented. Wood brown and useful for rough work.
362. C. gracile (Hook. f)	M. Atta karikka	Evergreen forests on river banks, 0—1000 ft.	Small tree. Leaves scented.
363. Machilus macrantha (Nees)	T. Kolla māvu M. Urvu. K. Ana kuru.	Evergreen forests, 0—4000 ft.	Large tree. Wood yellowish, light and soft, often used for common boats and tea boxes.
364. Phoebe lanceolata (Nees)		do.	Medium sized tree.
365. Alseodaphne semecarpifolia (Nees)		do.	Softy tree. Wood said to be good.
366. Actinodaphne Hookeri (Meisn)	T. Thali M. Iyelu : mala virinvi K. Neyarum : roanāli : pavaonha.	Evergreen forests, 0—3000 ft., common.	Small tree. Leaves used in washing.
367. A. angustifolia (Nees)		do.	do.
368. A. hirsuta (Hook. f)		do.	do.
369. Litsæa polyantha (Juss)		do.	do.
370. L. coriacea (Heyne)	T. Panni thali M. Maravetti thali.	do.	do. leaves used for washing.
371. L. glabrata (Wall)		Evergreen forests, 0—4000 ft.	Small tree.
372. L. Stocksii (Hook. f)		do.	do.
373. L. Wightiana (Wall)		do.	Medium sized tree.
374. L. zeylanica (C & F. Nees)	M. Vayana	do., 5000—5000 ft. do.	Small tree, wood yellow, strong, and sweet-scented.
N. O. PROTEACEÆ.			
375. Helicia travancorica		Evergreen forests about 4000 ft.	Medium sized tree.
N. O. THYMELEACEÆ.			
376. Lasiosiphon eriocephalus (Done)	K. Nanju ...	Evergreen forests 3000—6000 ft.	Small tree, bark used for poisoning fish.

Names of trees.	English and Vernacular names.	Habitat.	Remarks.
N. O. ELÆAGNACEÆ.			
377. Elæagnus latifolia (Linn) ...	...	...Evergreen forests 2000—4000 ft.	Small tree.
N. O. SANTALACEÆ.			
378. Santalum album (Linn)...	{ E. Sandalwood ... { T. and M. Chanthanam.	Open forests of the Anjinad valley only, 3000 ft.	Small tree. Wood highly scented and valuable.
379. Scleropyrum Wallichianum (Arn) ...	... ...	...Evergreen forests 3000—4000 ft.	Small thorny tree.
N. O. EUPHORBIACEÆ.			
380. Euphorbia Nivulia (Ham) ...	... ...	...Open rocky hills 0—1000 ft.	Small thorny tree with milky juice.
381. E. antiquorum (Linn) ...	T. and M. Kalli ...	do. 0—2000 ft.	do.
382. E. nerifolia (Linn) ...	do. ...	...Rocks 2000—4000 ft.	do.
383. Bridelia retusa (Spreng) ...	{ T. Mullu vengai: mullu marutha. { M. Mullangaynm.	Deciduous forest 0—3000 ft., common.	Large branching tree, wood strong, hard and heavy. Thorny when young.
384. B. sp	K. Olam : kadavu : asivel	Deciduous forest 3000—4000 ft., Cardamom Hills.	Branching tree, wood said to be good.
385. Cleistanthus patulus (Muell) ...	... ...	...Open forest in drier parts 0—1000 ft.	Small tree. Wood rose-coloured, hard and durable.
386. Actephila excelsa (Muell) ...	... ...	...Evergreen forests 0—5000 ft.	Small tree.
387. Phyllanthus emblica (Linn) ...	F. and M. Nelli	...Deciduous forests 0—3000 ft. abundant.	Small tree. Wood red hard and heavy.
388. P. indicus (Muell) ...	... ...	...Evergreen forests 0—3000 ft.	Small tree. Wood white and tough.
389. Glochidion lagifolium (Miquel) ...	{ M. Nīr vittil { K. Vayal nannal.	...Swampy places 0—2000 ft.	Small tree.
390. G. arboreum (Wight) ...	...	...Evergreen forests 2000—3000 ft.	Small tree.
391. Fluggea microcarpa (Blume) ...	...	Do.	Do.
392. Breynia rhamnoides (Muell) ...	...	Do.	Do.
393. Hemicyclia venusta (Thwaites) ...	{ T. Vellelamba { M. Vella katavu. { K. Palla ktui	...Evergreen forests 2000—4000 ft, common.	Small tree. Wood white, hard and heavy, used for posts.
394. Cyclostemon macrophyllus (Blume) ...	... ...	Evergreen forests 2000—4000 ft.	Large tree.
395. C. malabaricus (Bedd) ...	... ...	Do.	Do.
396. Bischofia javanica (Blume) ...	{ T. Malachithiyan : thondi { M. Nira. { K. Nannal : thiripa.	Evergreen forests 2000—5000 ft. common	A very large tree, wood red but not useful.

III

397.	Aporosa acuminata (Thwaites)	...		...Evergreen forests 0—3000 ft.	Small tree.
398.	A. Lindleyana (Baill) ...	{ T. and M. Vikili / K. Kodali.		...Evergreen forests 0—3000 ft.	Do. fruit edible.
399.	A. Bourdilloni (Stapf) ...	...	...	...	Do. 0—1000 ft. North Travancore.	Small tree.
400.	Daphniphyllum glaucescens (Blume)	...	...	...	Do. 5000—8000 ft.	Do.
401.	Antidesma Ghæsembilla (Gærtn)	...		...	Do. 0—2000 ft.	Medium sized tree. .
402.	A. Bunius (Spreng)	.	M. Cherukali	. .	Do.	Small tree.
403.	A. diandrum (Roth)	...		. .	Do.	Do.
404.	A. Menasu (Miquel)	...		...	Evergreen forests 0—2000 ft.	Small tree.
405.	Baccaurea Courtallensis (Muell)	...	M. Mula libiri : mulin kaipa		...Evergreen forests 1000—3000 ft.	Small tree. Fruit edible, wood hard and . heavy.
406.	Croton malabaricus (Bedd)	...	K. Anei kuru.		Evergreen forests 3000 4000 ft.	Small tree.
407.	C. scabiosus (Bedd)	...	. .	...	Do.	Do.
408.	C. Kleinzschianus (Wight)	...	...	...	Do.	Do.
409.	Givotia rottleriformis (Griff)	...	T. Vendálei...		...Dry forests 1000 ft. South Travancore.	Medium sized tree. Wood soft and very light.
410.	Trigonostemon nemoralis (Thwaites)		. .		...Evergreen forests 2000—4000 ft.	Small tree.
411.	Ostodes zeylanica (Muell)	...		. .	...	Do.	Do. Flowers large and handsome, yellow.
412.	Dimorphocalyx Lawianus (Hook. f) ...			...	...Evergreen forests 0—4000 ft. common.	Small tree.
413.	Agrostistachys longifolia (Benth)	...	{ T. Máochárei ... / K. Mulimpálei : erla pánei		Do. common 2000—4000 ft.	Small tree with a straight trunk, useful for posts. Wood hard.
414.	Adenochlæna indica (Bedd)	...		...	Evergreen forests 2000—5000 ft.	Medium sized tree.
415.	Cœlodepas calycinum (Bedd)	...	...	...	Do.	Small tree.
416.	Trewia nudiflora (Linn) ...		M. Pambara kumbil	...	Low country.	Large tree, wood soft and easily carved.
417.	Mallotus albus (Muell) ...	{ T. Mulla polava / K. Falchaparal		...Secondary and open forest 1000—4000 ft. common.	Small tree.
418.	M. muricatus (Bedd)	...	...	...	Evergreen forests 0—4000 ft.	Do. ,
419.	M. Beddomei (Hook)	...	...	...	Do.	Do.
420.	M. atrovirens (Muell)	...	...	...	Do.	Do.
421.	M. distans (Muell)	...	...	...	Do.	Do.
422.	M. repandus (Muell)	...	...	...	Do.	Do.
423.	M. philippinensis (Muell)	...	K. Manjana : ponni		...Secondary & open forests, 0—5000 ft.	Small tree, a red dye (kamila) is obtained from the fruit.
424.	Cladion javanicum (Blume)	...		...	Evergreen forests, 0—3000 ft.	Small handsome tree.
425.	Macaranga indica (Wight)	...	T. Vattathamarei		...Secondary and open forests, 0—4000 feet, common	Small tree.
426.	M. Roxburghii (Wight)...	{ T. Vattakanni ... / M. Vatta.		do.	do.
427.	Gelonium lanceolatum (Willd)	...	...	...	.. Evergreen forests.	do.
428.	Sapium indicum (Willd)	...	M. Komatsi	...	...Low country.	do. juice milky and poisonous.

Names of trees.	English and Vernacular names.	Habitat.	Remarks.
429. S. insigne (Benth)	—	Evergreen forests, 2000—5000 ft.	Small tree, juice milky and poisonous.
N. O. URTICACEÆ.			
430. Holoptelea integrifolia (Planch) ...	{ T. Ayā ... L. Arsl. H. Charcoal	Evergreen forests, 0—1000 ft.	Very large tree.
431. Trema orientalis (Blume) ...	{ T. Maddial : misl. M. Ama : potāgma. K. Ristthi : amberki : ayali.	Secondary and open forest, 0—5000 feet, common.	Small tree, wood white, soft and useless except for charcoal.
432. Gironniera rsticulata (Thwaites) ...	... ...	Evergreen forests, 0—3000 ft.	Large tree, timber hard and useful.
433. Phyllochlamys spinosa (Bureau) .	... ...	do.	Small thorny tree.
434. Streblus asper (Lour)	... ...	Dry forests, 0—1000 ft.	Small tree.
435. Piscospermum spinosum (Trecul) ...	... ...	Evergreen forests, 0—3000 ft., planted.	Immense tree, wood lasts under water.
436. Ficus bengalensis (Linn) ...	{ E. Banyan ... T. Al. M. Cherla. .		
437. B. tomentosa (Roxb) . .	...	do.	Large tree.
438. F. altissima (Blume) .	T. Kal atthi...	do.	do.
439. F. Benjamina (Linn) . .		do.	do.
440. F. religiosa (Linn) ...	T. Arasa	Much planted, doubtfully wild.	do.
441. F. Tsiela (Roxb) ...		Evergreen forests, 0—3000 ft.	do.
442. F. infectoria (Roxb) ...	... ...	do.	do.
443. F. asperrima (Roxb) ...	{ E. Sandpaper tree T. Irombaruthan. M. Thēragam.	Secondary and open forest 0—3000 feet.	Small tree. Leaves very rough.
444. F. hispida (Linn) ...	{ T. Chēna atthi : otta nâll M. Eruns nikku.	do.	do.
445. F. glomerata (Roxb) ...	T. Atthi ...		Large tree.
446. Antiaris toxicaria (Leschen) ...	M. Artojili : arenthal ...	Evergreen forests 0—3000 ft.	Immense tree. Wood soft and light, used for boats. Bark made into sacking.
447. Artocarpus hirsuta (Lamk) ...	T. and M. Anjili : ayani	do. Much planted.	Immense tree, yielding very valuable timber and edible fruit.
448. A. integrifolia (Linn) ...	{ S. Jack ... T. M. Pilāvu : pala	do.	Large tree, yielding excellent timber and good fruit.
449. A. Lakoocha (Roxb) ...		Evergreen forests, rare.	Large tree.
450. Laportea crenulata (Gaud) ...	{ E. Mauss . . T. Otta plavu. M. Ana choriya.	Evergreen forests 1000—3000 ft.	Small tree with very stinging leaves.

431. Villebrunea integrifolia (Gaud)		...Evergreen forests 1000—3000 ft.	Small tree.
N. O. SALICINEÆ.			
432. Salix tetrasperma (Roxb)		...On banks of streams 2000—6000 ft. common.	Small tree. Wood soft.
N. O. CONIFERÆ.			
433. Podocarpus latifolia (Wall)	T. Nerambali	...Evergreen forests 3000—4000 ft.	Large tree with aromatic wood.
N. O. CYCADACEÆ.			
434. Cyeas circinalis (Linn)	T. Chala ... / M. Entha. / K. Kaila.	...Deciduous forests 0—3000 ft.	Small tree. Kernels of fruit made into flour.
N. O. PALMÆ.			
455. Borassus flabelliformis (Linn)	E. Palmyra .. / T. Panoi.	...In the drier tracts of South Travancore 0—5000 ft., planted.	A lofty tree, yielding toddy and sugar. Wood tough and useful.
456. Corypha umbraculifera (Linn)	E. Talipot .... / T. Kondapanā. / M. Kodapana.	...Much planted, doubtfully indigenous but escaped.	Large tree with immense leaves used for thatching. Flour is obtained from the stem.
457. Caryota urens (Linn) ...	E. Bastard sago : fish tail palm / M. Chundapana.	...Evergreen forests 0—3000 ft.	Large tree yielding toddy. Sap can be obtained from the stem and also fibre.
458. Areca sp. ...	T. Kātiokanugu	...On rocks 2000—5000 ft.	Small tree.
459. A. Catechu (Linn) ...	T. and M. Kamugu	Planted.	do. yielding an edible nut.
460. Bentinckia condapana (Barry)		Evergreen forests.	do.
461. Cocos nucifera (Linn) ...	E. Cocoanut / T. and M. Thenga.	...Planted.	Large tree yielding toddy and fibre.
N. O. GRAMINEÆ.			
462. Bambusa arundinacea (Retz)	E. Bamboo ... / T. Mrangil. / M. Illi. / K. Mula.	..Dry forests 0—3000 ft.	Culms up to 6 inches in diameter and 80 ft. high. Flowers about every 30 years.
463. Oxytenanthora Thwaitesii (Munro) ...		...Evergreen forests 3000—6000 ft.	Reeds not exceeding one inch in diameter and 10 feet high. Flowers at long intervals.

Names of trees.	English and Vernacular names.	Habitat.	Remarks.
464. O. Bourdillonii (Gamble)	E. Thornless bamboo T. Pon mungil. M. Aramba. K. Kamba.	On rocks at 2000—4000 ft.	Culms 4 inches in diameter and 40 ft. high. Flowers at long intervals.
465. Teinostachyum Wightii (Munro)	K. Nanytva : meinetta : chittha	Evergreen forests 3000—4000 ft.	Reeds not exceeding one inch in diameter and 10 feet high. Flowers at long intervals.
466. Beesha Rheedii (Kunth)	M. Amma	Banks of rivers in the low country.	Reeds up to ½ inch in diameter and 10 ft. high Flowers annually.
467. B. Travancorica (Bedd)	T. Earal M. Eetta : vs. K. Kär eotta.	Evergreen forests 0—3000 ft.	Reeds up to 2 inches in diameter and 15 ft. high. Flowers every 7 years.
468. Dendrocalamus strictus (Nees)	E. Male bamboo T. Kal mungil	Anjuda valley 3000—4000 ft.	Culms up to 3 inches in diameter and 30 ft. high. Flowers gregariously and also sporadically.

APPENDIX IV.

List of Proclamations and notices connected with the Forests, appearing in the Government Gazette &c.

1st January 1837
16th Dhanoo 1013. — Prohibiting the collection of wax from the hills by private persons, and notifying that it will be sold by Government at certain places for ¼ rupee per lb.

16th February 1853
6th Koombom 1028. — Prohibiting the felling of kōl-teak except in the presence of a forest officer, fixing its price at 2 Rs. a candy, and notifying that no kōl-teak will be sold for mercantile transactions.

27th September 1858.
13th Kanni 1034 — Raising the price of kōl-teak from 4 to 6 rupees per candy.

11th April 1865.
31st Meenom 1040 — Saying that no one may take up land for coffee cultivation without applying to Government.

6th June 1865.
25th Edavom 1040. — Fixing the duty on planks and other materials.

13th June 1865.
1st Mithanam 1040. — Fixing the duty on timber at 10%.

20th June 1865.
8th Mithanam 1040 — Rules for the sales of waste land.

11th July 1865.
29th Mithanam 1040 — Notifying that instead of a fixed rate of 116 Rs. a candy for seeds and 240 Rs. for cardamoms, a share of ½ would be given to the cardamom ryots for their produce.

Do. — Prohibiting the felling of teak and blackwood and of jungle wood over 10 vannams in girth, by hill cultivators under fear of a penalty (not specified).

8th August 1865
25th Karkadagom 1040 — That on account of the trouble and delay in passing timber and getting it measured, seigniorage would in future be charged on the log, and not on the cubic contents as heretofore.

26th September 1865
12th Cunni 1041 — About measuring timber in blocks at the Chowkey and not piece by piece.

24th October 1865
9th Thulam 1041. — Offering a reward (but not saying how much) for giving information about the felling of teak and blackwood &c.

14th November 1865
1st Vrichegam 1041 — Abolishing the export duty on small articles of furniture &c.

21st November 1865.
8th Vrichegam 1041 — Abolishing the employment of watchers at watch stations on the rivers without pay.

2nd January 1866.
20th Dhanoo 1041. — Inviting applicants for the lease of the fruit trees and other forest produce of the Forest watch lands in South Travancore.

30th January 1866.
19th Magaram 1041 — Prohibiting the felling of teak, blackwood and ebony on private lands, and of all useful trees, and of trees near tanks and roads &c. " Jack, tamarind, and palmyra may be felled " by holders of tax paying lands, but only after getting leave, " and they can sell the timber after getting permission. " Holders of tax paying lands may sell or do what they like " with other trees, but, if taken by water, seigniorage will " have to be paid (no penalty is named)."

Do — Raising the price of kōl-teak from 4 Rs to 6 Rs. per candy.

25th December 1866 12th Dhanoo 1042.	Prohibiting the felling of forest land by hillmen if suitable for coffee cultivation.
13th August 1867. 30th Karkadagom 1042.	Raising the seigniorage on bamboos from 2 to 4 fanams per 100.
15th February 1868 5th Coombum 1043.	Imposing a charge on foreign cattle grazing on the hills.
2nd March 1869. 20th Coombum 1044.	Abolishing certain petty taxes hitherto levied on the hillmen near Thodupūra and Kārikōda.
14th October 1869. 3rd Thulam 1045.	Prohibiting the shooting of elephants.
29th December 1869 15th Dhanu 1045	Forbidding the sale of land draining into the Parali Aur.
12th April 1870. 1st Medam 1045.	Publishing rules for hill cultivation.
9th August 1870 26th Karkadagom 1045.	Abolishing certain dues paid by cardamom collectors and cultivators.
21st November 1871. 7th Kartigny 1047.	Cancelling the order about killing elephants in the neighbourhood of coffee estates in South Travancore, and giving permission to shoot any elephants within a certain area.
1st April 1873 21st Meenam 1048.	Levying taxes on firewood and eetas taken into British territory from Acchankōvil.
17th March 1874. 5th Pungoony 1049.	Allowing timber to be felled on the road near Punalūr, and permitting clearing.
15th September 1874. 1st Cunni 1050	About felling of kōl-teak, that applicants must go to the Taluq Cutcherries, and that a pillay will be deputed to fell and hand over the trees within six months.
17th August 1875. 2nd Chingom 1051.	Increasing the mēl-lāvam on teak and blackwood from 15 Rs. to 20 Rs. a candy (only on the road from Punalūr to Shencottah).
9th November 1875. 14th Tholam 1051.	Cancelling permission given to hill cultivators and others to dig pits and catch elephants.
15th February 1876. 5th Coombum 1051.	Prohibiting the felling of forest in the province of Thovalah.
2nd May 1876. 22nd Medam 1051.	Prohibiting the felling of forest near the keddah at Kōnniyūr.
22nd February 1881. 12th Coombum 1056	Increasing the reward for capturing smuggled cardamoms if the smugglers are caught to 22 Rs. and 12 Rs. per thulam, and if they are not caught to 20 Rs. and 10 Rs.
22nd March 1881. 11th Meenam 1056.	Defining the limits of hill cultivation.
2nd August 1881 19th Karkadagom 1056	Declaring that neglected cardamom gardens left untouched for 5 years will be given to others, and ordering the ryots to take up the cultivation of their gardens where neglected.
16th May 1882. 4th Edavom 1057	Conceding to ryots the right to collect, free of seigniorage, leaves for manure, firewood and wood for implements, and the right of grazing in the Kādukāval lands near Thovalah.
Do.	By the Conservator, offering rewards for giving information about the felling of teak without permission of ¼ of the price realized for the articles seized.

22nd June 1882. 10th Mithunam 1057.	Ordering the safe preservation and rescue of teak and other logs belonging to the Sirkar which had been carried away in the floods.
26th September 1882. 12th Cunni 1058.	About the felling of timber in the Shencottah forests, granting the right to cut timber on permit for one year.
3rd April 1883. 22nd Meenam 1058	Prohibiting the felling of forests or burning of grass land in the forests for fear of the fires spreading into Tinnevelly or Madura.
19th June 1883. 7th Mithanam 1058.	By the Conservator, ordering that all logs measured by the Forest Department should be stamped, and instructing the watchers and Police to examine them for the stamps.
3rd July 1893 21st Mithanam 1058.	By the Conservator, ordering that all receipts to merchants obtaining permits must be given on printed forms.
12th February 1884. 31st Magaram 1059.	Regarding the assessment on hill cultivation that on — Government lands 2 fanams per parrah of land be charged, and on Jenmies 1 fanam should be paid to Government, and 1 to the Jenmie.
26th August 1884. 12th Avani 1060.	By the Conservator, directing that all timber for which permits have been granted must be brought down within a year of this date, or it will be confiscated, and offering a reward of ⅓rd value of confiscated materials to any informant of cases of smuggling.
11th November 1884. 27th Thulam 1060.	By the Dewan, fixing the length of the measuring kole at 30 English inches.
19th May 1885. 24th Medom 1060.	Increasing the kuduyila on cardamoms from ⅓rd to ¾ths.
18th August 1885. 4th Chingam 1061.	By the Conservator, offering ¼ rewärd for information regarding the felling of teak on private lands.
22nd September 1885. 8th Cunni 1061.	By the Conservator, about renewing old permits and removing the timber covered by them within a certain date.
2nd March 1886. 21st Combum 1061.	By the Conservator, that old permits will be renewed from 30th Meenom 1061.
Do.	By the Conservator, that all claims for timber floated away by the floods should be preferred within a month.
31st May 1887. 19th Edavom 1062.	By the Dewan, that the Depôt system now obtaining in the Forest Department will be tried till Adi 1063, and its further continuance will be afterwards decided.
Do	By the Dewan, fixing the rate of assessment on waste — lands sown with gingelly and other produce than paddy @ 6 chs. per 128 perakoms.
6th December 1887. 22nd Vrichigam 1063.	The Forest Act. Regulation IV of 1063.
24th January 1889. 12th Magaram 1063.	By the Dewan, reintroducing the seigniorage system, and giving a list of the rates to be levied on all kinds of timber and produce.
Do.	By the Conservator, that permits will not be given to fell timber on the Calcollam and Neduvangâd hills for a time.
24th April 1888. 13th Madam 1063.	By the Dewan, reducing the seigniorage on mango-wood timber, logs or materials to 3 fanams a candy.
8th May 1898. 27th Madam 1063.	By the Quilon Peishcar, prohibiting the shooting of game in the neighbourhood of elephant pits at Acchankôvil.

22nd May 1888. 10th Edavom 1063.	By the Dewan, fixing the tax on Government lands planted with plantains and other produce at 2 fanams for 128 perukoms and 1½ and 1 fanam for Jenmi lands.
26th June 1888. 14th Mithunam 1063	By the Dewan, offering easy terms to natives of the country to take up low country land for tea.
10th July 1888. 28th Mithunam 1063.	By the Conservator, that when Proverthicarens give certificates for the removal of timber grown on private lands they must give full particulars of time and place &c.
18th September 1888. 4th Cunni 1064.	By the Trevandrum Peishcar, that the lands about Mukana mala are to be reserved.
9th October 1888. 25th Cunni 1064.	By the Quilon Peishcar, declaring a Reserve near Könniyür of 300 sq. miles.
13th November 1898. 29th Thulam 1064.	By the Conservator, that no timber may be felled under 2½ candies.
8th January 1889. 26th Dhanu 1064.	By the Dewan, remitting the duty on reeds cut on the hills for making mats.
5th February 1889 24th Magaram 1064	By the Quilon Peishcar, declaring a Reserve near Shencottah and Kulathurpura of 121 sq. miles.
26th February 1889 16th Coombum 1064.	To the Conservator from the Dewan directing him to notify auction sales of timber contracts in the Government Gazette.
19th March 1889. 7th Meenam 1064	By the Conservator, imposing a duty of 3½ fanams for every load of 56 lbs. of poonanpine brought from the hills.
26th March 1889 14th Meenam 1064.	Reserving lands near the Véli lake, 116 acres.
Do.	By the Dewan, that the duty on squared logs will in future be the same as on logs and not as on planks.
30th April 1889 19th Medam 1064	By the Conservator, that permits for the removal of timber from private property should hereafter be obtained from the Forest Department, and not from the Proverthicars.
12th November 1889. 28th Alpasy 1065	By the Dewan notifying a Reserve in the Kunnathunad Taluq of 345 sq. miles (Malayättür and Idiyara.)
19th November 1889. 5th Vrichigum 1065	By the Superintendent of Cardamom Hills, that the right to collect ginger and other forest produce had been sold to a merchant.
28th October 1890. 13th Alpasy 1066.	By the Dewan, notifying a Reserve of 1 sq. mile in Vykkeum Taluq (Manakuni Vadayär).
9th December 1890. 25th Kartigay 1066	By the Dewan Peishcar, notifying a Reserve in the Nöyyättinkarai Taluq of 539·51 acres (Mukuna Mala).
3rd February 1891. 22nd Tye 1066.	Regarding the reduction in the tariff rate on sawn timber.
18th August 1891. 3rd Avany 1067.	Directing, that confiscated articles should not be auctioned until 2 months have elapsed from the date of such orders.
19th January 1892. 7th Tye 1067.	By the Dewan, notifying a Reserve of 127 acres in the Kunnathunad Taluq (Maravanür 1st bit).
26th January 1892 14th Tye 1067.	By the Dewan, notifying a Reserve of 116 acres in the Trevandrum Taluq (Véli Reserve).
9th February 1892. 26th Tye 1067.	By the Dewan, notifying a Reserve of 87 acres in the Kunnathunad Taluq (Maravanür 2nd bit).
21st June 1892. 9th Auny 1067.	By the Dewan, notifying a Reserve of 100 sq. miles in the Thodupura Taluq.

APPENDIX V.

Receipts and expenditure of the Forest Department from 1047 to 1067 M. E.

Malayalam year	English year	Timber, chiefly teak, felled by Government and sold in logs.	Sequestrated timber felled by private parties.	Elephants &c. pads and mahouts depôts.	Quilon depôt.	Trevandrum depôt.	Neyyattinkara &c. Shali & Somali ers depôt.	Cumilly and Peermade depôts.	Other depôts.	Miscellaneous	Total	Salaries	Charges on felling timber.	Plantations	Miscellaneous	Total.	Net Revenue
1047	1871—2	82,368	24,444	18,222	...	...	...		..	21,037	1,64,211	No particulars.				55,161	88,030
1048	1872—3	53,760	22,474	39,370	...	...	...			17,361	1,32,504	...		...		40,148	92,735
1049	1873—4	62,017	24,038	45,789	...	...	...			22,916	1,66,780			...		61,448	94,316
1050	1874—5	50,142	21,725	42,917	...	...				20,801	1,35,543	...				42,351	93,192
1051	1875—6	18,465	30,241	51,608	...	...	...			19,051	99,365	..				28,238	71,127
1052	1876—7	54,459	30,850	82,898	..	...				39,160	1,48,267					45,492	1,02,775
1053	1877—8	58,575	41,765	32,846	..	...				27,470	1,60,886	...		...		48,419	1,12,237
1054	1878—9	52,944	22,778	41,555						34,542	1,51,819					49,371	1,01,716
1055	1879—80	81,787	18,540	36,747		...				24,801	1,62,785					47,915	1,14,870
1056	1880—1	1,06,004	26,316	43,541						28,558	2,09,190	12,784	55,874	3,501	5,491	37,080	1,92,110
1057	1881—2	1,36,042	22,733	53,803						10,107	2,62,653	13,025	46,383	10,765	4,010	74,183	1,48,519
1058	1882—3	1,64,650	20,802	26,881			5,566			40,726	2,78,834	No particulars.				63,310	2,15,044
1059	1883—4	1,33,543	17,507	57,608			9,478			25,138	2,61,114	27,649	26,717	14,581	8,650	79,073	1,14,411
1060	1884—5	2,14,690	8,345	27,310			12,486			45,896	3,12,727	26,160	22,818	7,145	7,865	1,23,983	1,83,742
1061	1885—6	2,05,248	8,275	79,125	18,677	39,291	11,068		6,917	34,977	4,03,545	29,877	1,44,069	4,345		1,79,31	2,24,211
1062	1886—7	1,00,235	7,655	89,086	40,916	29,291	18,210		15,391	68,294	3,90,496	38,338	1,84,750	2,971	14,008	2,15,158	1,80,360
1063	1887—8	1,60,795	31,427	85,207	35,275	36,381	20,079	1,368	24,281	19,556	4,34,128	37,607	1,92,254	4,178	8,607	2,12,643	2,21,485
1064	1888—9	1,54,879	65,335	1,30,248	40,515	35,548	44,960	4,129		11,177	4,72,721	42,435	1,70,101	4,633	29,032	2,45,601	2,27,120
1065	1889—90	1,61,671	61,345	1,89,306	65,403	53,229	58,841	15,008		21,518	5,74,587	47,478	7,27,789	9,344	22,292	3,16,839	2,57,719
1066	1890—1	2,34,475	35,192	80,484	41,748	45,488	50,271	23,758		15,367	5,59,912	47,068	3,14,362	5,066	23,270	2,99,966	2,59,316
1067	1891—2	2,36,490	76,783	69,058	44,379	38,814	42,268	28,968		30,960	5,72,292	50,763	1,85,635	4,610	20,498	2,61,680	3,11,806

NOTE—Up to 1056 the estimated value of timber in stock as well as what was sold was included in these totals, so that the figures are not exact. Up to 1057 the value of the elephants captured, and that of ivory, wax, cardamoms &c. collected by the Forest Department were included in these totals, but since 1057 the value of the elephants has not been credited to this Department, while the value of the ivory, wax, &c. have been credited to the Cardamom Department.

The large revenue shown in 1058 is due to a change in the manner of making up the account, rather than to any additional sales of timber &c.

The figures are taken from the Administration Reports.

Miscellaneous expenditure includes all sums advanced, and miscellaneous receipts all advances recovered.

APPENDIX VI.

Receipts and expenditure of the Cardamom Department from 1047 to 1067 M. E.

Malayalam year	English year	Cardamoms	Ivory	Wax	Dammer	Miscellaneous	Total	Charges	Profit or loss
1047	1871— 2	3,57,560	No particulars				20,725	3,78,585		
1048	1872— 3	2,54,702	..				18,642	2,73,394		
-1049	1873— 4	2,05,562	..				19,304	2,24,876		
1050	1874— 5	1,49,393	4,078	2,851	101	16,474	1,72,897		
1051	1875— 6	2,34,380	4,057	3,522	304	13,847	2,56,090		
1052	1876— 7	No particulars				...	94,729		
1053	1877— 8	2,22,526	3,871	1,795	17	23,196	2,57,405		
1054	1878— 9	3,21,950	3,667	3,918	229	21,866	3,51,530	No particulars	No particulars
1055	1879—80	4,87,520	4,025	3,187	248	21,531	5,16,523		
1056	1880— 1	3,43,922	3,004	3,156	166	18,817	3,69,967		
1057	1881— 2	2,29,985	4,353	4,075	501	16,518	2,55,332		
1058	1882— 3	1,15,422	3,094	3,781	571	17,051	1,39,319		
1059	1883— 4	3 14,112	5,873	3,181	1,139	3,937	3,28,242		
1060	1884— 5	3,78,925	3,012	3,663	454	12,053	3,98,307		
1061	1885— 6	1,02 906	3,450	2,984	524	17,021	1,21,893		
1062	1886— 7	92,379	6,770	3,607	415	12,122	1,15,293	73,084	+ 42,209
1063	1887— 8	1,57,171	3,764	4,240	477	16,030	1,81,682	1,39,509	+ 42,173
1064	1888— 9	1,65,210	3,454	2,805	627	11,966	1,84,062	1 18,834	+ 65,226
1065	1889—90	61,621	3,745	4,161	541	13,074	83,14	1,15,032	− 32,493
1066	1890— 1	2,29,23	2,538	3,303	438	13,751	2,49,468	1,43,078	+1,06,190
1067	1891— 2	2,31,301	4,959	4,174	755	11,828	2,53,017	1,89,144	+ 63,873

Note—Cardamoms, ivory, wax and dammer include the collections of the Forest Department from 1058 forward.

From 1062 forward the mèlvàram and other charges, besides the actual sums received by the sale of cardamoms, are included under the head of cardamoms.

Miscellaneous includes the value of sugar sold to the pagodas, generally about 10,000 Rs. a year, but not the value of elephants captured by this Department.

Charges do not include the pay of the Nair Brigade sepoys employed in protecting the cardamom gardens during the crop season.

The figures are taken from the Administration Reports.

APPENDIX VII.

Statement of timber sold on permits in 1066 and 1067 M. E.

Name of timber.	1066	1067	Remarks.
	Candies.	Candies.	
Anjili (*Artocarpus hirsuta*)	2,532	3,083	*Note*—A good deal of this timber was sold as sawn wood, and for this 33 per cent is added to show the quantity of timber in logs removed.
Kambagam(*Hopea parviflora*)	911	1,518	
Jack (*Artocarpus integrifolia*)	134	143	
Vênga (*Pterocarpus marsupium*)	914	753	
Agil (*Dysoxylum malabaricum*)	2,036	1,977	Almost all in logs.
Thèmbàvu (*Terminalia tomentosa*)	5,152	6,414	Do. Do.
Venteak (*Lagerstræmia lanceolata*)	8,203	9,160	Chiefly sawn wood.
Kadapilavu	204	158	
Irul (*Xylia dolabriformis*)	574	1,885	Chiefly in logs.
Thèvathàrum (*Cedrela toona*)	10	67	
Vàga (*Albizzia procera*)	—	104	
Mayila (*Vitex altissima*)	—	49	
Manimaruthu (*Lagerstræmia Flos reginæ*)	—	338	
Manjakadambu (*Adina cordifolia*)	—	294	
Ponna-pay (*Calophyllum tomentosum*)	186	45	For masts only.
Jungle wood of sorts	7,689	10,697	
Total	28,550	40,295	
Mango wood in planks (*Mangifera indica*)	1,840	2,858	Almost entirely from private lands.

Logs sold for boats.

Anjili (*Artocarpus hirsuta*) ...	152	135
Kambagam (*Hoped parviflora*)	457	581
Jungle wood	1,253	1,827
Total .	1,862	2,543

Free permits were given for 872 candies in 1066 and 1,515 in 1067.

Rate of seigniorage charged for the above timber &c. at the end of 1067.

Anjili ⎫
Kambagam ⎬ 2 rupees per candy in log, and 3½ rupees per candy for sawn
Jack ⎭ materials.

Venga : agil : thembāvu : ⎫
venteak : kadapilāvu : irūl : ⎬ 1½ rupees per candy in log, 2½ rupees per candy
thēvathāram : karinthakara ⎪ for sawn materials.
(vāga) ⎭

Poon spars per candy 1½ rupees.
Mango planks per candy ⅞ rupee.
Nedunār (*Polyalthia fragrans*) per log ¼ rupee.
All other timber ½ rupee per candy in log, and 1½ rupee per candy in sawn
materials.
Anjili boats ⅟₁₆ rupee, kambagom ⅛ rupee and other boats ⅛. of a rupee per
viral (1¼ inches) in diameter, irrespective of length.
Bamboos per 100, one rupee (but in South Travancore bamboos can only be
obtained from the depôts at varying and much higher rates).
Firewood brought to Quilon by cart ¼ rupee.
do. . do. in boats ½ to 1¼ rupee per boat.
Logs of firewood brought in rafts per 12 logs 1 rupee.
Figures supplied by the Forest Department.

APPENDIX VIII.

Sales of timber at the Land Depôts in 1067 M. E.

	Nagercoil	Trevandrum	Quilon.	Paliyam.	Kunili	Ramakal.	Total number of candies.
Teak	1,200	...	184	1,839	2,087	457	5,717
Blackwood ..	250	..	...	176	277	23	726
Thambagam or konga	450	156	3,077	2,292	....	..	5,975
Venga	240	793	742	255	1,504	2,079	5,613
Thombavu	108	1,407	100	...	..	...	1,705
Other kinds	662	11	...	10	204	4	891
Total.........	2,910	2,367	4,143	4,572	4,072	2,563	20,627

Note.—These figures are only approximate, the different species not having been kept separate in
all cases. This timber was delivered partly in logs, and partly as sawn wood. In the case of the latter
33 per cent. is added to bring all to one denomination.

Figures supplied by the Forest Department.

k

xlii

APPENDIX IX.

Statement of the Teak and Blackwood sold in logs at the River Depôts from 1058—1067 M. E.

	1058	1059	1060	1061	1062	1063	1064	1065	1066	1067	Total No. of candies.
Teak	11,145	8,186	12,897	16,280	14,262	12,098	11,914	13,487	20,821	22,968	1,44,058
Blackwood	101	948	98	350	255	716	637	708	556	419	4,808

Note —The River Depôts are Quilon, Vesapuram, Cottayam, Vettikattumukku and Virappura. Figures supplied by the Forest Department.

APPENDIX X.

Statement of the rates paid to the contractors for timber delivered at the depots and the selling prices at the end of 1067 M. E.

	River depots	Nagercoil	Trevandrum.	Quilon.	Pulyarah.	Kumili.	Ramakal.
Teak logs	$\frac{17}{5\frac{1}{4}}$ $\frac{13}{4\frac{1}{2}}$ $\frac{10}{3\frac{1}{4}}$	$\frac{13}{7\frac{1}{4}}$	...	...	$\frac{17}{7}$	5	$4\frac{1}{4}$
Sawn	...	$\frac{16}{12}$	...	$\frac{20}{10}$	$\frac{19}{7\frac{1}{4}}$ $\frac{18}{7\frac{1}{4}}$	6	5
Black wood logs	$\frac{18}{8}$ $\frac{15}{6}$	$\frac{15}{}$	$\frac{15}{8}$ $\frac{12}{6}$	...	$\frac{17}{7}$	5	$4\frac{1}{4}$
Sawn	...	$\frac{15}{12}$	...		$\frac{19}{7\frac{1}{4}}$ $\frac{18}{7\frac{1}{4}}$	6	5
Thambagam logs		$\frac{13}{6\frac{1}{2}}$	$\frac{14}{8}$	...	$\frac{17}{7}$	...	...
Sawn		$\frac{15}{12}$	$\frac{18}{10}$ $\frac{15}{9}$	$\frac{18}{9\frac{1}{4}}$ $\frac{17}{9\frac{1}{4}}$ $\frac{15}{9\frac{1}{4}}$	$\frac{19}{7\frac{1}{4}}$ $\frac{18}{7\frac{1}{4}}$	...	...
Venga logs		$\frac{13}{6\frac{1}{2}}$	$\frac{12\frac{1}{2}}{8}$	...	$\frac{17}{7}$	4	$2\frac{1}{4}$
Sawn		$\frac{15\frac{1}{2}}{12}$	$\frac{13\frac{1}{2}}{9}$	$\frac{14}{9\frac{1}{4}}$	$\frac{19}{7\frac{1}{4}}$ $\frac{18}{7\frac{1}{4}}$	5	3
Thembavu logs		$\frac{10}{5\frac{1}{4}}$	$\frac{11}{8}$	...	...	...	...
Sawn		$\frac{14}{12}$	$\frac{12}{9}$	$\frac{13}{9\frac{1}{4}}$	...	...	...
Jungle wood logs		$\frac{9}{5\frac{1}{4}}$	...	...	$\frac{13\frac{1}{4}}{6\frac{1}{4}}$	$2\frac{1}{4}$	2
Sawn		$\frac{13\frac{1}{4}}{12}$	..	...	$\frac{14}{7}$	3	2

Note.—The upper figure in each case represents the selling price, and the lower the price paid to the contractor per candy (15¼ cub. ft.)

At Kumili and Ramakal the contractors fell the timber and bring it to the depot, and remove it thence paying the mel-lavam shown above.

Figures supplied by the Forest Department.

APPENDIX XI:

Area and cost of the Teak Plantations.

Year.	Area planted.				Cost.			
	Koni.	Maliatar.	Ariankavu.	Total.	Koni.	Maliatur.	Ariankavu.	Total.
	Acres.	Acres	Acres	Acres.	Rs. 'ch ca	Rs. ch ca	Rs. ch ca	Rs. ch ca
1042	38 (26)	46	...	84	425 27 7	957 19 12	... .	1,383 19 3
1043	35 (17)	36	...	71	553 27 8	869 21 4	.. ...	1,423 20 12
1044	35 (21)	35	...	70	717 7 4	865 16 2	... .	1,562 23 6
1045	...	2	...	2	918 27 ...	1,522 15 12	... .	2,441 14 12
1046	30 (23)	16	...	46	622 5 9	1,098 15 .	... .	1,710 20 9
1047	.32 (12)	...	...	32	1,230 22 ..	800 5 8	... .. .	2,030 27 8
1048	...	...	...	...	1,577 22 ..	596 20 ...	... .	2,174 14 ...
1049	30 (11)	14	...	44	1,048 9 .	667 21 12	... ...	1,716 2 12
1050	..	...	...	...	492 9 8	694 15 ...	... .	1,186 24 8
1051	...	...	...	...	687 5 ...	807 17 ...	.. ...	1,494 22 ...
1052	...	...	...	..	724 7 .	815 22 ...	.. ..	1,540 1 ...
1053	...	...	...	...	806 5 ...	467 25 4	... ...	1,274 2 4
1054	50 (17)	...	...	50	1,110 18 ...	448 19 ...	... ...	1,559 9
1055	...	...	...	...	1,519 21 .	903 26 ...	.. ..	2,422 19 ...
1056	80 (46)	...	...	80	3,253 26 9	804 14 7	... ...	4,058 13 ...
1057	394 (208)	...	...	394	11,139 8 2	598 3 ..	... ...	11,737 11 2
1058	...	...	...	...	7,586 22 8	150 2 .	... .	7,736 24 8
1059	230 (136)	...	...	230	12.379 18 5	17 24 ..	... ...	12,397 14 5
1060	...	.	...	...	7,017 25 4	18 25 ..	... .	7,036 22 4
1061	...	...	...	...	5,129 21 12	597 2 ..	... ..	5,726 23 12
1062	...	...	...	...	2,887 13 9	315 20 8	... ...	3,203 5 11
1063	...	...	...	...	4,175 ... 7	6 4 .	... ...	4,181 4 7
1064	...	.	...	...	4,632 11 8	19 6 ...	... ...	4,651 17 8
1065	...	...	150 (0)	150	5,726 12 8	... .	2,905 9 13	8,631 22 5
1066	...	.	3¼(3¼)	3¼	4,501 12 6	480 3 8	564 21 2	5,546 10 ...
1067	...	...	36 (36)	36	3,314 10 15	79 6 11	381 4 8	3,774 22 2
Total......	954 (507)	149	180¼(39¼)	1,292¼ acres.	84,180 4 11	14,594 6 8	3,851 7 7	102635 18 10

Note—(1) The figures in brackets denote the present areas of the plantations as recently surveyed.

(2) The above figures are taken from our own accounts and are correct. They differ from those given in the Administration Reports owing to the system of preparing the accounts in the Huzur independently of us, as I have already explained.

APPENDIX XII.

Statement showing the number of Elephants captured from 1047 to 1067.

Places.	1047.	1048.	1049.	1050.	1051.	1052.	1053.	1054.	1055.	1056.	1057.	1058.	1059.	1060.	1061.	1062.	1063.	1064.	1065.	1066.	1067.	
Forest Department in pits	10	11	16	11	5	6	5	2	5	8	8	8	9	10	10	10	11	25	23	11	20	The elephants captured in "in-am" pits are also included in these returns. Capturing in this way was stopped in November 1875. M. E. 1051.
Forest Department in keddah			11	13	25	4	14	10	4	6	5	4										
Cardamom Dept. in pits	2	7	7	6	2									2	4	6	4	6				
Total	12	18	23	17	7	17	18	27	9	22	18	12	15	15	14	10	13	29	29	15	26	
Death	3	1	20	15	7	12	15	16	9	11	14	9	8	7	3	6	3	12	13	10	7	
Remainder	9	17	3	2		5	3	11	11	4	3	7	8	11	4	10	17	16	5	19		

Note—Figures supplied by the Elephant Department.

APPENDIX XIII.

Statement showing the value of timber exported between 1047 to 1067.

	1047.	1048.	1049.	1050.	1051.	1052.	1053.	1054.	1055.	1056.	1057.	1058.	1059.	1060.	1061.	1062.	1063.	1064.	1065.	1066.	1067.
Value of dutiable timber	1,10,900	1,08,700	96,438	1,16,500	1,92,800	2,14,400	2,69,600	2,02,600	1,87,800	1,93,000	1,74,700	2,54,270	3,05,161	2,55,784	2,81,691	2,88,501	2,57,406	2,86,452	2,83,634	2,91,781	2,53,313
Value of timber exported free.*	19,981	8,443	8,078	4,098	2,184	15,649	6,104	6,545	2,163	2,554	1,875	1,580	10,382	1,377	3,110	1,497	3,467	3,560	2,529	8,802	

Note.—The tariff valuation of timber is per candy for teak, blackwood and thěvathāram . .. .12/
Kole teak 5/
Anjili, jack, thembavu & karingali .. 6/
Jungle wood 4/
The duty is 10 per cent on the above valuations.
The above figures are taken from the Administration Reports.

* This does not include the large supplies of teak and blackwood to the value of more than 2 lacs a year exported by the Bombay firm, on which no duty is paid. No statistics seem available.

APPENDIX XIV.

LIST OF ELEVATIONS.

On the area drained by the Hanamanathi river. ‘Mâhênthragiri peak 5500. Panagudi gap 3600. Chûralvari 600.

The Palli or Vadasheri river. Waddamalei 3500. Tandaga malei 2500. Kalatthi mottei 1516. Vallât malei 600.

The Parali river. Valiya malei 4700. Balamore Rest-house 1500. Pandiyan dam 100.

The Kôtha river. Peaks above Eridge estate 5200. Mottachi peak 4500. Mutthu kuri vayal 4200. Vengalam malei 3000. Kalpadavu hill (Mauran mallay) 2700. Klâmalei (Cullan mallay) 2500. Kuniccha malei (Kautadi mallay) 2100. Thaccha malei (Succha mullay) 2000. Ariyanâda dam 210. Thripparpu fall 170.

The Ney river. Agasthiyar peak 6200. Nûrathôda pâra (Noortodia bluff rock) 5500. Kuru malei (Cooroo mallay) 1500. Kâttâdi malei 1300. Pêkkulattha malei (Awe mallay) 1250.

The Karamanay river. Sasthan kottei 3000. Uthal mala (Wodala mallay) 1400.

The Pâlôda or Vâmanapuram river. Chemmunji peak 4000. Pallipara hill 3510. Perambukotta station 3330. Sanitarium bungalow 3150. Kakkâd malei 600. Verliyan kunnu 600. Vithara 400. Palôda 200. Nedumungâda 120.

The Kalleda river. Alvâkkurichi peak 4500. Strathmore patenas 3920. Nâgamala peak 3500. Nedumpâra 2930. Alvâkkurichi gap 2600. Korakunnu 2000. Shânar pass 1940. Ariyankâvu pass 1210. Thên malei 1170. Eramalûr rock 800 Kôvarangudi 700, Camp Gorge bungalow 630, Angalam pâra 540, Kulatthûrpura 360, Koravanthâvalam 350, Karûr 160. Punalûr bridge 100, Kâlanyûr teak plantation 70.

The Thâmravarnni river. Râmakal rock 3600. Thûval malei (Coonnucal square rock) 3000. Acckankôvil pass 1470. Mêkkara 810, Kannupillai mettu 750. Puliyâra 720. Shenkotta 690.

The Acchankôvil river. Chêr malei (Shair malei) 4000. Karinkavala (Currin-cowly) 4000. Allapâda ridge 2960. Chompâlakkara 2420. Kûttampâra 1320. Pongam pâra 1285. Pâppankuli hill 1300. Palikka malei 1150. Kakkâri kudi 1050. Paligapara 620. Acchankôvil 250. Konni Forest bungalow 220. Thora 200.

The Itâni river. Amaratha mêda 4500. Manammutti mala 3970. Koretti mudi (Corayty moody) 3750. Puliyar kôvil 3700. Davis thâvalam 3630. Chittambalam mala (Chitumblum mala) 3540 Nallathanni pâra 3400. Chennâtta kara 3370. Kanakan mala (Conaca mole) 3210. Koretti mudi camp 3200. Tholapa thâvalam (Tolapay tavalam) 3100. Chemmani kotta (Chemmung cottay) 3050. Perambu kotta (Perunby cotay) 3050. Kombukotta 3050. Dighton's Dingle 3030. Ilamba-cheddi 2980. Perumpâra mala 2500. Thêvara mala 2000. Gopura mala 1800. Nânnâttapâra 1750. Mêlpâra thâvalam 1720. Sabari mala pagoda (Choura malay) 1680. Vampali ala pâra (Bumbalaly para) 1540. Nâladi pâra 1500. Râjampâra 1500. Kombukutti 1240. Nellikal 1100. Shellikal 1000. Vilangapâra 700. Junc-tion of Kalâr with Maniâr 500. Junction with Pâkkalâr 410. Mûngapâra 420. Kunam 360. Shêttakal 250. Keddah bungalow 180. Peranthôn aruvi 90. Pera-nâda 40. Kumaram pêrûr 30.

The Manimala river. Karuvâli kâda (Curvâlicad hill) 800. Manittûkka peruva mala 800. Palippura 710. Edakkônam hill 700. Pârathôda 370. Chera vayal 320. Pêrathoda 265. Alapâra 200. Mundakkayam 150. Erumêl 150. Edakkônam 150.

The Pâlâyi river. Kallâla mêda 4000. Kudamurutti cliff 3500. Mâvadi mala 1000. Vettimala 1000. Addakkam 1000. Thallanâda 1000. Ayyampâra 1000. Kallilapâra 650. Thêvara mala 600. Nellappâra 450.

The Múvāttapura river. Nagarampāra camp 2550. Uppu kunnu 2500. Neddiyattha (Midietthu mallay hill station) 2000. Thoppi mala 2000 Padikkānam peak (Kydhapāra hill) 1500. S. Nādagāni 1500. Thūmbipāra 1200. Pūcchappara 1000. Perambukkāda 1000. Udambannūr 200. Arakkulam 150.

The Periyār river. (a) *On the river itself.* Mlāppāra (Sangany tavalam) 3000. Junction with Varukkapara thōda 2950. Mullayār thāvalam 2880. The Dam 2840. Bandy road crossing 2700. Chenkara crossing 2540. Thodupura crossing 2300. Madatthin kadavu 1650. Junction with Mothirapura river 800. Uttharam 300. Neriyamangalam 200.

(b) *Between Vandanvatta and Mlāppara.* Kōtta mala 6400. Paccha mottei 5500. Varayāttin mudi 5170. Chōttūr (Shaitoor) hill 5100. Chingammala 5100. Kūttukal 5000. Kathira mudi 4250. Cheyitthan mala (Shaitan) 4250. Kūttan mēda 4200. Karintīra mala (Autmode) 4200. Puliyan mala (Poolimulla)4200. Kumārikulam 4000. Bāla kumāri mēda (Walum coomry mode) 4000. Pālkācchiya mēda 4000. Mangalam dēvi 4000. Ottatthala mēda 3800. Vandanvatta camp 3750. Uppakulam hill 3710. Māvadimala 3700. Bālarāma kotta (Balrunguddy) 3700. Chōta pāra 3090. Kumili 3000.

(c) *On the Peermerd hills and the country to the north.* Muppura mudi (Kolekote peak) 4900. Vanjūram padi (Pearmoode hills) 4200. Bāgamannān hill 4020. Pālkulam mēda 3960. Heeman kunnu 3810. Varei aut mala 3300. Residency 3300. Chakkakol mēda 3240. Idukka vetta (Cullyka mulla) 2600.

(d) *On the Cardamom Hills.* Muracchu mala (Moolchindy hill) 4510. Bodi nāyakanūr mettu 4490. Kaliyāna pāra 4245. Mottayam mēda (Moolēar mudi) 4200. Nayāndi mala (Ninar mala) 3950. Perumānkayam thāvalam 3780. Elayakād thavalam (Poolcuddy pāra) 3660. Kambam mettu (Ramagerry colum) 3600. Kāmakchi mēda (Nachambulla hill) 3480. Chāntha pāra bungalow 3450. Udanbanchōla bungalow 3420. Tingam mēda 3310. Thēvāram mettu 3230. Tambarakan pāra 3250. Rāmakal mettu 3200. Kalkunthal 3150. Idokki 3150. Pichatthi kulam thāvalam 2580. Pirinyan kūtti 1800.

(e) *To the west of the Cardamom hills and High Range.* Thēra thandu 3210. Chokkan mudi 3150. Kadakay (Kaddacky) thāvalam 2600. Mothirapura camp 2465. Mannān kandam 1910. Kolapura hill 1900. Kuthira kutti thāvalam 1800. Chennāyi pāra 1500. Junction of Mothirapura and Kalār 1340. Malayāttūr hill 1040. Pēndi mād kutthu 840. Junction of Mothirapura and Periyār 800. Vāda muri 420.

(f) *On the High Range and in the Anjinada Valley.* Anei mudi 8837. Kāttumala 8100. Vaga varei 8000. Pāmbādichōla 8000. Karumpāra 7900. Eravi mala 7881. Chenda varei 7664. Kumārikal 7540. Karinkulam 7480. Hamilton's Plateau 7400. Payrat mala 7400. Chokkan mudi 7300. Thēvi mala 7250. Periya varei 7200. Chemman peak 7100. Nīlakal ōda 7000. Korka kombu 7000. Aliyar mala 6900. Kuriyan mala 6800. Vattavadi 6000. Kottakombu 5950. Thēvikulam 5710. Perumola 5570. Puthūr 5400. Perumpattikal 5380 Kundali 5370. Perumāl mala 5350. Kāndel 5200. Pālakadavu 5100 Gudalūr ala 5060. Kalandūr 4850. Mūnār flāt 4700. Pannipāra kuppu 3850. Karūr 3800. Maravūr 3500. Nāchivayal 3000.

On the Travancore-Cochin Boundary. Pannimāda trijunction station 3800. Kaliyāli pāra 3490. Payiram pāra 3490. Pālan mudi 3350. Kallithōda 3210. Anei adi pāra 2900. Malampāra 2880. Chūral variccha pāra 2865. Kilithōda fall 2750. Kārapilla thōda 2560. Pannimāda cairn 2550. Vīramudi 2450. Ayira kunnu 2390. Aneimādan fall 2360. Mukkampāra 2340. Ulchēri hill 1890. Kāleikal 1770. Mēsapola 1750. Manimāda fall 1420. Chennāyi pāra 750. Karadi pāra 685. Mīnāri thōda 610. Athirapura fall 400.

Additions and Corrections.

Page 5 line 53 For " has always been" read "is."
Page 6 line 35 After "east" put a full stop.
Page 7 line 32 For " Mūnā " read " Mūnār."
Page 8 line 38 For " Konniyūr" read " Kōnniyūr."
Page 9 line 9 For " Neduvengaud" read " Nedumangād."
     line 11 Since this was written, Plumbago mining has become a large industry.
Page 10 line 39 For " precepitation " read " precipitation."
     line 41 For " in " read " on."
Page 12 line 37 For " develope" read " develop."
     line 51 margin. For " apon " read " upon."
Page 13 line 29 For " in " read " on."
Page 14 line 20 For " our " read " one "
Page 17 line 22 For " Magiltunni " read " Mayilunni."
Page 19 line 5 For " Karrimia " read " Kurrimia."
     line 15 For " Holoptelia " read " Holoptelea."
     line 38 For " Carya " read " Careya."
Page 21 line 27 The returns are more than double this amount now, or upwards of 2000 rupees.
     line 56 An Act has since been passed.
Page 24 line 4 Add " and near Puliyara."
Page 25 line 13 The returns have exceeded half a lac in some years. See Appendix V.
Page 27 line 15 This depôt has since been closed again.
Page 28 line 6 For " Katār" read " Kalār."
Page 31 line 9 A reserve has since been declared there.
     line 25 For " other " read " others."
Page 32 line 45 A reserve has since been chosen here.
Page 34 line 1 A reserve has since been chosen here.
     line 49 For " higer " read " higher."
Page 35 Note. For " a chain or half a chain " read " half a chain or a chain."
     line 51 For " kind " read " kinds."
Page 36 line 35 margin. For " sperations " read " operations."
     line 53 A reserve has since been chosen.
Page 43 line 27 For " occassionally." read " occasionally."
     line 44 This idea has been abandoned.
     last line do. do. do.
Page 44 line 26 The Kōtta vāthal ridge where these sandalwood trees are growing is not the boundary between the Shencottah and Kottarakara taluqs which is farther west.
Page 48 bottom of page. The rates for charcoal in Travancore have now been raised, and there is no longer any trouble with the British Forest Dept. through inequality of rates.
Page 49 line 1 A reserve has since been declared in the Shencottah taluq.
     line 18 For " decending " read " descending."
Page 50 line 4 For " kunan " read " kuman."
     line 35 For " pantagyna " read " pentagyna."
Page 53 line 2 Mr. Thomas was Assistant Conservator at Konniyūr at the time of my visit.
     line 11 The forests on the southern bank have also since been reserved.
Page 55 line 12 margin. For " Valia " read " Valiya."
Page 57 line 6 For " Rāgan " read " Rājam."
Page 64 line 28 A reserve has been recently gazetted on this river and southeast of Mūndakkayam.
.Page line 48 For " acquired " read " required."
Page 67 line 35 An Amindar is now stationed at Pālāyi as suggested.

Page 70 line 28 This Chowkey has since been abolished.
Page 71 line 4 For "above" read "about."
Page 72 line 10 For "is" read "in."
line 38 For "Nèthalampàra" read "Mèthalampàra."
Page 73 line 16 This valley has since been reserved.
line 49 The work of the Thodupura Amindar has since been reduced.
Page 75 line 31 Kathira mudi is only 4250 ft. high.
Page 76 line 6 For "Màradi" read "Màvadi."
line 40 And a little blackwood.
Page 79 line 25 Take the comma from after "river" and place it after "point."
Page 83 line 58 For "obstructien" read "obstruction."
Page 85 line 47 For "builders" read "boulders."
Page 86 line 2 For "centre" read "middle."
Page 87 line 34 For "centre" read "middle."
Page 88 line 16 For "centre" read "middle."
Page 92 line 46 For "where" read "whence."
Page 101 line 10 The Forest Department collections are not likely to be in-
creased so long as the suitable land is all supervised by the
Cardamom Department.
Page 103 line 17 For "Chàlakkada" read "Chàlakkudi."
line 45 For "in" read "it."
Page 105 line 5 For "one" read "our."
Page 111 line 4 For the second "thus" read "this,"
Page 112 line 33 For "old" read "ala."
Note line 3 For "boundary" read "boundaries."
line 10 For "to" read "too."
Page 113 line 16 The Southern boundary is the latitude of the Karutha-urutti
fall, but much of the land south of the Kòrakkunna ridge
has been given to others, so that the ridge is the accepted
boundary now.
Page 114 Note "Chèrikkal" lands are of different kinds, but I am told they are
all registered in the names of persons, whereas ordinary
Virippu lands are not so registered.
Page 115 line 3 For "of" read "off."
line 53 For "brunt" read "burnt."
Page 119 line 46 Under the Rules passed under the Forest Act the occupation
of waste lands will be regulated.
Page 121 line 30 Note For "distruction" read "destruction."
Page 131 line 43 Omit the second "from."
Page 132 line 4 For "Pàlayai" read "Pàlàyi."
Page 133 line 50 Some of the trees in the plantations at Ariyankàvu have
attained a height of 15 ft. in one year and more than 20 ft.
in eighteen months.
Page 134 Footnote. For $\frac{1 \times w}{b \times d}$ read $\frac{1 \times w}{b \times d^2}$
Page 136 line 38 For "750" read "500" and for "12" read "20."
Page 139 footnote Correct 1 Kings X. 11.
Page 143 line 9 For "founded" read "found."
line 35 For "course" read "coarse."
Page 147 line 14 For "mahoganny" read "mahogany."
15 For "excallant" read "excellent."
Page 150 line 18 For "been" read "be."
Page 154 line 11 "Chokala" I have since ascertained to be *Aglaia Roxburghiana*
line 49 After "others" insert "is."
Page 156 line 13 For "Actinodaphue" read "Actinodaphne."
line 16 For "Actinodophue" read "Actinodaphne."
line 22 For "Odallam" read "Odollam."
Page 157 line 27 For "crest" read "crests."
Page 158 line 19 For "tranquility" read "tranquillity."
Note line 6 For "censequence" read "consequence."
Page 167 line 46 This fifth reserve turned out to be less than one square mile
in area when surveyed.

169 line 48 This new bungalow has since been built.
170 line 3 This has since been built.
line 37 The Rules to be passed under the new Forest Act will greatly help us in this respect.
Page 171 line 44 A longer experience has taught us that a clearing can only be planted and kept in good order during the first year for 36 rupees if all the conditions are favourable, and 50 rupees is a safer estimate.
Page 178 line 19 And since then to 12 rupees a candy.
Page 179 line 48 The rates for boats have since been raised.
Page 180 line 2 For "fined" read "paid."
line 8 For "authorised" read "required."
line 41 For "depôt" read "station."
Page 181 line 39 For "piels" read "plies."
Page 187 line 58 Since the passing of the new Forest Act beeswax has ceased to be a monopoly.
Page 188 line 17 Lac is no longer a monopoly.
line 20 Dammer is not a monopoly now.
line 25 All charges on firewood have been abolished except in the Shencottah taluq and at Panagudi.
Page 191 Note: line 3 For "Prussion" read "Prussian."
Page 194 line 33 A new Forest Act has since been passed.
Page 195 line 17 For "cunnivance" read "connivance."
line 40 For "is" read "are."
Page 204 line 43 Reserves have since been proclaimed in S. Travancore.
line 50 For "maher" read "matter."
line 57 Reserves have since been gazetted for settlement in the localities mentioned.
Page 209 line 42 That is the total number of all the Hill men.
Page 216 line 13 For "increasing" read "increase."
line 38 This firm has since entered into a new contract.
Page 217 line 16 For "one quarter" read "one third."
Page 222 line last but 4 Omit the second "to."
Page 223 Note § The net Forest revenue of the Madras Presidency was in
1891-2...........3,80,901 rupees.
1892-3...........3,74,000 „
1893-4...........2,76,000 „ (estimate).

Appendix page XII For "Stercoliaceœ" read "Sterculiaceœ."
„ page XXX Insert 393 bis. Hemicyclia elata.........evergreen forests 1000—3000 ft. medium sized tree with white hard wood.
„ page XXXVIII line 10 Omit the long mark over "mâla."
„ page XLI line 23 All charges in firewood have been abolished since writing this.
„ page XLV line 6 For 1516 put 1560.